Handbook of Irrigation System Selection for Semi-Arid Regions

Handbook of Irrigation System Selection for Semi-Arid Regions

Edited by
Mohammad Albaji
Saeid Eslamian
Abd Ali Naseri
Faezeh Eslamian

CRC Press
Taylor & Francis Group
Boca Raton London New York

CRC Press is an imprint of the
Taylor & Francis Group, an **informa** business

First edition published 2020
by CRC Press
6000 Broken Sound Parkway NW, Suite 300, Boca Raton, FL 33487-2742

and by CRC Press
2 Park Square, Milton Park, Abingdon, Oxon, OX14 4RN

© 2020 Taylor & Francis Group, LLC

CRC Press is an imprint of Taylor & Francis Group, LLC

Library of Congress Cataloging-in-Publication Data

Names: Albaji, Mohammad, editor. | Eslamian, Saeid, editor. | Naseri, Abd Ali, editor. | Eslamian, Faezeh, editor.
Title: Handbook of irrigation system selection for semi-arid regions /
edited by Mohammad Albaji, Saeid Eslamian, Abd Ali Naseri, Faezeh Eslamian.
Description: First edition. | Boca Raton, FL : CRC Press/Taylor & Francis
Group, [2020]
Identifiers: LCCN 2020007246 (print) | LCCN 2020007247 (ebook) | ISBN
9780367505363 (hardback ; acid-free paper) | ISBN 9781003050261 (ebook)
Subjects: LCSH: Irrigation. | Arid regions agriculture. | Irrigation
engineering.
Classification: LCC S613 .H35 2020 (print) | LCC S613 (ebook) | DDC
631.7--dc23
LC record available at https://lccn.loc.gov/2020007246
LC ebook record available at https://lccn.loc.gov/2020007247

ISBN: 978-0-367-50536-3 (hbk)
ISBN: 978-0-367-51877-6 (pbk)
ISBN: 978-1-003-05026-1 (ebk)

Typeset in Times
by Lumina Datamatics Limited

Contents

SECTION I   Introduction

SECTION II   Strategies for Irrigation Development and Management

SECTION III   Case Studies: Irrigation Method Selection for the Crossroads of Central and Southeast Europe and South Asia

SECTION IV   Case Studies: Irrigation System Selection in Western Asia

Preface

Irrigation, the addition of water to lands via artificial means, is essential to profitable crop production in arid climates. Irrigated agriculture makes a major contribution to food security in the world. It uses more than 60%–70% of the water withdrawn from the world's water resources. The irrigated area at a global level represents around 20% of the total agricultural land and contributes 40% of the total food production. However, the performance of the majority of existing irrigation systems remains low. With increasing competition from other sectors, irrigation is under pressure to reduce its share of water use. The available world's water resources may not be able to meet various demands that will inevitably result in the irrigation of additional lands in order to achieve a sustainable food security. Due to the scarce water supply in the arid and semiarid climates, maximizing water use efficiency is critical. A shift from surface irrigation to high-tech irrigation technologies, e.g., sprinkle and drip irrigation systems, may yield significant water savings. Also, in lands where irrigation is introduced as a new practice, and in lands that currently irrigate, there must be irrigation method choices. Alternative irrigation methods may become more advantageous as the land organization, objectives, price, and cost change. The selection of irrigation methods will often be influenced by costs, energy, crop, soil texture, land topography, water supply, climate parameters, labor availability, etc. In other words, irrigation application method and system selection should result in optimum use of available water. The selection should be based on a full awareness of management considerations, such as water source and cost, water quantity and quality, irrigation effects on the environment, energy availability and cost, farm equipment, product marketability, and capital for irrigation system installation, operation, and maintenance. The purpose of this book is to bring together and integrate in a single text the subject matter that deals with the irrigation selection methods in arid and semiarid regions. The book is divided into four major sections dealing with the subject mentioned above and is intended for students, researchers, and professionals working on various aspects of irrigation method selection. Each section is composed of some chapters from various research groups and individuals working separately. Various case studies have been discussed in the chapters to present a general scenario of irrigation method selection management.

The first section highlights the parametric evaluation system for irrigation purposes. The parametric approach in the evaluation of land characteristics consists of a numeral rating of the different limitation levels of the land characteristics in a numerical scale from a maximum (normally 100) to a minimum value. If a land characteristic is optimal for the considered land utilization type, the maximum rating of 100 is attributed; if the same land characteristic is unfavorable, a minimal rating is applied. The aim of the parametric evaluation system is to provide a method that permits evaluation for irrigation purposes and that is based on the standard granulometrical and physicochemical characteristics of a soil profile. The section concludes with a review of the analytical hierarchy process integrated with GIS in an arid region for site selection of different irrigation systems.

Section two focuses on strategies for irrigation development and management. The deficit irrigation and partial root-zone drying irrigation systems are introduced in an arid area. Deficit irrigation (DI) and partial root-zone drying system (PRD) are strategies that involve this need that express irrigating the root zone with less water than the maximum evapotranspiration without loss in yield. Also, the solute leaching modeling under different irrigation regimes, soil moisture conditions, and organic fertilizer application in an arid area is introduced. This study examines the experimental and numerical modeling of soil solute leaching under different irrigation regimes, soil moisture conditions, and vermicompost application. Finally, this section concludes with an evaluation of the BUDGET model in simulating different strategies of irrigation in the west of Asia.

Section three highlights the multi-criteria decision-making for irrigation management in Europe. Decision-making related to irrigation management needs to be made in the context of multiple, usually conflicting, criteria that are coming from economic, environmental, and social domains. Multi-criteria decision analysis (MCDA) is the most often used approach for this type of decision-making, and different (MCDA) methods have been developed in the past. Hence, the first goal of this section was to make MCDA methods more intelligible (compared with the current level of understanding) to novice users within the field of irrigation management. Therefore, basic ideas as well as the main steps of selected (MCDA) methods are presented. Also, this section focuses on appraisal of agricultural lands for irrigation in the hot, subhumid region of Jayakwadi Command Area, Parbhani District, Maharashtra, India. The appraisal of basaltic clays soils under the hot, subhumid Marathwada region of Maharashtra is to tackle the recurring droughts and to explore answers to two vital questions such as water to every field and more crop per drop. The detailed land resource inventory is very essential for sustainable agriculture under irrigation with the help of the FAO and Sys parametric approach. This study showed the limits of soil–topography-yield variations in interpreting site-specific data for irrigation.

The final section starts with an evaluation of the soils delta of the Wadi Horan within the province of the Upper Euphrates, Iraq, for some technologies of irrigation systems. The results showed that drip irrigation proved more appropriate than the surface irrigation system in the study area. However, the main determinants of both methods were the characteristics of the physical environment and the topography of the study area. Drip irrigation research has proved that the conservation of soil and keeping it within the water field capacity will be more useful for application of the irrigation method, especially in climates and arid areas; therefore, it is recommended to exercise as the best way appropriate for the study area. Also, this section focuses on analysis and comparison of three irrigation systems (surface, sprinkle, and drip) in six major Iran watersheds by taking into account various soil and land characteristics. Details are given for the analysis of field data to compare the suitability of the land for surface, sprinkle, and drip irrigation systems. The analyzed parameters included soil and land characteristics. In this research, the land is not being "improved." The lands are just being ranked according to defined criteria to establish which irrigation method (surface, sprinkle, or drip) is best suited to the characteristics of the land. That is, the land suitability for irrigation methods (surface, sprinkle, or drip) is being compared while taking into consideration the characteristics of the land. The results obtained showed that sprinkle and drip irrigation methods are more suitable than a surface or gravity irrigation method for most of the soils tested. Moreover, because of the insufficiency of surface and groundwater resources, and the aridity and semiaridity of the climate in these areas, sprinkle and drip irrigation methods are highly recommended for a sustainable use of this natural resource; hence, the changing of current irrigation methods from gravity (surface) to pressurized (sprinkle and drip) in the study area are proposed.

It is evident but nevertheless worth mentioning that all the chapters have been prepared by individuals who are experts in their fields. The views expressed in the book are those of the authors, and they are responsible for their statements. An honest effort has been made to check the scientific validity and justification of each chapter through several iterations. We, the editors, publisher, and the authors of the chapters, have put together a comprehensive reference handbook on irrigation method selection in arid and semiarid regions with a belief that this book will be of immense use to present and future colleagues who teach, study, research, and/or practice in this particular field.

Dr. Mohammad Albaji
Department of Irrigation and Drainage
Faculty of Water Science Engineering
Shahid Chamran University of Ahvaz, Ahvaz, Iran

Professor Saeid Eslamian
Department of Water Engineering, College of Agriculture
Center of Excellence in Risk Management and Natural Hazards
Isfahan University of Technology, Iran

Professor Abd Ali Naseri
Department of Irrigation and Drainage
Faculty of Water Science Engineering
Shahid Chamran University of Ahvaz, Ahvaz, Iran

Dr. Faezeh Eslamian
Bioresource Engineering Department
McGill University
Montreal, Quebec, Canada

Editors

Mohammad Albaji is a faculty member at Shahid Chamran University of Ahvaz, Iran. He earned his PhD in Irrigation and Drainage Engineering from Shahid Chamran University of Ahvaz, Iran. Dr. Albaji is active in the fields of irrigation and drainage, precision irrigation, agricultural water management, water productivity, water & soil salinity and alkalinity, soil science, land suitability for irrigation, land suitability for crops, and has more than 70 publications in reputed journals, books, and refereed conferences. He has been working as a reviewer for many reputed journals, such as *Agricultural Water Management, Computers and Electronics in Agriculture, Environment, Development and Sustainability, Transaction of the Royal Society of South Africa, Clean Soil, Air, Water,* etc. He had several executive posts such as dean of Jundishapur's Water and Energy Research Institute (Shahid Chamran University of Ahvaz; 2016–2018), head of Soil Science Department (Khuzestan Water & Power Authority; 2000–2012), etc. He has membership in the International Center for Biosaline Agriculture (ICBA), American Society of Civil Engineering (ASCE), etc.

Saeid Eslamian is a full professor of environmental hydrology and water resources engineering in the Department of Water Engineering at Isfahan University of Technology, where he has been since 1995. His research focuses mainly on statistical and environmental hydrology in a changing climate. In recent years, he has worked on modeling natural hazards, including floods, severe storms, wind, drought, pollution, water reuses, sustainable development and resiliency, etc. Formerly, he was a visiting professor at Princeton University, New Jersey, and ETH Zürich, Switzerland. On the research side, he started a research partnership in 2014 with McGill University, Canada. He has contributed to more than 600 publications in journals, books, and technical reports. He is the founder and chief editor of both the *International Journal of Hydrology Science and Technology* (IJHST) and the *Journal of Flood Engineering* (JFE). Professor Eslamian is now associate editor of four important publications: *Journal of Hydrology* (Elsevier), *Ecohydrology & Hydrobiology* (Elsevier), *Journal of Water Reuse and Desalination* (IWA), and *Journal of the Saudi Society of Agricultural Sciences* (Elsevier).

Professor Eslamian is the author of approximately 35 books and 180 chapter books. Dr. Eslamian's professional experience includes membership on editorial boards, and he is a reviewer of approximately 100 Web of Science (ISI) journals, including the ASCE *Journal of Hydrologic Engineering*, ASCE *Journal of Water Resources Planning and Management*, ASCE *Journal of Irrigation and Drainage Engineering*, *Advances in Water Resources, Groundwater, Hydrological Processes, Hydrological Sciences Journal, Global Planetary Changes, Water Resources Management, Water Science and Technology, Ecohydrology, Journal of American Water Resources Association, American Water Works Association Journal,* etc. UNESCO also nominated him for a special issue of the *Ecohydrology & Hydrobiology* journal in 2015. Professor Eslamian was selected as an outstanding reviewer for the *Journal of Hydrologic Engineering* in 2009 and received the EWRI/ASCE Visiting International Fellowship in Rhode Island (2010). He was also awarded outstanding prizes from the Iranian Hydraulics Association in 2005 and Iranian Petroleum and Oil Industry in 2011. Professor Eslamian has been chosen as a distinguished researcher of Isfahan University of

Technology (IUT) and Isfahan Province in 2012 and 2014, respectively. In 2016, he was a candidate for national distinguished researcher in Iran.

He has also been the referee of many international organizations and universities. Some examples include the U.S. Civilian Research and Development Foundation (USCRDF), the Swiss Network for International Studies, the Majesty Research Trust Fund of Sultan Qaboos University of Oman, the Royal Jordanian Geography Center College, and the Research Department of Swinburne University of Technology of Australia. He is also a member of the following associations: American Society of Civil Engineers (ASCE), International Association of Hydrologic Science (IAHS), World Conservation Union (IUCN), GC Network for Drylands Research and Development (NDRD), International Association for Urban Climate (IAUC), International Society for Agricultural Meteorology (ISAM), Association of Water and Environment Modeling (AWEM), International Hydrological Association (STAHS), and UK Drought National Center (UKDNC).

Professor Eslamian finished Hakimsanaei High School in Isfahan in 1979. After the Islamic Revolution, he was admitted to IUT for a BS in water engineering and graduated in 1986. After graduation, he was offered a scholarship for a master's degree program at Tarbiat Modares University, Tehran. He finished his studies in hydrology and water resources engineering in 1989. In 1991, he was awarded a scholarship for a PhD in civil engineering at the University of New South Wales, Australia. His supervisor was Professor David H. Pilgrim, who encouraged him to work on "Regional Flood Frequency Analysis Using a New Region of Influence Approach." He earned a PhD in 1995 and returned to his home country and IUT. In 2001, he was promoted to associate professor and in 2014 to full professor. For the past 24 years, he has been nominated for different positions at IUT, including university president consultant, faculty deputy of education, and head of the department. Professor Eslamian is now director for the Center of Excellence in Risk Management and Natural Hazards (RiMaNaH).

Professor Eslamian has made three scientific visits to the United States, Switzerland, and Canada in 2006, 2008, and 2015, respectively. In the first, he was offered the position of visiting professor by Princeton University and worked jointly with Professor Eric F. Wood at the School of Engineering and Applied Sciences for one year. The outcome was a contribution in hydrological and agricultural drought interaction knowledge by developing multivariate L-moments between soil moisture and low flows for northeastern U.S. streams. Recently, Professor Eslamian has published the editorship of 12 handbooks published by the Taylor & Francis Group (CRC Press): the three-volume *Handbook of Engineering Hydrology* in 2014, *Urban Water Reuse Handbook* in 2016, *Underground Aqueducts Handbook* (2017), the three-volume *Handbook of Drought and Water Scarcity* (2017), *Constructed Wetlands: Hydraulic Design* (2020), and the three-volume *Flood Handbook* (2020). He has also published the two-volume *Handbook of Water Harvesting and Conservation* (2020) by Wiley-Blackwell and the New York Academy of Sciences. In addition, *An Evaluation of Groundwater Storage Potentials in a Semiarid Climate* by Nova Science Publishers is also his joint book publication in 2019.

Abd Ali Naseri is a full professor of irrigation and drainage in the *Faculty of Water Science Engineering at Shahid Chamran University of Ahvaz, Iran.* He earned his PhD in irrigation and drainage engineering from the University of Southampton, England, 1998. His research focuses mainly on irrigation and drainage, soil physics, leaching, precision drainage, agricultural water management, and modeling. He has contributed to more than 350 publications in journals, books, and technical reports. He has published more than 150 articles in many reputed journals, such as *Agricultural Water Management, Journal of Irrigation and Drainage Engineering (ASCE), Irrigation and Drainage (ICID) Chemical Engineering Journal, Journal of Cleaner Production, Journal of Powder Technology, International Journal of Applied Earth Observation and*

Geoinformation, Journal of Ecological Engineering, Quarterly of RS & GIS for Natural Resources, Transaction of the Royal Society of South Africa, etc.

Professor Abd Ali Naseri is a member of journal editorial board of Agricultural Research, Journal of Applied Sciences, Asian Journal of Scientific Research, Journal of Agronomy and Irrigation Engineering and Sciences. He had supervised more than 120 MSc and doctoral dissertations and research projects. He earned four scientific awards (The top researcher of 2010 in Shahid Chamran University of Ahvaz; The top researcher of 2001 from the Sugarcane & by Products Development Company; The top researcher of 2000 in Hamedan Province, Iran; The winner of the second top paper in the 2d Conference of River Engineering, Ahvaz, Iran).

Professor Abd Ali Naseri has held several executive posts such as head of irrigation and drainage department in the *Faculty of Water Science Engineering at Shahid Chamran University of Ahvaz, Iran* (2009–2011; 2015–2017), member of expert commission of *Shahid Chamran University* Press Council (June 2011 to till date), member of the technical and expert committee on Irrigation and Drainage Research of Khuzestan Water and Power Authority (June 2009 to till date), member of regional committee of irritation and drainage in Khuzestan province (September 2011 to till date), head of the Drainage Researches Center (2015–2019) and director manager of Sugarcane and By Products Development Company (2019 to till date), etc.

Faezeh Eslamian holds a PhD in Bioresource Engineering from McGill University. Her research focuses on the development of a novel lime-based product to mitigate phosphorus loss from agricultural fields. Faezeh completed her bachelor's and master's degrees in Civil and Environmental Engineering from Isfahan University of Technology, Iran, where she evaluated natural and low-cost absorbents for the removal of pollutants such as textile dyes and heavy metals. Furthermore, she has conducted research on the worldwide water quality standards and wastewater reuse guidelines. She is an experienced multidisciplinary researcher with interest in soil and water quality, environmental remediation, water reuse, and drought management.

Contributors

Mohammad Albaji
Department of Irrigation and Drainage
Faculty of Water Science Engineering
Shahid Chamran University of Ahvaz
Ahvaz, Iran

Farhan J. Mohamed Althyby
Department of Soil and Water Science
Faculty of Agriculture
Anbar University
Anbar, Iraq

AbdulKarem Ahmed Al-Alwany
Center for Desert Studies
Anbar University
Anbar, Iraq

Hossein Bagheri
Department of Water Engineering
Faculty of Agriculture
Bu Ali Sina University
Hamedan, Iran

Amin Behmanesh
College of Water Engineering
College of Agriculture
Isfahan University of Technology
Isfahan, Iran

Atila Bezdan
Department of Water Management
Faculty of Agriculture
University of Novi Sad
Novi Sad, Serbia

Jovana Bezdan
Department of Water Management
Faculty of Agriculture
University of Novi Sad
Novi Sad, Serbia

Bhaskara Phaneendra Bhaskar
Regional Centre, ICAR-National Bureau of Soil
    Survey and Land Use Planning
Bangalore, India

Boško Blagojević
Department of Water Management
Faculty of Agriculture
University of Novi Sad
Novi Sad, Serbia

Saeed Boroomand Nasab
Department of Irrigation and Drainage
Faculty of Water Science Engineering
Shahid Chamran University of Ahvaz
Ahvaz, Iran

Niaz Ali Ebrahimi Pak
Department of Irrigation and Soil Physics, Soil
    and Water Research Institute
Agricultural Research, Education and
    Extension Organization (AREEO)
Karaj, Iran

Faezeh Eslamian
Bioresource Engineering Department
McGill University
Montreal, Canada

Saeid Eslamian
Department of Water Engineering
College of Agriculture
Center of Excellence in Risk Management and
    Natural Hazards
Isfahan University of Technology
Isfahan, Iran

Shahrokh Fatehi
Soil and Water Research Department
Kermanshah Agricultural and Natural
    Resources Research and Education Center
Kermanshah, Iran

Mona Golabi
Department of Irrigation and Drainage
Faculty of Water Science Engineering
Shahid Chamran University of Ahvaz
Ahvaz, Iran

Rajendra Hegde
Regional Centre, ICAR-National Bureau of Soil
    Survey and Land Use Planning
Bangalore, India

Majid Heydari
Department of Water Engineering
Faculty of Agriculture
Bu Ali Sina University
Hamedan, Iran

Azizallah Izady
Water Research Centre
Sultan Qaboos University (SQU)
Muscat, Oman

Mehdi Jovzi
Soil and Water Research Department
Kermanshah Agricultural and Natural
    Resources Research and Education Center
Agricultural Research, Education and
    Extension Organization (AREEO)
Kermanshah, Iran

Sampura Chinnappa Ramesh Kumar
Regional Centre, ICAR-National Bureau of Soil
    Survey and Land Use Planning
Bangalore, India

Abd Ali Naseri
Department of Irrigation and Drainage
Faculty of Water Science Engineering
Shahid Chamran University of Ahvaz
Ahvaz, Iran

Lamya Neissi
Department of Irrigation and Drainage
Faculty of Water Science Engineering
Shahid Chamran University of Ahvaz
Ahvaz, Iran

Venkataramappa Ramamurthy
Regional Centre, ICAR-National Bureau of Soil
    Survey and Land Use Planning
Bangalore, India

Reza Sadegh Mansouri
Department of Irrigation and Drainage
Faculty of Water Science Engineering
Shahid Chamran University of Ahvaz
Ahvaz, Iran

Maasomeh Salehi
National Salinity Research Center
Agricultural Research, Education and
    Extension Organization (AREEO)
Yazd, Iran

Adel K. Salemn
Center for Desert Studies
Anbar University
Anbar, Iraq

Radovan Savić
Department of Water Management
Faculty of Agriculture
University of Novi Sad
Novi Sad, Serbia

Milica Vranešević
Department of Water Management
Faculty of Agriculture
University of Novi Sad
Novi Sad, Serbia

Hamid Zare Abyaneh
Department of Water Engineering
Faculty of Agriculture
Bu Ali Sina University
Hamedan, Iran

Section I

Introduction

1 Introduction of Parametric Evaluation System for Irrigation Purposes

Mohammad Albaji, Saeid Eslamian, Abd Ali Naseri, and Faezeh Eslamian

CONTENTS

1.1 INTRODUCTION

In arid and semiarid climates, where the study area has been carried out, according to the results of the climate classification, the most relevant method to improve agriculture production is irrigation. An adequate supply of water is important for plant growth. When rainfall is not sufficient, crops must be provided with additional water from other sources. To decide where to apply irrigation and to choose the appropriate method, natural conditions, previous experience with irrigation, required labor inputs, type of crops and technology, costs and benefits, and other factors should be considered. As a matter of fact, irrigation practices can be very expensive and may cause negative phenomena, such as soil erosion and salinization. For this reason, evaluation systems for irrigation purposes must be developed. The prevailing semiarid climatic conditions and an insufficient amount of precipitation are a major limitation to crop yields in the study area. Agricultural fields are often required to be irrigated if food security is to be ensured, especially with vegetable crops. The purpose of land suitability for irrigation was therefore aimed at visualizing parcels of land that may be conveniently irrigated for agricultural production based on the physical and chemical parameters of the underlying soil (IAO 2007).

On the other hand, food security and stability in the world greatly depend on the management of natural resources. Due to the depletion of water resources and an increase in population, the extent of irrigated area per capita is declining, and irrigated lands now produce 40% of the food supply (Hargreaves and Mekley 1998). Consequently, available water resources will not be able to meet various demands in the near future, and this will inevitably result in the seeking of newer lands for irrigation in order to achieve sustainable global food security. Land suitability, by definition, is the natural capability of a given land to support a defined use. The process of land suitability classification is the appraisal and grouping of specific areas of land in terms of their suitability for a defined use.

According to FAO methodology (1976), land suitability is strongly related to "land qualities" including erosion resistance, water availability, and flood hazards, which are in themselves immeasurable qualities. Since these qualities are derived from "land characteristics," such as slope angle and length, rainfall, and soil texture—which are measurable or estimable—it is advantageous to use the latter indicators in the land suitability studies and then use the land parameters for determining the land suitability for irrigation purposes. Sys et al. (1991a) suggested a parametric evaluation system for irrigation methods that was primarily based upon physical and chemical soil properties (Dengiz 2006; Liu et al. 2006; Naseri et al. 2009; Albaji 2010; Albaji et al. 2012).

1.2 INTRODUCTION OF PARAMETRIC EVALUATION SYSTEM FOR IRRIGATION PURPOSES

1.2.1 GENERAL PRINCIPLES

The parametric approach in the evaluation of land characteristics consists of a numeral rating of the different limitation levels of the land characteristics in a numerical scale from a maximum (normally 100) to a minimum value. If a land characteristic is optimal for the considered land utilization type, the maximum rating of 100 is attributed; if the same land characteristic is unfavorable, a minimal rating is applied.

The successful application of the system implies the respect of the following rules:

1. The number of land characteristics to consider has to be reduced to a strict minimum to avoid repetition of related characteristics in the formula, leading to a depression of the land index. Therefore, all land qualities expressed by one characteristic should be rated together. As such, the single rating of texture should be done with regard to the capacity to retain nutrients, water availability, permeability, and one should avoid introducing separate ratings for these single qualities.
2. An important characteristic is rated in a wide scale (100–25), a less important characteristic in a narrower scale (100–60). This introduces the concept of a weighting factor. Example: studying the suitability for irrigation the very important factor of texture is rated from 100 to 25, the less important factor of calcium carbonate content from 100 to 80.
3. The rating of 100 is applied for optimal development or maximum appearance of a characteristic. If, however, some characteristics are better than the usual optimal, the maximum rating can be chosen higher than 100. Example: if the most common organic carbon content of the top 15 cm in a specific area varies from 1% to 1.5%, the rating of 100 is applied for that carbon level. Soils with more than 1.5% O.C are attributed a rating of more than 100 for organic matter.
4. The depth to which the land index has to be calculated must be defined for each land utilization type. If one considers that for a specific land utilization type all horizons have a similar importance, the weighted average of the profile section until the considered depth is calculated for each characteristic. If, on the other hand, one considers that the importance of a horizon becomes greater when this position is nearer to the surface, a different proportional rating can be given to the depth sections of the profile in such a way that they increase when approaching the surface. Therefore, the profile can be subdivided into equal sections; to each of these sections one attributes a "depth correction index" (weighting factor) starting with a minimum value in depth and becoming gradually greater when approaching the surface section.

The depth to be considered should coincide with the normal depth of the root system in a deep soil. The weighting factors or depth correction indices suggested are given in Table 1.1.

TABLE 1.1
Number of Sections and Weighting Factors for Different Depths

Depth (cm)	Number of Equal Sections	Weighting Factors
125–150	6	2.00,1.50,1.00,0.75,0.50,0.25
100–125	5	1.75,1.50,1.00,0.50,0.25
75–100	4	1.75,1.25,0.75,0.25
50–75	3	1.50,1.00,0.50
25–50	2	1.25,0.75
0–25	1	1.00

A land suitability classification for surface, sprinkler, and drip irrigation was provided. It was carried out by a parametric system according to the methodology proposed by Sys et al. (1991a). The aim of this parametric evaluation system (Sys et al. 1991a) is to provide a method that permits evaluation for irrigation purposes, and that is based on the standard granulometrical and physico-chemical characteristics of a soil profile. It has been estimated that the soil as a medium for plant growth under irrigation should in the first place provide the necessary water and plant nutrients in an available form, and in the most economic way. A method to improve the agriculture production can be the application of irrigation; the decision regarding if it is possible, how and in which area, should be taken according to economic and agronomic factors. In order to avoid bad effects such as salinization or water stagnating and in order to estimate the impact of agronomic parameters, an evaluation of irrigation suitability has to be applied (IAO 2002).

In the system proposed by Sys et al. (1991b), the factors affecting soil suitability for irrigation purposes can be subdivided into four groups:

1. PHYSICAL PROPERTIES determine the soil-water relationship in the solum such as permeability and available water content both related to texture, structure, and soil depth; also, $CaCO_3$ status could be considered here.
2. CHEMICAL PROPERTIES interfere in the salinity/alkalinity status, such as soluble salts and exchangeable Na.
3. DRAINAGE PROPERTIES.
4. ENVIRONMENTAL FACTORS, such as slope.

The soil drainage, depth, slope, texture, electric conductivity, and $CaCO_3$ content were the considered parameters. The parametric method is based on the concept that index rating (capability index for irrigation) is obtained by multiplying all the parameters involved in the evaluation. Thus, any factor may dominate or control the final rating. Recall that according to the specific suitability evaluation, this multiplication is applied considering all parameters involved in the evaluation.

The different land characteristics that influence the soil suitability for irrigation are rated, and a capability index for irrigation (Ci) is calculated according to the formula:

$$Ci = A \times \frac{B}{100} \times \frac{C}{100} \times \frac{D}{100} \times \frac{E}{100} \times \frac{F}{100} \qquad (1.1)$$

where:
Ci = Capability index for irrigation
A = Soil texture rating
B = Soil depth rating
C = Calcium carbonate content rating
D = Electrical conductivity rating
E = Drainage rating
F = Slope rating.

TABLE 1.2

Suitability Classes for the Irrigation Capability Indices (Ci) Classes

Capability Index	Definition	Symbol
>80	Highly suitable	S_1
60–80	Moderately suitable	S_2
45–59	Marginally suitable	S_3
30–44	Currently not suitable	N_1
<29	Permanently not suitable	N_2

The suitability classes are defined according to the value of the capability (or suitability) index (Ci) (Table 1.2).

The classes S_2 to N_2 can have the following subclasses with regard to the nature of the limiting factors:

s = Limitations due to physical soil properties (A, B, C)
n = Limitations due to salinity/alkalinity (D)
w = Wetness limitations (E)
t = Topographic limitations (F).

The land suitability classification consists of assessing and grouping the land types in orders and classes according to their aptitude (FAO methodology 1976). The order defines the suitability and is expressed by:

- S (suitable) characterizes a land were sustainable use giving good benefits is expected.
- N (not suitable) indicates a land with qualities that do not allow the considered type of use or are not enough for sustainable outcomes.

The classes (S_1, S_2 and S_3 for suitable order; N_1 and N_2 for unsuitable order) express the degrees of suitability or unsuitability. Thus, there are five classes according to Table 1.3 for irrigation.

1.2.2 Factors Influencing the Soil Suitability for Irrigation

1.2.2.1 Texture (A)

Texture is rated (Table 1.4) with regard to permeability and available water content, and weighted average is calculated for the upper 1.5 m.

TABLE 1.3

Land Suitability Classes

Order	Class	Description
Suitable	S_1 (Highly suitable)	Land having no or insignificant limitations for irrigation
	S_2 (Moderately suitable)	Land having minor limitations for irrigation
	S_3 (Marginally suitable)	Land having moderate limitations for irrigation
Not Suitable	N_1 (Currently not suitable)	Land having severe limitations for irrigation but can be improved by specific management
	N_2 (Permanently not suitable)	Land having such severe limitations for irrigation that are very difficult to overcome

TABLE 1.4
Rating of Textural Classes for Irrigation

Tex[a]	Rating for Surface Irrigation					Rating for Sprinkler Irrigation					Rating for Drip Irrigation				
	Fine Gravel (%)			Coarse Gravel (%)		Fine Gravel (%)			Coarse Gravel (%)		Fine Gravel (%)			Coarse Gravel (%)	
	<15	15–40	40–75	15–40	40–75	<15	15–40	40–75	15–40	40–75	<15	15–40	40–75	15–40	40–75
CL[b]	100	90	80	80	50	100	90	80	80	50	100	90	80	80	50
SiL	100	90	80	80	50	100	90	80	80	50	100	90	80	80	50
SCL	95	85	75	75	45	95	85	75	75	45	95	85	75	75	45
L	90	80	70	70	45	90	80	70	70	45	90	80	70	70	45
SiL	90	80	70	70	45	90	80	70	70	45	90	80	70	70	45
Si	90	80	70	70	45	90	80	70	70	45	90	80	70	70	45
SiC	85	95	80	80	40	85	95	80	80	40	85	95	80	80	40
C	85	95	80	80	40	85	95	80	80	40	85	95	80	80	40
SC	80	90	75	75	35	95	90	80	75	35	95	90	85	80	35
SL	75	65	60	60	35	90	75	70	70	35	95	85	80	75	35
LS	55	50	45	45	25	70	65	50	55	30	85	75	55	60	35
S	30	25	25	25	25	50	45	40	30	30	70	65	50	35	35

a Tex: Textural Classes.
b CL: Clay Loam; SiL: Silty Loam; SCL: Sandy Clay Loam; L: Loam; SiL: Silty Loam; Si: Silty; SiC: Silty Clay; C: Clay; SC: Sandy Clay; SL: Sandy Loam; LS: Loamy Sand; S: Sandy.

1.2.2.2 Soil Depth (B)

Soil depth is defined as the thickness of the loose soil above a limiting layer, which is impenetrable for roots or percolating water. The most common types of such limiting layers are:

- An unconsolidated gravelly or stony horizon with at least 75% coarse fragments (by weight)
- A continuous, more or less consolidated, calcium carbonate or gypsiferous layer with a minimum thickness of 30 cm, and containing at least 75% calcium carbonate or gypsum (or both together)
- A continuous hard rock or hardpan more than 10 cm thick.

Table 1.5 gives the soil depth ratings used for the suitability classification for irrigation.

1.2.2.3 Calcium Carbonate Status (C)

The presence of free lime in the soil has not only an effect on the structural arrangement of the soil mass, interfering thus directly in water infiltration rate and evaporation processes, but also plays a role in the soil reaction and the physicochemical constitution of the solum as a whole. Thus, the calcium carbonate status influences at the same time the soil-water relationship of the soil and its available nutrient supply for plant growth.

A moderate $CaCO_3$ content has a favorable effect on soil suitability for irrigation. Table 1.6 gives the $CaCO_3$ ratings used in the system. The $CaCO_3$ content of the profile represents the weighted average over the superficial 150 cm.

1.2.2.4 Soil Salinity (D)

The unfavorable effect of salinity hazards depends on soil texture. Ratings are given in Table 1.7. The values for electrical conductivity (Ec) are weighted averages for the upper 150 cm.

1.2.2.5 Drainage (E)

Imperfect or poor drainage is an evident limiting factor. The drainage problems for irrigation are related to soil texture and to the depth and salinity status of the groundwater. Ratings are given in Table 1.8.

TABLE 1.5
Rating of Soil Depth for Irrigation

Soil Depth (cm)	Rating for Surface Irrigation	Rating for Sprinkler Irrigation	Rating for Drip Irrigation
<20	25	30	35
20–50	60	65	70
50–80	80	85	90
80–100	90	95	100
>100	100	100	100

TABLE 1.6
Rating of $CaCO_3$ for Irrigation

$CaCO_3$ (%)	Rating for Surface Irrigation	Rating for Sprinkler Irrigation	Rating for Drip Irrigation
<0.3	90	90	90
0.3–10	95	95	95
10–25	100	100	95
25–50	90	90	80
>50	80	80	70

TABLE 1.7
Rating of Salinity for Irrigation

EC (ds. m⁻¹)	Rating for Surface Irrigation		Rating for Sprinkler Irrigation		Rating for Drip Irrigation	
	C, SiC, SiCL, S, SC Textures[a]	Other Textures	C, SiC, SiCL, S, SC Textures	Other Textures	C, SiC, SiCL, S, SC Textures	Other Textures
<4	100	100	100	100	100	100
4–8	90	95	95	95	95	95
8–16	80	50	85	50	85	50
16–30	70	30	75	35	75	35
>30	60	20	65	25	65	25

[a] C: Clay; SiC: Silty Clay; SiCL: Silty Clay Loam; S: Sand; SC: Sandy Clay.

TABLE 1.8
Rating of Drainage Classes for Irrigation

Drainage Classes	Rating for Surface Irrigation		Rating for Sprinkler Irrigation		Rating for Drip Irrigation	
	C, SiC, SiCL, S, SC Textures[a]	Other Textures	C, SiC, SiCL, S, SC Textures	Other Textures	C, SiC, SiCL, S, SC Textures	Other Textures
Well Drained	100	100	100	100	100	100
Moderately Drained	80	90	90	95	100	100
Imperfectly Drained	70	80	75	85	80	90
Poorly Drained	60	65	65	70	70	80
Very Poorly Drained	40	65	45	65	50	65
Drainage Status Not Known	70	80	70	80	70	80

[a] C: Clay; SiC: Silty Clay; SiCL: Silty Clay Loam; S: Sand; SC: Sandy Clay.

TABLE 1.9
Rating of Slope for Irrigation

Slope Classes (%)	Rating for Surface Irrigation		Rating for Sprinkler Irrigation		Rating for Drip Irrigation	
	Non-terraced	Terraced	Non-terraced	Terraced	Non-terraced	Terraced
0–1	100	100	100	100	100	100
1–3	95	95	100	100	100	100
3–5	90	95	95	100	100	100
5–8	80	90	85	95	90	100
8–16	70	80	75	85	80	90
16–30	50	65	55	70	60	75
>30	30	45	35	50	40	55

1.2.2.6 Slope (F)

The dominant topographic factor that influences the irrigation suitability concerns the slope. Rating the overall slope can be considered as sufficient. It is also estimated that a difference should be made between terraced and nonterraced slopes. Ratings are given in Table 1.9.

TABLE 1.10

Some of Physicochemical Characteristics for Reference Profiles of Soil Series Coded 1

Soil Depth (cm)	Sand (%)	Silt (%)	Clay (%)	CaCO$_3$ (%)	ECe (ds.m^{-1})	Drainage Classes	Slope (%)
0–15	20	64	16	40.50	1.00	Moderately Drained	2–5
15–52	18	55	27	44.30	1.30	Moderately Drained	2–5
52–150	15	54	31	43.60	1.50	Moderately Drained	2–5

Example

Calculating the Ci and land suitability classes of different irrigation methods based on the parametric evaluation in Dasht Bozorg Plain, Soil series coded 1.

Some of the physicochemical characteristics for reference profiles of soil series coded 1 are shown in Table 1.10.

1. The textural rating of the profile calculates as follows:

First section (0–25 cm)	$15 \times 2 \times 20$	= 600
	$10 \times 2 \times 18$	= 360
Second section (25–50 cm)	$25 \times 1.5 \times 18$	= 675
Third section (50–75 cm)	$2 \times 1 \times 18$	= 36
	$23 \times 1 \times 15$	= 345
Fourth section (75–100 cm)	$25 \times 0.75 \times 15$	= 281.25
Fifth section (100–125 cm)	$25 \times 0.5 \times 15$	= 187.50
Sixth section (125–150 cm)	$25 \times 0.25 \times 15$	= 93.75
	Sum	= 2,578.50

Use six sections of 25 cm with weighting factors: 2.00, 1.50, 1.00, 0.75, 0.50, and 0.25.
The amount of sand particle in the profile calculates as follows:
The amount of sand particle in the profile: (2,578.5)/(150) = 17/19%
The amount of silt particle in the profile calculates as follows:

First section (0–25 cm)	$15 \times 2 \times 64$	= 1,920
	$10 \times 2 \times 55$	= 1100
Second section (25–50 cm)	$25 \times 1.5 \times 55$	= 2,062.50
Third section (50–75 cm)	$2 \times 1 \times 55$	= 110
	$23 \times 1 \times 54$	= 1,242
Fourth section (75–100 cm)	$25 \times 0.75 \times 54$	= 1,012.50
Fifth section (100–125 cm)	$25 \times 0.5 \times 54$	= 675
Sixth section (125–150 cm)	$25 \times 0.25 \times 54$	= 337.50
	Sum	= 8,459.50

The amount of silt particle in the profile: (8,459.5)/(150) = 56.39%
The amount of sand particle and silt particle in the profile are 17/19% and 56.39%, respectively; therefore, by referring to the triangular soil texture, the soil texture of profile is **silty loam**.
The rating of silty loam textures for different irrigation systems are as follows (Refer to Table 1.4):

Soil Texture	Rating for Surface Irrigation	Rating for Sprinkler Irrigation	Rating for Drip Irrigation
Silty Loam	100	100	100

2. The soil depth ratings for different irrigation systems are as follows (Refer to Table 1.5):

Soil Depth (cm)	Rating for Surface Irrigation	Rating for Sprinkler Irrigation	Rating for Drip Irrigation
150	100	100	100

3. The $CaCO_3$ (%) in the profile calculates as follows:

First section (0–25 cm)	$15 \times 2 \times 40.50$	= 1215
	$10 \times 2 \times 44.30$	= 886
Second section (25–50 cm)	$25 \times 1.5 \times 44.30$	= 1661.25
Third section (50–75 cm)	$2 \times 1 \times 44.30$	= 88.60
	$23 \times 1 \times 43.60$	= 1002.80
Fourth section (75–100 cm)	$25 \times 0.75 \times 43.60$	= 817.50
Fifth section (100–125 cm)	$25 \times 0.5 \times 43.60$	= 545
Sixth section (125–150 cm)	$25 \times 0.25 \times 43.60$	= 272.50
	Sum	= 6488.65

The $CaCO_3$ (%) in the profile: (6488.65)/(150) = 43.25%
The ratings of $CaCO_3$ (%) for different irrigation systems are as follows (Refer to Table 1.6):

$CaCO_3$ (%)	Rating for Surface Irrigation	Rating for Sprinkler Irrigation	Rating for Drip Irrigation
43.25	90	90	80

4. The electrical conductivity (ds.m^{-1}) in the profile calculates as follows:

First section (0–25 cm)	$15 \times 2 \times 1.00$	= 30
	$10 \times 2 \times 1.30$	= 26
Second section (25–50 cm)	$25 \times 1.5 \times 1.30$	= 48.75
Third section (50–75 cm)	$2 \times 1 \times 1.30$	= 2.60
	$23 \times 1 \times 1.50$	= 34.50
Fourth section (75–100 cm)	$25 \times 0.75 \times 1.50$	= 28.12
Fifth section (100–125 cm)	$25 \times 0.5 \times 1.50$	= 18.75
Sixth section (125–150 cm)	$25 \times 0.25 \times 1.50$	= 9.37
	Sum	= 198.10

The electrical conductivity (ds.m^{-1}) in the profile: (198.10)/(150) = 1.32%
The ratings of electrical conductivity (ds.m^{-1}) for different irrigation systems are as follows (Refer to Table 1.7):

ECe (ds.m^{-1})	Rating for Surface Irrigation	Rating for Sprinkler Irrigation	Rating for Drip Irrigation
1.32	100	100	100

5. The ratings of drainage classes for different irrigation systems are as follows (Refer to Table 1.8):

Drainage Classes	Rating for Surface Irrigation	Rating for Sprinkler Irrigation	Rating for Drip Irrigation
Moderately Drained	90	95	100

6. The ratings of slope for different irrigation systems are as follows (Refer to Table 1.9):

Slope (%)	Rating for Surface Irrigation	Rating for Sprinkler Irrigation	Rating for Drip Irrigation
2–5	97.50	97.50	100

7. Capability index for surface, sprinkler, and drip irrigation systems is calculated according to the formula:

$$Ci = A \times \frac{B}{100} \times \frac{C}{100} \times \frac{D}{100} \times \frac{E}{100} \times \frac{F}{100}$$

7.1. The capability index for surface irrigation system is:

$$Ci = 100 \times (100/100) \times (90/100) \times (100/100) \times (90/100) \times (97.5/100) = 74.92$$

The suitability class for surface irrigation system is (Refer to Table 1.2): S_2
The subclasses for surface irrigation system are:
s = Limitations due to CaCO$_3$ (%) content
w = Drainage limitations.
Finally, the land suitability class for surface irrigation system is: S_{2sw}

7.2. The capability index for sprinkler irrigation system is:

$$Ci = 100 \times (100/100) \times (90/100) \times (100/100) \times (95/100) \times (97.5/100) = 83.36$$

The suitability class for sprinkler irrigation system is (Refer to Table 1.2): S_1

7.3. The capability index for drip irrigation system is:

$$Ci = 100 \times (100/100) \times (80/100) \times (100/100) \times (100/100) \times (100/100) = 80.00$$

The suitability class for drip irrigation system is (Refer to Table 1.2): S_1

1.3   LAND SUITABILITY MAPS

In order to develop land suitability maps for different irrigation methods, a semi-detailed soil map (for all plains) was used, and all the data for soil characteristics were analyzed and incorporated into the map using ArcGIS 9.2 software.

The digital soil map base preparation was the first step toward the presentation of a GIS module for land suitability maps for different irrigation systems (Albaji et al. 2010, 2014). The soil map was then digitized and a database prepared. The total different polygons or soil series were determined in the base map. Soil characteristics were also given for each soil series. These values were used to generate the land suitability maps for surface, sprinkle, and drip irrigation systems using geographic information systems (Albaji et al. 2008; Landi et al. 2008; Rezania et al. 2009; Jovzi et al. 2012).

Suitability classes are defined by considering the value of the capability indices and presented in Table 1.2. Each of the land and soil characteristics with associated attribute data are digitally encoded in a GIS database to eventually generate six thematic layers.

Data are digitally encoded in a GIS database to eventually generate six thematic layers. The diagnostic factors of each thematic layer were assigned values of factor rating identified in Tables 1.3 through 1.8. The parametric model is defined using the value of factor rating as formula (1.1). These six layers were then spatially overlaid to produce resultant layers. In Figure 1.1, a schematic chart of GIS application for land suitability map for different irrigation methods is shown.

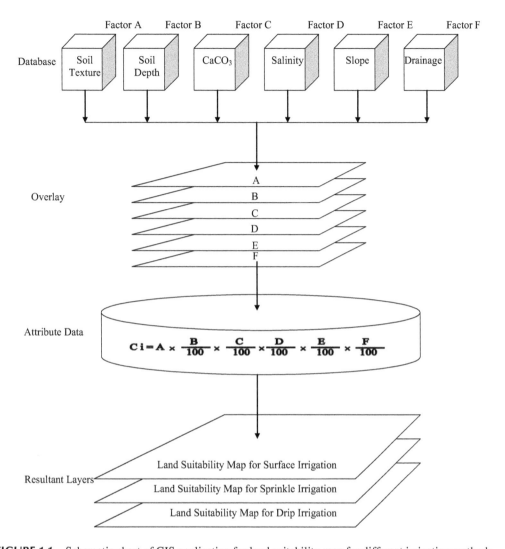

FIGURE 1.1   Schematic chart of GIS application for land suitability map for different irrigation methods.

1.4   CONCLUSIONS

The parametric approach in the evaluation of land characteristics consists of a numeral rating of the different limitation levels of the land characteristics in a numerical scale from a maximum (normally 100) to a minimum value. If a land characteristic is optimal for the considered land utilization type, the maximum rating of 100 is attributed; if the same land characteristic is unfavorable, a minimal rating is applied. The successful application of the system implies the respect of the following rules: (a) The number of land characteristics to consider has to be reduced to a strict minimum to avoid repetition of related characteristics in the related formula; (b) An important characteristic is rated in a wide scale (100–25), a less important characteristic in a narrower scale (100–60); (c) The rating of 100 is applied for optimal development or maximum appearance of a characteristic; (d) The depth to which the land index has to be calculated must be defined for each land utilization type; and (e) The depth to be considered should coincide with the normal depth of the root system in a deep soil.

In the proposed system, the factors affecting soil suitability for irrigation purposes can be subdivided into four groups: physical properties, chemical properties, drainage properties, and environmental properties. The soil drainage, depth, slope, texture, electric conductivity, and $CaCO_3$ content were the considered parameters. The different land characteristics that influence the soil suitability for irrigation are rated, and a capability index for irrigation (Ci) is calculated according to the formula.

In order to develop land suitability maps for different irrigation methods, a semi-detailed soil map was used, and all the data for soil characteristics were analyzed and incorporated into the map using ArcGIS 9.2 software. The soil map was then digitized and a database prepared. The total different polygons or soil series were determined in the base map. Soil characteristics were also given for each soil series. These values were used to generate the land suitability maps for surface, sprinkle, and drip irrigation systems using geographic information systems.

ACKNOWLEDGMENTS

We are grateful to the Research Council of Shahid Chamran University of Ahvaz for financial support (GN:SCU.WI98.280).

REFERENCES

Albaji, M. 2010. *Land Suitability Evaluation for Sprinkler Irrigation Systems*. Khuzestan Water and Power Authority (KWPA), Ahvaz, Iran (in Persian).
Albaji, M., Boroomand Nasab, S., Kashkuli, H.A., Naseri, A.A., Sayyad, G., Jafari, S. 2008. Comparison of different irrigation methods based on the parametric evaluation approach in North Molasani Plain, Iran. *Journal of Agronomy* 7 (2), 187–191.
Albaji, M., Boroomand Nasab, S., Kashkoli, H.A., Naseri A. 2010. Comparison of different irrigation methods based on the parametric evaluation approach in the plain west of Shush, Iran. *Irrigation and Drainage* 59 (5), 547–558.
Albaji, M., Golabi, M., Egdernejad, A., Nazarizadeh, F. 2014. Assessment of different irrigation systems in Albaji Plain. *Water Science and Technology: Water Supply* 14 (5), 778–786.
Albaji, M., Papan, P., Hosseinzadeh, M., Barani, S. 2012. Evaluation of land suitability for principal crops in the Hendijan region. International *Journal of Modern Agriculture* 1 (1), 24–32.
Dengiz, O. 2006. A comparison of different irrigation methods based on the parametric evaluation approach. *Turkish Journal of Agriculture Forestry* 30, 21–29.
FAO (Food Agriculture Organization of the United Nations). 1976. *A Framework for Land Evaluation*. Soil Bulletin No. 32. FAO, Rome, Italy, 72pp.
Hargreaves, H.G., Mekley, G.P. 1998. *Irrigation Fundamentals*. Water Resource Publication, LLC, 200p.
IAO (Istituto Agronomico Per L'Oltremare). 2002. Land evaluation in the province of larache, morocco. *22nd Course Professional Master. Geometric and Natural Resources Evaluation*. November 12, 2001–June 21, 2002. In: Istituto Agronomico per l'Oltremare, IAO, Florence, Italy. 17, 29–48.
IAO (Istituto Agronomico Per L'Oltremare). 2007. Land evaluation in the Essaouira province, Morocco. *27th Course Professional Master. Geometric and Natural Resources Evaluation*. November 6, 2006–June 22, 2007. In: Istituto Agronomico per l'Oltremare, IAO, Florence, Italy.
Jovzi, M., Albaji, M., Gharibzadeh, A. 2012. Investigating the suitability of lands for surface and under-pressure (drip and sprinkler) irrigation in Miheh Plain. *Research Journal of Environmental Sciences* 6 (2), 51–61.
Landi, A., Boroomand-Nasab, S., Behzad, M., Tondrow, M.R., Albaji, M., Jazaieri, A. 2008. Land suitability evaluation for surface, sprinkle and drip irrigation methods in Fakkeh Plain, Iran. *Journal of Applied Sciences* 8 (20), 3646–3653.
Liu, W., Qin, Y., Vital, L. 2006. Land evaluation in Danling county, Sichuan province, China. *26th Course Professional Master. Geometric and Natural Resources Evaluation*. November 7, 2005–June 23, 2006, IAO, Florence, Italy, 26, 33–64.
Naseri, A.A., Albaji, M., Boroomand Nasab, S., Landi, A., Papan, P., Bavi, A. 2009. Land suitability evaluation for principal crops in the Abbas Plain, Southwest Iran. *Journal of Food, Agriculture & Environment* 7 (1), 208–213.
Rezania, A.R., Naseri, A.A., Albaji, M. 2009. Assessment of soil properties for irrigation methods in North Andimeshk Plain, Iran. *Journal of Food, Agriculture & Environment* 7 (3&4), 728–733.
Sys, C., Van Ranst, E., Debaveye, J. 1991a. *Land Evaluation*, Part II, Methods in Land Evaluation. International Training Centre for Post-graduate Soil Scientists, Ghent University, Belgium, 247pp.
Sys, C., Van Ranst, E., Debaveye, J. 1991b. *Land Evaluation*, Part I, Principles in Land Evaluation and Crop Production Calculations. International Training Centre for Post-graduate Soil Scientists, Ghent University, Belgium, 265pp.

2 Using Analytical Hierarchy Process Integrated with GIS in an Arid Region for Site Selection of Different Irrigation Systems

Lamya Neissi, Saeed Boroomand Nasab, and Mohammad Albaji

CONTENTS

2.1 INTRODUCTION

Multi-criteria decision-making combines technically feasible, economically viable, socially acceptable, and environmental friendly criteria with respect to their importance of suitability. Then these criteria are analyzed by using GIS and treated spatially to generate regions susceptibility maps (Anane et al. 2012).

Multi-criteria decision-making (MCDM) methods are recommended for considering different parameters that can affect the appropriate irrigation system site selection. Integrating MCDM with a geographical information system (GIS) has been used in site selection assessment for proper decision-making to increase the potential of environmental resources in arid and semiarid regions.

Saaty (1997) defined analytical hierarchy process (AHP) as one of the MCDM methods that changes a complex multi-criteria decision problem into a simple hierarchy.

AHP has several advantages including overspecification of judgment, built-in consistency tests, and use of appropriate measurement scales and applicability in elicitation of utility functions (Chen and Huang 2004).

Site selection by using AHP integrated with GIS has been applied for many environmental resources (Assefa et al. 2018; Pan and Xu 2018; Garcia et al. 2014; Akinci et al. 2013; Anane et al. 2012; Chandio et al. 2011; Montazar and Zadbagher 2010; Gilliams et al. 2005). AHP solved a variety of complex environmental problems including land suitability (Feizizadeh and Blaschke 2013; Mendas and Delali 2012; Cengiz and Akbulak 2009), selecting strategies (Abdollahzadeh et al. 2016), ranking different alternatives (Baffoe 2019; Sharma et al. 2018), selecting best crop pattern (Dekamin et al. 2018; Werner et al. 2014), and flood hazard assessment (Phonphoton and Pharino 2019; Seejata et al. 2018). Due to the successful application of AHP in solving environmental and agricultural problems, this method was adopted to evaluate different irrigation systems for the most suitable regions for water-saving and higher irrigation system efficiency in the context of this study.

Maps contain valuable spatial information from a general point of view. In the case of water management for best irrigation system site selection in arid regions, using AHP integrated with GIS can be helpful. AHP assigns appropriate weight to each irrigation system according to considered criteria under drought conditions in arid regions; then GIS maps show the most suitable sites for each irrigation system due to assigned weights calculated by AHP.

The present study aims to select the most suitable sites for different traditional and modernized irrigation systems according to analytical hierarchy process and GIS maps in arid regions. The methodology uses easy-to-get data from the official institution of the Khuzestan Water and Power Authority (KWPA) and available satellite images.

2.2   MATERIALS AND METHOD

2.2.1   CHARACTERIZATION OF THE STUDY AREA

The Izeh Plain is located in Khuzestan Province at the southwestern part of Iran, 49°45′ to 49°59′ E and 31°46′ to 31°57′ N with 11,080.5 km² area (Figure 2.1). The climate is arid with 1,685 mm as annual average evaporation at Izeh (city). The average temperature is 30°C. The Izeh Plain is considered a flat area with a slope that varies from 2% to 5% and loamy soil texture. Agriculture is the basic economic activity in the Izeh Plain. Over much of the Izeh Plain, the use of surface irrigation systems has been applied specifically for field crops to meet the water demand of both summer and winter crops. The major irrigated broad-acre crops grown in this area are wheat, barley, and maize, in addition to fruits, melons, watermelons, and vegetables such as tomatoes and cucumbers. But like other plains in Khuzstan Province, the important major crops in the Izeh Plain are wheat and barley (Albaji and Alboshokeh 2017; Behzad et al. 2009a, 2009b; Naseri et al. 2009; Albaji et al. 2012b).

Surface water used for irrigation is classified in the C3S1 water quality class (Anonymous 2017). Surface irrigation is the most important irrigation in the Izeh Plain, as well as other plains in Khuzstan Province; there are very few instances of sprinkle and drip irrigation on large area farms in this area (Albaji et al. 2008, 2012a, 2014a, 2014b, 2016; Boroomand Nasab et al. 2010; Rezania et al. 2009; Landi et al. 2008). For this research, irrigation systems such as solid set irrigation system, wheel move irrigation system, low-pressure irrigation system, surface irrigation system, and drip irrigation system were considered in selecting the most suitable sites for different irrigation systems in the Izeh Plain.

Two wetlands, Miangaran and Bondoun, are located in the Izeh Plain. The Miangaran wetland is located 1.5 km north of Izeh City with a 2,500 ha area, and the Bondoun wetland is located 3 km south of Izeh City with a 1,300 ha area. These two wetlands have important role in environmental conservation and surface water saving. Drought conditions of the Izeh Plain caused these wetlands to become dry over the years.

The Zagros Mountains are located in the north and south parts of the Izeh Plain. Highest elevation of the plain is 342 m in the mountainous area, which can affect site selection of the best irrigation system.

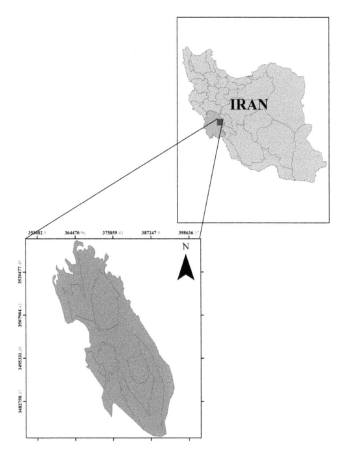

FIGURE 2.1 Location map of the study area.

2.2.2 METHODOLOGY OVERVIEW

To locate the suitable sites for irrigation systems, an analytic hierarchy process (AHP) integrated with GIS was used. The methodology involved the following steps:

- Determine the different main criteria, criteria, and subcriteria.
- Develop a decision hierarchy structure and identify priorities by using a pairwise comparison matrix.
- Check the accuracy of weights by applying a consistency ratio.
- Use GIS for extracting the geographic layers corresponding to each subcriterion.

Socioeconomic and physical criteria were the two main criteria selected to determine suitable sites for each irrigation system.

The proposed criteria and subcriteria indicators include:

1. Socioeconomic: Acceptability of an irrigation system (R_{as}), technical support requirements (T_{sr}), system costs (S_{ec}), and labor skills (L_{ls})
2. Physical: Topography, water, climate, soil, crop; subcriteria of physical criteria are classified as:

Topography: Elevation difference (L_{ad}), land slope (L_{so})
Water: Suspended materials (W_{sm}), sodium concentration (W_{na}), biological materials (W_{bm}), availability of water (W_{aw}), EC, PH

Climate: Climate (C_{re}), wind speed (W_{ws})
Soil: Infiltration rate (I_{ir}), available water in the soil (AW)
Crop: Crop density (C_{cd}), crop type (P_{pk}), and crop pest (P_{pd}).

2.2.3   DATA ANALYSIS

2.2.3.1   Analytical Hierarchy Process and Weighting

In order to evaluate the suitability of a given site, a weight for each subcriterion is assigned. Weighting expresses the criterion degree of relevance or preference relative to the others. The process is achieved through the pairwise comparison between the elements for each hierarchical level (Saaty 1980). The pairwise comparison employed a semantic 9-point scale for the assignment of priority values: 1, 3, 5, 7, and 9 correspond respectively to equally, moderately, strongly, very strongly, and extremely important criterion, when compared with other numbers. The intermediate values are 2, 4, 6, and 8. The assignment of preference values is based on experts consulting and reviewing technical documents and published international guidelines.

Figure 2.2 show the analytical hierarchy process (AHP) structure for selecting suitable irrigation systems.

2.2.3.2   Consistency Ratio (CR)

Decision consistency is very important in decision-making. Controlling decision consistency is always amenable to computation and evaluation. For each matrix, the quotient of consistency index to inconsistency index of a stochastic matrix of the same vector is taken as the criterion to judge the decision inconsistency, i.e., the consistency ratio. In cases where the ratio is less than 0.1, the system has an acceptable consistency (Saaty 1980). Otherwise, judgments must be repeated and a new comparison matrix is solicited. In the present study, evaluation of decision consistency was performed for each of the matrices developed based on research by Saaty (1980).

Each matrix consistency is checked out through the calculation of consistency ratio (CR), which is defined as the quotient between the consistency index (CI) and the random index (RI) as Equation (2.1):

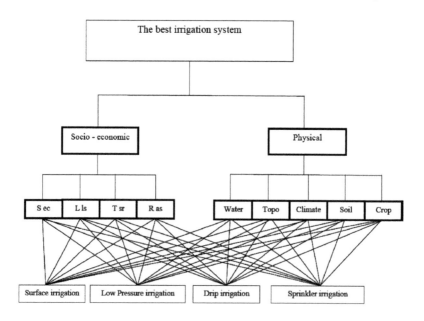

FIGURE 2.2   AHP structure for selecting suitable irrigation systems.

TABLE 2.1
Random Index (R.I)

R.I	1	2	3	4	5	6	7	8	9	10	11	12	13	14	15
n	0.00	0.00	0.58	0.90	1.12	1.24	1.32	1.41	1.45	1.49	1.51	1.48	1.56	1.57	1.59

$$CR = (CI/RI) \tag{2.1}$$

The consistency index (CI) is determined using the following Equation (2.2):

$$CI = (\lambda_{max} - n)/(n - 1) \tag{2.2}$$

where n is the criteria number and λ_{max} is the maximum value of Eigenvector.

The random index (RI) is obtained from a Table 2.1 established by Oak Ridge National Laboratory (Saaty 1980). The main idea of using RI is that the CR is a normalized value, since it is divided by an arithmetic mean of random matrix consistency indices (Alonso and Lamata 2006).

Consistency ratio less than 0.1 is considered acceptable. If the CR was higher than 0.1, the generated weights have to be revised until the CR is less than 0.1.

2.2.3.3 Subcriterion and Constraints Layering by GIS

Identifying suitable regions for irrigation systems starts with representing each selected subcriterion by a thematic layer. In order to map all the considered criteria, data are gathered from satellite images and official sources for maps of DEM, slope, climate, etc. Afterwards, the gathered data are analyzed and treated using GIS and geostatistical tools. Each layer is obtained in the raster data model. Spatial data on water characteristics, topography, and climate are obtained from the "water and power authority" of the Khuzestan district, which is the Iranian official source of an agricultural spatial database (Anonymous 2017).

2.3 RESULTS AND DISCUSSION

2.3.1 CALCULATED WEIGHTS BY ANALYTICAL HIERARCHY PROCESS (AHP)

In order to select the most suitable sites for different irrigation systems, the AHP method was combined with GIS. Due to Saaty's scale for comparative judgments by using pairwise comparison of different criteria, weight of each criterion was calculated. Table 2.2 shows the results of computations and weights of the considered irrigation systems in this research.

According to Table 2.2, the cost subcriterion of the socioeconomic criterion has the highest weight compared with other socioeconomic subcriteria. The cost of emitters and the necessary equipment for installing and maintaining a drip irrigation system make this subcriterion more influenced by the cost criterion. Farm land leveling operations could be costly for low-pressure and surface irrigation systems, but most agricultural lands of the Izeh Plain consist of flat lands. Low-pressure and surface irrigation system alternatives could cost less when implementing operations in flat areas, which can reduce the final weight of the cost criterion comparing with other irrigation system alternatives.

Skilled labor and acceptability of irrigation system criteria have higher weight for solid set irrigation systems because modern and automated systems have not been installed in this plain. Due to drought conditions, modern water-saving irrigation systems are recommended for the study area.

The physical main criterion could influence the site selection more than the socioeconomic criterion due to climate conditions of the study area.

TABLE 2.2
The Results of Computations and Weights of Irrigation Systems

Solid Set	Low Pressure	Wheel Move	Surface	Drip	Subcriterion		Criterion
0.0182	0.0282	0.0420	0.0258	0.0313	Tsr		Socioeconomic
0.0166	0.0399	0.0425	0.0365	0.0313	Lls		
0.0960	0.0441	0.0425	0.0403	0.1563	cost		
0.0390	0.0255	0.0425	0.0233	0.0313	Ras		
0.1523	0.0827	0.1015	0.1888	0.0375	L ad	Topography	Physical
0.0508	0.2481	0.1015	0.2832	0.0375	Slope		
0.0508	0.1034	0.0508	0.1368	0.0563	C re	Climate	
0.1523	0.0345	0.1523	0.0456	0.0188	W ws		
0.0194	0.0197	0.0194	0.0077	0.0235	Wna	Water	
0.0073	0.0197	0.0073	0.0188	0.0123	Waw		
0.0073	0.0197	0.0073	0.0114	0.1036	Wsm		
0.0055	0.0197	0.0055	0.0114	0.1036	Wbm		
0.0194	0.0197	0.0194	0.0094	0.1036	EC		
0.0194	0.0197	0.0194	0.0094	0.1036	PH		
0.1015	0.0345	0.1015	0.0126	0.0188	I ir	Soil	
0.1015	0.1034	0.1015	0.0631	0.0563	AW		
0.0356	0.0345	0.0356	0.0189	0.0250	C cd	Crop	
0.0224	0.0345	0.0224	0.0189	0.0250	P pk		
0.0847	0.0689	0.0847	0.0379	0.0251	P pd		

Different irrigation systems can operate appropriately in special physical conditions. Site selection could find the most suitable place for each irrigation system. Physical conditions of the Izeh Plain differ spatially.

Calculating the weights and matrix of each criterion and subcriterion was done for site selection of the appropriate irrigation system.

The final step is to calculate the consistency ratio (CR). For this set of judgment using the CI for the corresponding value from large samples of matrices of purely random judgments using Table 2.1, the upper row is the order of the random matrix, and the lower is the corresponding index of consistency for random judgments (Coyle 2004). Consistency ratios were less than 0.1, and the pairwise comparison matrix and the weights are acceptable.

2.3.2 SPATIAL DISTRIBUTION OF THE CALCULATED WEIGHTS

Spatial distribution for the most appropriate irrigation systems in each site can be achieved by using GIS. Different regions of the study area related to a different weight, which was calculated by AHP, and then the weights were classified into three different classes. Class 4 is related to weak quality for the considered irrigation system; 5 and 6 are related to moderate and good quality, respectively.

Figures 2.3 through 2.7 show site selection for different irrigation systems. Wetland, urban, and mountain were shown in the maps and these parts weren't considered for evaluation of the irrigation systems.

Figure 2.3 shows the site selection of drip irrigation system.

When the physical main criterion was considered, according to Figure 2.3, most of the mountainous area of the Izeh Plain was classified as high value and acceptable for implementing a drip irrigation system. Socioeconomic conditions of the study area are important for drip irrigation systems; therefore, all regions around the urban area are classified as high value for drip irrigation systems.

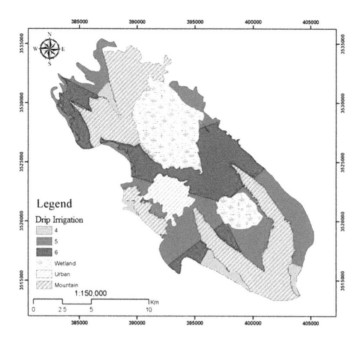

FIGURE 2.3   Site selection of drip irrigation system.

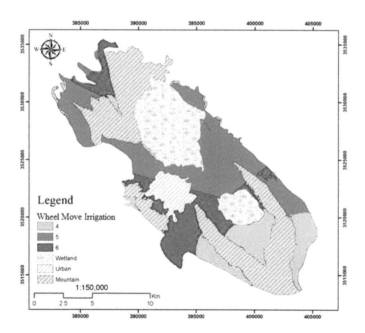

FIGURE 2.4   Site selection of wheel move irrigation system.

Small mountainous areas with high slope are considered in the low class due to the far distance from Izeh (city) for equipment support.

Figure 2.4 shows the site selection of a wheel move irrigation system.

Physical conditions of the study area can affect wheel move irrigation system efficiency. Due to site selection of the most appropriate area for a wheel move irrigation system, mountainous areas are classified as low value for this irrigation system. Also, by considering socioeconomic conditions

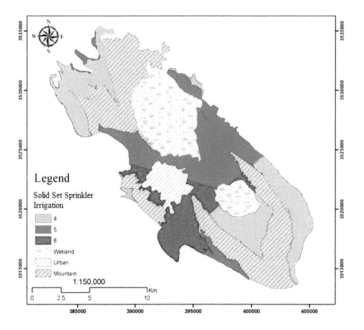

FIGURE 2.5   Site selection of solid set irrigation system.

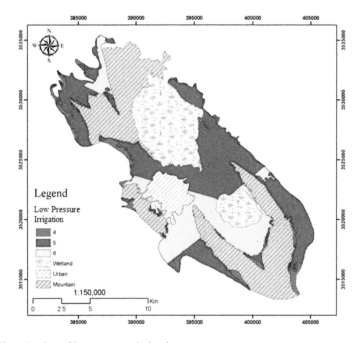

FIGURE 2.6   Site selection of low-pressure irrigation system.

especially equipment support for a wheel move irrigation system, far distance regions from the city are classified as low-value class and regions around the city classified as high value.

Wetland regions are classified by less value compared with urban regions; this makes it clear that physical conditions of the study area have a higher effect on site selection compared with socioeconomic conditions.

Figure 2.5 shows site selection of a solid set irrigation system.

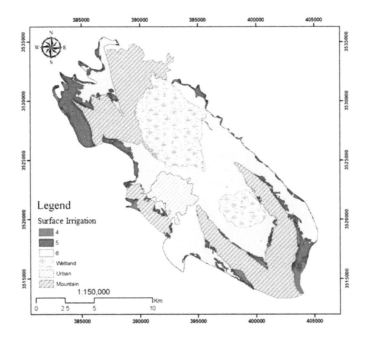

FIGURE 2.7 Site selection of surface irrigation system.

Selection of a solid set irrigation system is influenced by physical conditions because all- around mountainous regions are classified as low value. The southern part of the Izeh Plain, between the mountains and wetlands, is classified as low value due to critical physical (slope and topography) conditions. Regions around Izeh (city) have acceptable socioeconomic conditions for implementing solid set irrigation systems and are classified as high value.

Figure 2.6 shows the site selection of a low pressure irrigation system.

The southern part of the Izeh Plain between the wetland and mountain areas is classified as high value for implementing low-pressure irrigation system. This is proof that this irrigation system is not affected by physical conditions of the Izeh Plain. Regions around the city are classified as high value, too, which show that this irrigation system is less affected by socioeconomic conditions. Low-pressure irrigation systems convey water by pipe, so the regions far from water reservoirs are classified as less valuable.

Figure 2.7 shows the site selection of a surface irrigation system.

A surface irrigation system is selected as the most appropriate irrigation system in the study area especially around the urban and wetland areas. In the regions around the mountains, due to physical criteria of the study area, surface irrigation systems are categorized in class 4, which reveals that these regions are less suitable for implementing surface irrigation systems. In case of appropriate water management in a surface irrigation system at the arid regions, a surface irrigation system can extend to high water use efficiency. But irrigation system management was not considered as a criterion because modern irrigation systems were not installed and implemented in the study area, and the experts did not have any background in the modernized irrigation system management. This could be a disadvantage of the AHP method; when there was no background for an alternative, the experts could not value the criteria for the considered alternative. Table 2.3 represents the percentage of covering area for each irrigation system. Table 2.3 reveals that surface irrigation systems and low-pressure irrigation system have the highest percentage of suitable site in classes 6 and 5, respectively. Due to arid conditions of the study area, drip irrigation systems had the highest percentage in classes 5 and 6 and can be selected as the most suitable modernized irrigation system.

TABLE 2.3
Final Results of Irrigation Systems

Irrigation Method Value	4	5	6
Drip	3.45	52.26	44.29
Wheel move	18.32	59.1	22.58
Solid set	43.52	34.54	21.94
Surface	0.53	8.36	91.11
Low-pressure	1.36	84.15	14.49

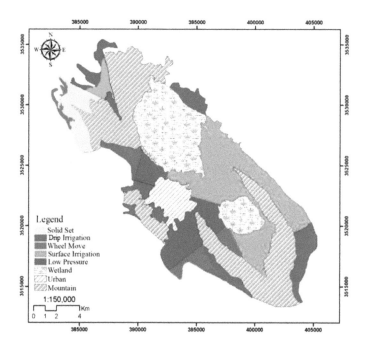

FIGURE 2.8   Result map of suitable sites for different irrigation systems.

Figure 2.8 shows the result map of suitable sites for different irrigation systems after integrating the calculated weights of considered criteria by using GIS.

Figure 2.8 shows that regions around the mountains and city are suitable for drip irrigation systems. Most flat areas and around wetlands are suitable for surface irrigation systems and low-pressure irrigation systems because even in drought conditions there is enough water for operating surface irrigation systems. Other considered irrigation systems allocated less suitability areas, which could be replaced with drip irrigation systems. A drip irrigation system is the most suitable modern system for water savings, but it is better to operate common irrigation systems of the study area (surface irrigation system and low-pressure irrigation system) for economic justification.

2.4   CONCLUSIONS

Different climate conditions can affect water consumption of irrigation systems. Changing water consumption patterns influence irrigation system efficiency. By considering climate conditions, the most appropriate irrigation system must be installed.

Multi-criteria decision-making can consider different factors that can have an effect on site selection for the most appropriate irrigation system in a specified region. By using this approach, two main criteria were considered for the study area: physical and socioeconomic criteria. A single-objective AHP integrated with a GIS was carried out to identify suitable sites for irrigation systems in the Izeh Plain.

Evaluation of suitable sites for different irrigation systems, using AHP integrated into a GIS, reveals that surface irrigation systems, drip irrigation systems, and low-pressure irrigation systems are the most suitable irrigation systems due to the results of site selection.

ACKNOWLEDGMENTS

We are grateful to the Research Council of Shahid Chamran University of Ahvaz for financial support (GN: SCU.WI98.280).

REFERENCES

Abdollahzadeh, G., Damalas, C.A., Sharifzadeh, M.S., Ahmadi-Gorgi, H. 2016. Selecting strategies for rice stem borer management using the analytic hierarchy process (AHP). *Crop Protection* 84, 27–36.

Akinci, H., Ozalp, A.Y., Turgut, B. 2013. Agricultural land use suitability analysis using GIS and AHP technique. *Computers and Electronics in Agriculture* 97, 71–82.

Albaji, M., Alboshokeh, A. 2017. Assessing agricultural land suitability in the Fakkeh region, Iran. *Outlook on Agriculture* 46 (1), 57–65.

Albaji, M., Boroomand Nasab, S., Kashkuli, H.A., Naseri, A.A., Sayyad, G., Jafari, S. 2008. Comparison of different irrigation methods based on the parametric evaluation approach in North Molasani Plain, Iran. *Journal of Agronomy* 7 (2), 187–191.

Albaji, M., Boroomand Nasab, S., Hemadi, J. 2012a. Comparison of different irrigation methods based on the parametric evaluation approach in West North Ahvaz Plain. In: *Problems, Perspectives and Challenges of Agricultural Water Management*, M. Kumar, (ed.), InTech, Croatia, pp. 259–274.

Albaji, M., Papan, P., Hosseinzadeh, M., Barani, S. 2012b. Evaluation of land suitability for principal crops in the Hendijan region. *International Journal of Modern Agriculture* 1 (1), 24–32.

Albaji, M., Golabi, M., Boroomand Nasab, S., Jahanshahi. M. 2014a. Land suitability evaluation for surface, sprinkler and drip irrigation systems. *Transactions of the Royal Society of South Africa* 69 (2), 63–73.

Albaji, M., Golabi, M., Piroozfar, V.R., Egdernejad, A., Nazari Zadeh, F. 2014b. Evaluation of agricultural land resources for irrigation in the Ramhormoz plain by using GIS. *Agriculturae Conspectus Scientificus* 79 (2), 93–102.

Albaji, M., Golabi, M., Hooshmand, A.R., Ahmadee, M. 2016. Investigation of surface, sprinkler and drip irrigation methods using GIS. *Jordan Journal of Agricultural Sciences* 12 (1), 211–222.

Alonso, J.A., Lamata, M.T. 2006. Consistency in the analytic hierarchy process: A new approach. *International Journal of Uncertainty, Fuzziness and Knowledge-Based Systems* 14 (4), 445–459.

Anane, M., Bouziri, L., Limam, A., Jellali, S. 2012. Ranking suitable sites for irrigation with reclaimed water in the nabeul-hammamet region (Tunisia) using GIS and AHP-multicriteria decision analysis. *Resources, Conservation and Recycling* 65, 36–46.

Anonymous. 2017. Meteorology Report of Izeh Plain, Iran. Khuzestan Water and Power Authority (KWPA). http://www.kwpa.com. (in Persian).

Assefa, T., Jha, M., Reyes, M., Srinivasan, R., Worqlul, A.W. 2018. Assessment of suitable areas for home gardens for irrigation potential, water availability, and water-lifting technologies. *Water Journal* 10 (4), 495.

Baffoe, G. 2019. Exploring the utility of analytic hierarchy process (AHP) in ranking livelihood activities for effective and sustainable rural development interventions in developing countries. *Evaluation and Program Planning* 72, 197–204.

Behzad, M., Albaji, M., Papan, P., Boroomand Nasab, S., Naseri, A.A., Bavi, A. 2009a. Qualitative evaluation of land suitability for principal crops in the Gargar Region, Khuzestan Province, Southwest Iran. *Asian Journal of Plant Sciences* 8 (1), 28.

Behzad, M., Albaji, M., Papan, P., Boroomand Nasab, S. 2009b. Evan region qualitative soil evaluation for wheat, barley, alfalfa and maize. *Journal of Food, Agriculture & Environment* 7 (2), 843–851.

Boroomand Nasab, S., Albaji, M., Naseri, A.A. 2010. Investigation of different irrigation systems based on the parametric evaluation approach in Boneh Basht plain, Iran. *African Journal of Agricultural Research* 5 (5), 372–379.

Cengiz, T., Akbulak, C. 2009. Application of analytical hierarchy process and geographic information systems in land-use suitability evaluation: A case study of Dumrek village. *International Journal of Sustainable Development & World Ecology* 16 (4), 286–294.

Chandio, I.A., Matori, A.N., Lawal, D.U., Sabri, S. 2011. GIS-based land suitability analysis using AHP for public parks planning in Larkana City. *Modern Applied Science* 5 (4), 177–189.

Chen, C., Huang, C. 2004. A multiple criteria evaluation of high tech industries for the science-based industrial parks in Taiwan. *Information and Management* 41, 839–851.

Coyle, G. 2004. *The Analytic Hierarchy Process (AHP). Practical Strategy. Open Access Material.* Glasgow: Pearson Education Ltd.

Dekamin, M., Barmaki, M., Kanooni, A. 2018. Selecting the best environmental friendly oilseed crop by using life cycle assessment, water footprint and analytic hierarchy process methods. *Journal of Cleaner Production* 198, 1239–1250.

Feizizadeh, B., Blaschke, T. 2013. Land suitability analysis for Tabriz County, Iran: A multi criteria evaluation approach using GIS. *Journal of Environmental Planning and Management* 56 (1), 1–23.

Garcia, J.L., Alvarado, A., Blanco, J., Jimenez, E., Maldonado, A.A., Corte, G. 2014. Multi-attribute evaluation and selection of sites for agricultural product warehouses based on an analytic hierarchy process. *Computers and Electronics in Agriculture* 100, 60–69.

Gilliams, S., Raymaekers, D., Muys, B., Van Orshoven, J. 2005. Comparing multiple criteria decision methods to extend a geographical information system on afforestation. *Computers and Electronics in Agriculture* 49, 142–158.

Landi, A., Boroomand-Nasab, S., Behzad, M., Tondrow, M.R., Albaji, M., Jazaieri, A. 2008. Land suitability evaluation for surface, sprinkle and drip irrigation methods in Fakkeh Plain, Iran. *Journal of Applied Sciences* 8 (20), 3646–3653.

Mendas, A., Delali, A. 2012. Integration of multi-criteria decision analysis in GIS to develop land suitability for agriculture: Application to durum wheat cultivation in the region of Mleta in Algeria. *Computers and Electronics in Agriculture* 83, 117–126.

Montazar, A., Zadbagher, E. 2010. An analytical hierarchy model for assessing global water productivity of irrigation networks in Iran. *Water Resource Management* 24, 2817–2832.

Naseri, A.A., Albaji, M., Boroomand Nasab, S., Landi, A., Papan, P., Bavi, A. 2009. Land suitability evaluation for principal crops in the Abbas Plain, Southwest Iran. *Journal of Food, Agriculture & Environment* 7 (1), 208–213.

Pan, H., Xu, Q. 2018. Quantitative analysis on the influence factors of the sustainable water resource management performance in irrigation areas: An empirical research from China. *International Journal of Sustainable Future for Human Security* 10(1), 264.

Phonphoton, N., Pharino, C. 2019. Multi-criteria decision analysis to mitigate the impact of municipal solid waste management services during floods. *Resources, Conservation and Recycling* 146, 106–113.

Rezania, A.R., Naseri, A.A., Albaji, M. 2009. Assessment of soil properties for irrigation methods in North Andimeshk Plain, Iran. *Journal of Food, Agriculture & Environment* 7 (3&4), 728–733.

Saaty, T.L. 1977. A scaling method for priorities in hierarchical structure. *Journal of Mathematical Psychology* 15, 228–234.

Saaty, T.L. 1980. *The Analytic Hierarchy Process.* McGraw Hill, New York.

Seejata, K., Yodying, A., Wongthadam, T., Mahavik, N., Tantanee, S. 2018. Assessment of flood hazard areas using analytical hierarchy process over the lower Yom Basin, Sukhothai Province. *Procedia Engineering* 212, 340–347.

Sharma, Y.K., Yadav, A.K., Mangla, S.K., Patil, P.P. 2018. Ranking the success factors to improve safety and security in sustainable food supply chain management using fuzzy AHP. *Materials Today: Proceedings* 5(5), 12187–12196.

Werner, A., Werner, A., Wieland, R., Kersebaum, K.C., Wiggering, H., 2014. Ex ante assessment of crop rotations focusing on energy crops using a multi-attribute decision-making method. *Ecological Indicators* 45, 110–122.

Section II

Strategies for Irrigation
Development and Management

3 Crop Yield Response to Partial Root Drying Compared with Regulated Deficit Irrigation

*Hamid Zare Abyaneh, Mehdi Jovzi, Niaz Ali
Ebrahimi Pak, and Mohammad Albaji*

CONTENTS

3.1 INTRODUCTION

In many agroecosystems worldwide, water availability for agriculture is anticipated to decline as a consequence of global climate change, environmental pollution, and growing demand for other uses (Mossad et al., 2018). Moreover, the scarcity of water in arid and semiarid regions such as Iran is a restrictive element for the agricultural sector. Hence, water scarcity is emerging as the limiting factor for sustainable agricultural production; therefore, there is an urgent need to develop initiatives to save water in this particular sector. Among the proposed solutions is the use of deficit irrigation. Deficit irrigation is a method that irrigates the entire root zone with an amount of water less than that of the potential evapotranspiration with induced minor stress, which has minimal effects on the yield (English and Raja, 1996). Deficit irrigation strategies help to control excessive vegetative growth and thus bring about substantial water savings (Fernández et al., 2013; Hernandez-Santana et al., 2017; Padilla-Díaz et al., 2016).

N-fertilizer is considered a limiting factor in the obtaining of high yield and quality. An adequate supply of nitrogen is essential for optimum yield (Kiymaz and Ertek, 2015a). Furthermore, N-fertilizer is among the high-ranking chemical products and has become very expensive. Thus, the sustainable use of water and N-fertilizer is becoming a priority for agriculture, particularly in water-scarce regions (Wang et al., 2013).

Sugar beet is a plant adapted to deficit irrigation (FAO, 2002). The cultivated area under cultivation of sugar beet is estimated at about 4.35 million hectares globally with an average yield of 56,912 kg ha^{-1}. The cultivation area for this crop in Iran is about 82,516 hectares with an average yield of 42,021 kg ha^{-1} (FAOSTAT, 2013). The water requirements for this crop with growth period of approximately 150 days are about 350 (humid areas) up to 1,150 mm (arid areas) (Allen et al., 1998).

Regulated deficit irrigation (RDI) and partial root drying (PRD) are among the most important deficit irrigation methods, aimed at limiting the irrigation water volume. PRD is a new approach for irrigation management and improving crop quality while preventing yield loss (Albaji et al., 2011a; Posadas et al., 2008; Shahnazari et al., 2007; Steduto et al., 2012; Zegbe and Serna-Pérez, 2011). Abboud et al. (2019) state that under water restriction conditions, new approaches for irrigation management are required to reduce water consumption and make more efficient use of water by maximizing water saving with a minimum impact on crop productivity, PRD irrigation being one of the most promising techniques to attain this objective. The basic principle underlying PRD irrigation is to alternately let one part of the root system be exposed to soil drying, while the other part is irrigated, which in itself induces more soil water dynamics with dry/wet cycles in the soil profile relative to the RDI by which the water is delivered evenly to the soil surface (Wang et al., 2013). In PRD irrigation, the root sectors that grow on the irrigated side absorb enough water to maintain high shoot water potential, and the root sectors that grow on the non-irrigated side produce abscisic acid (ABA). ABA affects the regulating of plants' physiological processes such as the decreasing of leaf growth and the reducing of stomatal conductance. This mechanism optimizes the water use of the plant (Ahmadi et al., 2010; Davies et al., 2002; FAO, 2002; Kang and Zhang, 2004; Steduto et al., 2012).

PRD irrigation can create different physiological responses that allow plants to cope with water-deficit stress (Tahi et al., 2008). In this regard, Santos et al. (2003) and de Souza et al. (2005) investigated PRD irrigation on grapevines. Their results showed that the PRD irrigation decreases stomatal conductance compared with full irrigation without decreases in photosynthesis rate, which led to an increase in the water use efficiency. However, the decrease in photosynthesis under PRD has been reported by Yuan et al. (2013). The results of Aganchich et al. (2009) showed that split-root pot experiments trials of PRD irrigation on olive induced a reduction in leaf water potential, shoot growth, and stomatal conductance; however, relative water content and photosynthetic capacity are maintained.

The effects of PRD irrigation have been extensively investigated utilizing different fruit trees and field crops especially on the crop yields of apples (Leib et al., 2006), grapes (Kusakabe et al., 2016), olives (Dbara et al., 2016), potatoes (Ahmadi et al., 2014; Xie et al., 2012; Yactayo et al., 2013), sunflowers (Albaji et al., 2011b), tomatoes (Savić et al., 2008), and maize (Kang and Zhang, 2004). PRD irrigation and nitrogen levels have also been researched utilizing crops such as wheat (Sepaskhah and Hosseini, 2008), potato (Jovanovic et al., 2010), tomato (Bogale et al., 2016; Wang et al., 2013), sunflower (Sezen et al., 2011), cotton (Du et al., 2008), and maize (Li et al., 2010). The research results indicated the effectiveness of PRD irrigation on the fruit trees' and field crops' performance. The PRD irrigation can wet the lower parts of the root zone, and this could affect the N uptake from the subsoil, which is important especially under field conditions (Kuhlmann et al., 1989). Abboud et al. (2019) studied differential agro-physiological responses induced by partial root-zone drying irrigation in olive cultivars grown in semiarid conditions. Their results showed that reducing irrigation volumes by 25% (75% PRD) and 50% (50% PRD) compared to the control (100% ET$_C$) increased oil yield and water productivity mainly for olive cultivar without significant reductions on yield components.

In recent years, many studies on the effects of the RDI method on sugar beet cultivation have been carried out by various researchers (Esmaeili and Yasari, 2011; Fabeiro et al., 2003; Kiymaz and Ertek, 2015b; Mahmoodi et al., 2008; Topak et al., 2011; Uçan and Gençoglan, 2004; Winter, 1980). Hassanli et al. (2010) and Ghamarnia et al. (2012) studied the influence of drip and furrow irrigation on sugar beet yield. Haghverdi et al. (2017) studied the impact of irrigation, surface residue cover, and plant population on sugar beet growth, in addition to yield, irrigation water use efficiency, and soil water dynamics. They looked into the effects of the RDI technique by taking into account furrow and sprinkler irrigation on the sugar beet yield. Kiymaz and Ertek (2015a) studied the influence of RDI and fertilizer-nitrogen levels with drip irrigation methods on sugar beet yield. A few studies tested the effects of the PRD technique on sugar beet using furrow (Sepaskhah and Kamgar-Haghighi, 1997) and drip irrigation methods (Sahin et al., 2014). Topak et al. (2016) investigated the effects of PRD, RDI, and fertilizer-nitrogen levels on sugar beet using the drip irrigation method. Nevertheless, there is still little understanding of the mechanisms of PRD irrigation in sugar beet grown under furrow irrigation with different N-fertilizer levels.

Water and nitrogen are important resources for sugar beet cultivation, and their availability affects the quantity and quality of sugar beet yield. To the best of our knowledge, based on the above studies, no field study has yet been undertaken to investigate the impact of PRD, RDI, and nitrogen levels on sugar beet crop using the furrow irrigation method. Therefore, the present study has been carried out to evaluate the sugar beet grown and yield, under RDI and PRD irrigation and different levels of N-fertilizer.

3.2 MATERIALS AND METHODS

3.2.1 Experimental Site

The research was conducted on farmland in the Karafs plain, which is located in the Hamedan Province, in the west of Iran at latitude 35° 18′ N, longitude 49°20′ E, and altitude 1,915 m above sea level. The study was conducted over two growing seasons from October to June 2013–2014. According to the meteorological statistics for 1997 to 2012 in Table 3.1, the average annual rainfall and temperature were 337.7 mm and 11.1°C (Hamedan Meteorological Office, 2013).

3.2.2 Soil and Water Properties

Some physical and chemical properties of soil are given in Table 3.2. Water supply was groundwater, which was supplied from a well. The water quality was based on the Wilcox diagram and ranked as a class C2S1. Chemical properties of the water used in this study are presented in Table 3.3.

TABLE 3.1
Mean Air Temperature, Relative Humidity, and Total Monthly Rainfall (1997–2012) at Karafs Plain in the Hamedan Province

Parameter	Jan	Feb	Mar	Apr	May	Jun	Jul	Aug	Sep	Oct	Nov	Dec	Average
Temperature (°C)	−2.8	−0.9	4.6	9.8	14.6	21.0	24.2	24.1	19.3	13.1	5.2	0.7	11.1
Relative humidity (%)	75.6	71.9	53.8	46.4	38.3	28.2	26.6	26.6	29.3	37.7	59.3	68.3	46.8
													Total
Rainfall (mm)	49.1	46.5	40.3	56.5	24.9	4.2	7.0	2.9	1.2	27.8	36.4	40.8	337.7

TABLE 3.2

Physical and Chemical Properties of Soil

Soil Depth (cm)	Clay (%)	Siltc (%)	Sandc (%)	Soil Texture	θ_{FC} (%)	θ_{PWP} (%)	Bulk Density (g cm^{-3})	pH	EC (dS m^{-1})	Organic Carbon (%)	Total N (%)
0–30	24	41	35	Loam	21.33	10.36	1.11	7.9[A]–7.5[B]	0.7[A]–0.6[B]	0.34[A]–0.8[B]	0.03[A]–0.08[B]
30–60	41	22	37	Clay	21.52	11.15	1.10	7.8–7.5	0.7–0.7	0.51–0.7	0.05–0.07
60–100	28	47	25	Clay Loam	22.36	11.80	1.07	7.9–7.7	0.6–0.6	0.38–0.3	0.04–0.03

Note: A, soil properties in year of 2013 and B, 2014; θ_{FC}, Field capacity; θ_{PWP}, Permanent wilting point; EC, electrical conductivity; N, nitrogen.

TABLE 3.3
Physical and Chemical Properties of Water

Year	Cations	Ca^{2+}	Na^+	Mg^{2+}	Anions	SO_4^{2-}	Cl^-	HCO_3^-	CO_3^{2-}	pH	EC (10^{-3}dS m^{-1})	TDS (mg l^{-1})	SAR
					(meq l^{-1})								
2013	2.9	0.7	0.7	1.5	2.9	0.9	0.5	1.5	0	8.9	293	188	0.67
2014	2.8	0.6	0.6	1.6	2.8	0.8	0.6	1.4	0	8.8	284	182	0.57

Note: SAR, sodium absorption ratio; EC, electrical conductivity; TDS, total dissolved solids.

3.2.3 EXPERIMENTAL DESIGN, SUGAR BEET PLANTATION, IRRIGATION, AND N-FERTILIZER TREATMENTS

The study was conducted in a split block with a randomized complete block design in order to investigate the effects of partial root drying, regulated deficit irrigation, and nitrogen fertilizer levels on the sugar beet characteristics (*Canaria* hybrid). The main block included: full irrigation (FI); three levels of partial root drying: 85 (PRD_{85}), 75 (PRD_{75}), and 65% (PRD_{65}); and regulated deficit irrigation at three levels: 85 (RDI_{85}), 75 (RDI_{75}), and 65% of the crop water requirements (RDI_{65}); and the sub block encompassed two levels of 100 (f_{100}) and 75% of N-fertilizer requirement (f_{75}). Furrows spacing was 0.55 m and a length of 20 m, and the ends of the furrows were closed. In each treatment, there were five rows. The crops were hand sown on June 6, 2013 and 2014; plant-plant spacing was 0.2 m. All treatments were irrigated under full irrigation conditions until plants were established (plants with 6–8 leaves). Once the plants were well established, the FI treatment was irrigated with 100% of the water requirements (calculated using the soil moisture content). Then the RDI_{85}, RDI_{75}, and RDI_{65}, and also the PRD_{85}, PRD_{75}, PRD_{65} treatments were irrigated at 85%, 75%, and 65% of the water requirements, respectively. Figure 3.1 shows the layout of the treatments at experimental blocks.

In all treatments, two rows of furrows were created for each plant row. In each irrigation, the FI and RDI treatments were watered on both sides (both furrows) of each plant row; however, in the PRD treatment, both sides of the plant row were alternatively watered. The volume of water delivered in each treatment was controlled utilizing a volume method (cylindrical tank). Figure 3.2 shows the layout of the irrigation system used on the farm.

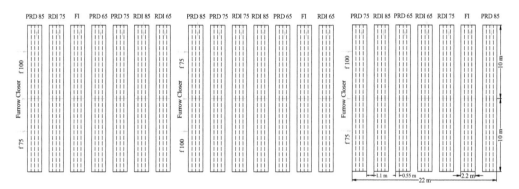

FIGURE 3.1 Layout of the treatments at experimental blocks.

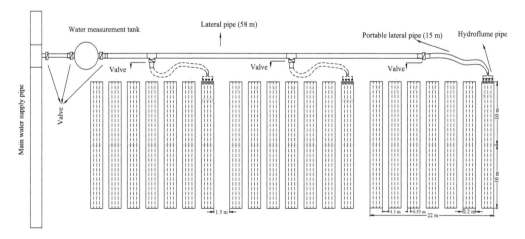

FIGURE 3.2 Layout of the irrigation system used on the farm.

Nitrogen fertilizer requirements in 2013 and 2014 were measured using a soil analysis method and were found to be 184 and 138 kg ha^{-1}, respectively. One-third and two-thirds of the fertilizer requirements were applied to all experimental plots in the form of urea fertilizer (46% N) 48 and 71 days after planting.

The irrigation interval was 7 days. The irrigation water depth was calculated based on soil moisture depletion from the root zone on the day before irrigation as follows:

$$dn = \sum\nolimits_{i=1}^{k} \left[\left(\theta_{FCi} - \theta_{BIi} \right) \times \rho_{bi} \times D_{RZi} \right] \tag{3.1}$$

where dn is the net irrigation requirement (mm), i is the number of soil layers to which the root has penetrated, θ_{FCi} is the soil moisture in the field capacity in each soil layer, θ_{BIi} is the soil moisture on the day before irrigation in each soil layer, ρ_{bi} is the bulk density per each soil layer (g cm^{-3}), and D_{RZi} is the development depth of plant roots in each soil layer (mm).

Irrigation efficiency was considered as being 90% due to the short length and closed end of the furrows (Waller and Yitayew, 2016).

3.2.4   MEASUREMENTS OF QUANTITY AND QUALITY PARAMETERS OF SUGAR BEET

At the end of the growing season, the plants of each treatment were harvested from the three center rows at a level equivalent of 6.6 m^2. After separating the leaves from the roots, the leaves' fresh weight was measured with a digital scale. The roots of the plants that had received treatment were washed and then measured. Pulp samples, taken from the roots of each treatment type, were analyzed qualitatively using a Betalyser. In the qualitative analysis, sugar content (SC) was defined using the polarimetric method; moreover, the amount of potassium (K) and sodium (Na) were obtained using a flame photometer and the amount of alpha-amino-nitrogen (α-amino-N) was measured using a blue number (Kernchen, 1997). Sugar molasses (MS) was calculated from the empirical Equation (3.2) (Reinefeld et al., 1974).

$$MS = 0.343 \left(K + Na \right) + 0.094 \left(\alpha \, amino\text{-}N \right) + 0.29 \tag{3.2}$$

where MS is the sugar molasses (%), K is the potassium (mmol 100 g^{-1} root), Na is the sodium (mmol 100 g^{-1} root), and α-amino-N is the alpha-amino-nitrogen (mmol 100 g^{-1} root).

Values for white sugar content (WSC) and white sugar yield (WSY) were obtained via the following equations, respectively.

$$WSC = SC - MS \tag{3.3}$$

$$WSY = RY \times WSC \tag{3.4}$$

where WSC is the white sugar content (%), SC is the sugar content (%), MS is the molasses sugar (%), RY is the root yield (kg ha^{-1}), and WSY is the white sugar yield (kg ha^{-1}).

3.2.4.1   Water Productivity

Water productivity indexes were determined to evaluate the productivity of irrigation in the treatments. Water productivity indexes for yield (biomass, root, and white sugar yield) were calculated by the following equations (Tavakkoli and Oweis, 2004; Zhang, 2003):

$$WP = Y/Ig \tag{3.5}$$

where WP is the water productivity for yield (kg m^{-3}); Y is the biomass, root, and white sugar yield (kg ha^{-1}); and Ig is the amount of the seasonal irrigation (m^3 ha^{-1}).

3.2.4.2 Data Analysis

An SAS 9.1 software program (SAS Institute, 2006) was used for the mean comparison of the parameters using the Tukey test at a significance level of 5% and 1%.

3.3   RESULTS AND DISCUSSION

3.3.1   RAINFALL AND IRRIGATION WATER APPLIED

The total rainfall during the growing season was 18 mm in 2013, which mainly occurred in the last week of the planting period, and 17 mm in 2014 of which 11 and 6 mm of the rainfall occurred in the first and last weeks of the planting period, respectively. Thus, the amount of precipitation was similar in both years.

The total amounts of irrigation water applied for the FI, RDI$_{85}$, PRD$_{85}$, RDI$_{75}$, PRD$_{75}$, RDI$_{65,}$ and PRD$_{65}$ treatments were 9,761; 8,715; 8,715; 8,017; 8,017; 7,319; and 7,319 m^{-1} ha^{-1} in 2013 and 10,272; 9,278; 9,278; 8,616; 8,616; 7,954; and 7,954 m^3 ha^{-1} in 2014, respectively. The results of the study show that the amount of irrigation water applied in the second year was higher than the first year. As portrayed in Figure 3.3, this phenomenon can be attributed to the increase in the average daily temperature in the second year as compared to the first year. Sepaskhah and Kamgar-Haghighi (1997) reported that sugar beet irrigation water applied through furrow irrigation was 10,680 m^3 ha^{-1} in similar weather conditions. Fabeiro et al. (2003) reported that sugar beet irrigation water applied via drip irrigation was 8,965 m^3 ha^{-1} in the FI treatment. Kiymaz and Ertek (2015b) stated that sugar beet irrigation water applied during 91.5 mm rainfall in the growing season was 6,136 m^3 ha^{-1} combined with drip irrigation in the FI treatment. Topak et al. (2016) reported that sugar beet irrigation water applied through drip irrigation was 8,542 m^3 ha^{-1} in the FI treatment. Haghverdi et al. (2017) stated that sugar beet irrigation water applied utilizing sprinkler irrigation in Nebraska, U.S., with an annual precipitation of 394 mm was 6,900 m^3 ha^{-1} in the FI treatment. It is of note that the irrigation water applied in this study is in accordance with a study by Sepaskhah and Kamgar-Haghighi (1997).

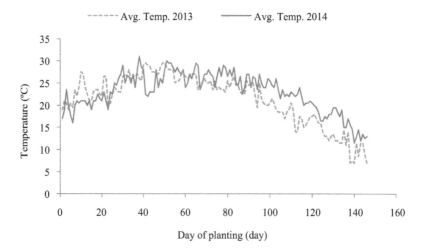

FIGURE 3.3   Daily average temperatures in cultivation seasons at 2013–2014.

3.3.2 Biomass and Root Yield

The combined analysis of the variance effect of irrigation and fertilizer treatments on the quantity of sugar beet tested over two years (2013 and 2014) is presented in Table 3.4. According to Table 3.4, the experiment shows differences in biomass and root yield significantly at 5% over the years of the study. The results show irrigation and fertilizer treatment effects are significant on biomass and root yield at 1% probability level. The results also show that the interaction of the irrigation and fertilizer treatments was nonsignificant on biomass and root yield. The average comparison tests for biomass and root yield for both years and for each individual year separately are given in Table 3.5. According to the average comparison test in Table 3.5, the biomass and root yield in the second year at 94,409 and 48,664 kg ha^{-1}, respectively, is higher than the first year at 74,946 and 38,286 kg ha^{-1}, respectively. As seen in Figure 3.3, the average daily temperature in the second year was 1.5°C less than the first year for 50 days after planting; however, after day 50 to the end of the growing period, the average daily temperature in the second year was on average 3.1°C higher than the first year. The temperature increase in the second year brought about the early growth period due to a lower temperature compensation by which the growth rate increases and results in an improved product performance. Nelson (1978) and Ghamarnia et al. (2012) found that the quantity yield of sugar beet was greater in years with higher temperatures.

According to the average comparison of both years utilizing various irrigation treatments (Table 3.5), FI and PRD$_{65}$ treatments had the highest and lowest biomass yields at 11,4638 and 60,066 kg ha^{-1} respectively. FI treatment accounted for the highest biomass yield among irrigation treatments for 2013 and 2014 at about 103,255 and 126,022 kg ha^{-1} respectively. The lowest biomass yields were attributed to RDI$_{65}$ (51,318 kg ha^{-1}) and PRD$_{65}$ (67,926 kg ha^{-1}) treatments, respectively. It seems that water use via an FI treatment can increase the activity of the leaves, photosynthesis, and thus increase the biomass yield. By contrast, drought stress lowers the activity of the leaves, decreases leaf area, and causes the leaves to fall thus decreasing photosynthesis, which in itself causes the biomass yield to become reduced. Hernandez-Santana et al. (2017) state that plant water stress causes reduction of

---

TABLE 3.4
Results of Combined Variance Analysis (Mean Square) the Effects of Irrigation and Nitrogen Fertilizer Treatments on Biomass and Root Yield of Sugar Beet at Karafs Region (2013–2014)

S.O.V	df	Biomass	Root Yield
Year	1	7955096420[a]	2261561207[a]
Irrigation	6	4961622015[b]	906622868[b]
Year × Irrigation	6	33470253 ns	11566571 ns
error	24	31879195	5132937
Fertilizer	1	733956237[b]	173706633[b]
Year × Fertilizer	1	424284781 ns	4881325 ns
error	4	1117293	2117414
Irrigation × Fertilizer	6	6587703 ns	2313684 ns
Year × Irrigation × Fertilizer	6	2215183 ns	965159 ns
Residual error	24	4889827	2142494
Generalized error	83	486277853	101572979
Coefficients variance (C.V)	—	2.61	3.37

Note: ns, nonsignificant.

[a] Significant at the 5% of probability level (P < 0.05).

[b] Significant at the 1% of probability level (P < 0.01).

TABLE 3.5

Mean Comparison of the Effects of Irrigation and Nitrogen Fertilizer Treatments on Biomass and Root Yield of Sugar Beet at Karafs Region (2013–2014)

Year	Factor	Treatment	Biomass	Root Yield
			(kg ha^{-1})	
Avg. 2013–2014	Irrigation	FI	114638 a	54235 a
		RDI_{85}	99932 b	48770 b
		RDI_{75}	80673 c	40952 c
		RDI_{65}	60831 d	30588 e
		PRD_{85}	96479 b	51113 ab
		PRD_{75}	80124 c	44047 c
		PRD_{65}	60066 d	34621 d
	Fertilizer	f_{100}	87634 a	44913 a
		f_{75}	81722 b	42037 b
	Year	Y_{2013}	74946 b	38286 b
		Y_{2014}	94409 a	48664 a
2013	Irrigation	FI	103255 a	47667 a
		RDI_{85}	88058 ab	42505 abc
		RDI_{75}	71821 b	36258 cd
		RDI_{65}	51318 c	26136 e
		PRD_{85}	85619 b	45369 ab
		PRD_{75}	72345 b	39606 c
		PRD_{65}	52206 c	30465 de
	Fertilizer	f_{100}	77676 a	39483 a
		f_{75}	72216 b	37090 b
2014	Irrigation	FI	126022 a	60804 a
		RDI_{85}	111806 b	55036 b
		RDI_{75}	89524 c	45647 c
		RDI_{65}	70343 d	35040 d
		PRD_{85}	107340 b	56857 ab
		PRD_{75}	87902 c	48488 c
		PRD_{65}	67926 d	38777 d
	Fertilizer	f_{100}	97591 a	50343 a
		f_{75}	91227 b	46985 b

Note: Means within the same column and factors, followed by the same letter using Tukey's Studentized Range (HSD) Test are not significantly different.

stomatal conductance and photosynthesis rate, which results in the decline of the leaf area. In this regard, Abboud et al. (2019) stated that a coordinated adjustment in stomatal responses may represent an adaptive advantage in conditions of water deficit induced by PRD irrigation. Mehrabi and Sepaskhah (2019) reported the leaf photosynthesis rate was not significantly lowered in PRD in comparison with that obtained in FI. The PRD strategy reduced the stomatal conductance about 12% to 7% in comparison with that obtained in FI that was statistically significant. Cooke and Scott (1993) also indicate that when drought stress occurs, the leaves change; and these changes involve a delay in the emergence, slow spread, accelerated aging, and reduction of the photosynthetic production. Adu et al. (2018) reported the observed lower yields under RDI and PRD compared to FI can be due to insufficient water supply at critical growth stages (especially during yield formation).

Based on the two-year average comparison test in Table 3.5, biomass yield in the f_{100} fertilizer treatment with 87,634 kg ha^{-1} is more than the f_{75} treatment (81,722 kg ha^{-1}). What is more, as in 2013 and 2014 for fertilizer treatments, the f_{100} treatment showed more biomass yield (77,676 and 97,591 kg ha^{-1}) than the f_{75} treatments (72,216 and 91,227 kg ha^{-1}), respectively. The low nitrogen fertilizer reduces the amount of chlorophyll and the photosynthesis rate of older leaves and older leaves' loss (before maturity), thereby reducing nitrogen fertilizer and causing it to reduce the biomass yield. Cooke and Scott (1993) believe that with the reduction of the nitrogen fertilizer application, the biomass yield is reduced.

Based on the average comparison of both years among the irrigation treatments for root yield (Table 3.5), the maximum root yield averaged 54,235 kg ha^{-1} and was associated with FI treatment. The RDI_{65} treatment averaged 30,588 kg ha^{-1} and was related to the minimum root yield. The highest root yield among the irrigation treatments for 2013 and 2014 was recorded in the FI treatment with 47,667 and 60,804 kg ha^{-1}, respectively. By contrast, the RDI_{65} treatment with 26,136 and 35,040 kg ha^{-1} showed the lowest root yield. In similar weather conditions Topak et al. (2011) reported 77,300 and 28,100 kg ha^{-1} of sugar beet root yield for full irrigation and RDI_{25} treatments using the drip irrigation method in the Mid-Anatolian semiarid climate of Turkey. Adu et al. (2018) stated crops under RDI produce similar yields as PRD, but yields under both are typically lower than yields of FI. Also, when PRD was compared with FI, yield reductions under PRD were large for tomatoes, grapes, apples, and pepper but not orange and potato. Esmaeili and Yasari (2011) reported that the sugar beet root yield for the treatments without stress and the RDI_{80} treatment using basin irrigation method were 62,540 and 47,540 kg ha^{-1}, respectively, in the east of Iran. Stevens et al. (2015) reported 61,150 kg ha^{-1} root yield using sprinkler irrigation in Montana in the U.S. In Nebraska, Haghverdi et al. (2017) reported the average sugar beet root yield was equal to 71,110 kg ha^{-1} over six years of their study on full irrigation treatment using a sprinkler irrigation method while applying 75% and 50% of the ET_c, which on average caused a 9%–11% reduction in the yield. In general, the results of the current study are in accordance with the abovementioned research findings.

The results show that with the lessening of irrigation water, the root yield is reduced. It seems that root yield losses under deficit irrigation are due to reduced leaf area, photosynthetic material, and consequently reduced storage materials in the root. Some researchers believe that the increase in drought stress causes a reduction in the turgescence of plant cells and increases the mechanical strength of the soil, in as such that root growth is reduced (Bondok, 1996; Cooke and Scott, 1993; Tognetti et al., 2003). These researchers also believe that the main cause of root growth reduction is the reduction in the supplying of carbohydrates from leaves. A large number of studies also indicate that with the increasing of drought stress, root growth and root yield are reduced (Baigy et al., 2012; Chołuj et al., 2014; Haghverdi et al., 2017; Kiymaz and Ertek, 2015a, 2015b; Sepaskhah and Kamgar-Haghighi, 1997; Tognetti et al., 2003; Topak et al., 2016).

According to results portrayed in Table 3.5, root yield in all PRD treatments when compared to RDI treatments is more at similar irrigation levels. The average two-year root yield difference between PRD and RDI at 85% and 75% of the plant irrigation requirements was nonsignificant; however, with the intensification of tension up to 65% of plant irrigation requirements, this difference became significant. The average root yield under the PRD_{65} and RDI_{65} treatment was 34,621 and 30,588 kg ha^{-1}, respectively; thus, the root yield of PRD_{65} treatment increased 13.2% as compared to the RDI_{65} treatment. This result corresponded with the study carried out by Topak et al. (2016) who reported sugar beet root yield increasing up to 10.73% in the PRD_{50} treatment as compared to the RDI_{50} treatment. This result could be attributed to more water penetrating in the soil from furrow irrigation in the PRD treatments. This causes less soil mechanical resistance against root growth and better use of soil nutrients by the root to be more influential. Previous studies showed that PRD irrigation extends the root system to greater soil depths (Dry et al., 2000; Kuhlmann et al., 1989; Wang et al., 2013), in addition to increasing root mass (Kang et al., 2000; Mingo et al., 2004) and soil nutrients absorption (Wang et al., 2009). Hu et al. (2006) and Li et al. (2007) have expressed that drying and wetting cycles via PRD irrigation may cause uneven availability of nutrients in soil, leading to uneven nutrient absorption by the roots in different root zones. In this regard, Abboud et al. (2019) stated that moderate water stress generated by PRD irrigation (75% PRD) produced an effective control of tree vigor compared to well-irrigated trees. However, a severe water deficit (50% PRD) led to a reduction in vegetative growth.

The average two-year comparison test in Table 3.5 showed that root yield in the f_{100} fertilizer treatment with 44,913 kg ha^{-1} is more than the f_{75} treatment (42,037 kg ha^{-1}). In addition, in terms of fertilizer treatments as in 2013 and 2014, the f_{100} treatment showed greater root yield (39,483 and 50,343 kg ha^{-1}) than the f_{75} treatment (37,090 and 46,985 kg ha^{-1}). The proper use of nitrogen fertilizer therefore improves the growth of leaves, increases production and photosynthesis, and thus increases the root yield. Other studies show that increasing use of nitrogen fertilizer causes an increase in leaf area, the number of leaves, canopy cover, and the root yield of sugar beet (Abdel-Motagally and Attia, 2009; El-Gizawy et al., 2014; Loomis and Nevins, 1963; Martin, 1986).

3.3.3 SUGAR BEET QUALITATIVE TRAITS

The combined variance analysis of the effect of irrigation and fertilizer treatments on the quality characteristics of sugar beet, which was studied over two consecutive years (2013 and 2014), is presented in Table 3.6. According to Table 3.6, the two experimental years show differences in sugar content, K, Na, and white sugar yield at 5% probability level and α-amino-N, molasses sugar, which are significant at the 1% probability level; however, the effect on white sugar content is non-significant. The effect of irrigation treatments on all quality characteristics studied is significant at 1% probability level. Fertilizer treatment effects are significant on the sugar content, white sugar content, and white sugar yield at 1% probability level. The impact of fertilizer treatments on other

TABLE 3.6

Results of Combined Variance Analysis (Mean Square) the Effects of Irrigation and Nitrogen Fertilizer Treatments on Sugar Content, α-amino-N, K, Na, Molasses Sugar, with Sugar Content and with Sugar Yield of Sugar Beet at Karafs Region (2013–2014)

S.O.V	df	Sugar Content	α-amino-N	K	Na	Molasses Sugar	White Sugar Content	White Sugar Yield
Year	1	20.88[a]	125.7[b]	43.9[a]	17.2[a]	22.6[b]	0.03 ns	51991351[a]
Irrigation	6	5.18[b]	1.35[b]	1.88[b]	0.57[b]	0.70[b]	2.63[b]	16218064[b]
Year × Irrigation	6	0.03 ns	0.08[b]	0.05 ns	0.002 ns	0.006 ns	0.03 ns	214934 ns
error	24	0.07	0.0196	0.04	0.007	0.007	0.07	101855
Fertilizer	1	0.92[b]	0.117 ns	0.02 ns	0.06 ns	0.03 ns	1.24[b]	2310393[b]
Year × Fertilizer	1	0.02 ns	0.007 ns	0.003 ns	0.02 ns	0.001 ns	0.03 ns	52431 ns
Error	4	0.02	0.134	0.06	0.014	0.024	0.03	40175
Irrigation × Fertilizer	6	0.06 ns	0.016 ns	0.002 ns	0.008 ns	0.001 ns	0.05 ns	39654 ns
Year × Irrigation × Fertilizer	6	0.03 ns	0.013 ns	0.003 ns	0.004 ns	0.002 ns	0.04 ns	39625 ns
Residual error	24	0.03	0.022	0.006	0.004	0.003	0.30	46963
Generalized error	83	0.74	1.65	0.71	0.27	0.33	0.30	1951487
Coefficients variance (C.V)	—	1.01	6.32	1.52	4.79	2.39	1.16	3.34

Note: ns, nonsignificant; K, potassium; Na, Sodium; α-amino-N, alpha-amino-nitrogen.
[a] Significant at the 5% of probability level (P < 0.05).
[b] Significant at the 1% of probability level (P < 0.01).

TABLE 3.7

Mean Comparison of the Effects of Irrigation and Nitrogen Fertilizer Treatments on Sugar Content, α-amino-N, K, Na, Molasses Sugar, with Sugar Content and with Sugar Yield of Sugar Beet at Karafs Region (2013–2014)

Year	Factor	Treatment	Sugar Content (%)	α-amino-N	K	Na	Molasses Sugar (%)	With Sugar Content (%)	White Sugar Yield (kg ha⁻¹)
				(mmol 100 g⁻¹ root)					
Avg. 2013–2014	Irrigation	FI	16.82 f	1.97 d	4.66 e	1.04 f	1.83 f	14.39 c	7803 a
		RDI$_{85}$	17.15 ef	2.14 dc	5.11 dc	1.20 de	2.05 de	14.50 c	7064 b
		RDI$_{75}$	17.80 dc	2.37 b	5.40 bc	1.38 bc	2.24 bc	14.95 b	6125 c
		RDI$_{65}$	18.30 ab	2.90 a	5.84 a	1.68 a	2.54 a	15.17 b	4635 e
		PRD$_{85}$	17.52 de	2.11 dc	4.91 de	1.13 ef	1.96 ef	14.96 b	7644 a
		PRD$_{75}$	18.05 bc	2.32 bc	5.24 bc	1.29 cd	2.15 cd	15.30 b	6734 b
		PRD$_{65}$	18.71 a	2.70 a	5.55 ab	1.47 b	2.35 a	15.76 a	5446 d
	Fertilizer	f$_{100}$	17.66 b	2.40 a	5.26 a	1.34 a	2.18 a	14.88 b	6659 a
		f$_{75}$	17.87 a	2.32 a	5.23 a	1.29 a	2.14 a	15.13 a	6327 b
	Year	Y$_{2013}$	18.26 a	3.58 a	5.97 a	1.76 a	2.68 a	14.98 a	5706 b
		Y$_{2014}$	17.27 b	1.14 b	4.52 b	0.86 b	1.64 b	15.02 a	7280 a
2013	Irrigation	FI	17.38 e	3.09 d	5.52 d	1.52 e	2.39 e	14.38 c	6850 a
		RDI$_{85}$	17.66 de	3.30 dc	5.83 bcd	1.64 cde	2.56 cde	14.50 c	6150 abc
		RDI$_{75}$	18.25 bcd	3.66 b	6.09 b	1.84 bc	2.76 bc	14.90 bc	5397 cd
		RDI$_{65}$	18.78 ab	4.19 a	6.51 a	2.12 a	3.04 a	15.14 abc	3947 e
		PRD$_{85}$	17.94 cde	3.25 d	5.66 cd	1.60 de	2.48 de	14.86 bc	6731 ab
		PRD$_{75}$	18.58 abc	3.57 bc	5.95 bc	1.73 bcd	2.66 bcd	15.32 ab	6065 bc
		PRD$_{65}$	19.25 a	4.02 a	6.2 ab	1.91 b	2.85 ab	15.80 a	4803 d
	Fertilizer	f$_{100}$	18.17 b	3.61 a	5.98 a	1.78 a	2.69 a	14.88 a	5847 a
		f$_{75}$	18.35 a	3.55 a	5.94 a	1.75 a	2.66 a	15.09 a	5565 b
2014	Irrigation	FI	16.27 d	0.85 c	3.81 d	0.56 e	1.27 e	14.40 c	8755 a
		RDI$_{85}$	16.64 d	0.98 bc	4.38 bcd	0.75 cde	1.54 cd	14.50 c	7977 ab
		RDI$_{75}$	17.34 c	1.08 bc	4.70 abc	0.93 bc	1.72 bc	15.02 b	6853 cd
		RDI$_{65}$	17.83 ab	1.61 a	5.17 a	1.24 a	2.04 a	15.20 b	5323 e
		PRD$_{85}$	17.09 c	0.97 c	4.16 cd	0.67 de	1.44 de	15.06 b	8558 a
		PRD$_{75}$	17.51 bc	1.07 bc	4.53 bc	0.85 bcd	1.64 bcd	15.27 ab	7404 bc
		PRD$_{65}$	18.17 a	1.38 ab	4.89 ab	1.03 ab	1.85 ab	15.72 a	6088 de
	Fertilizer	f$_{100}$	17.14 b	1.18 a	4.53 a	0.90 a	1.66 a	14.88 b	7470 a
		f$_{75}$	17.39 a	1.09 a	4.51 a	0.82 a	1.62 a	15.17 a	7089 b

Note: Means within the same column and factors, followed by the same letter using Tukey's Studentized Range (HSD) Test are not significantly different.

quality characteristics of sugar beet was nonsignificant. Except for the interactions of year and irrigation for α-amino-N, none of the factors' interactions in this study had any significant effect on the quality characteristics of sugar beet (Table 3.6). Due to the significant interaction of the effects of the year and the irrigation for α-amino-N traits, the average comparisons of each separate year for the abovementioned traits and other traits are presented in Table 3.7.

Based on the comparison of the averages for both years in Table 3.7, the sugar content, α-amino-N, K, Na, and molasses sugar in 2013 show an average of 18.26%, 3.58, 5.97, 1.76 mmol 100 g⁻¹ root, and 2.68% respectively, which is in itself more than 2014 with 17.27%, 1.14, 4.52, 0.86 mmol 100 g⁻¹ root, and 1.64%, respectively. However, the average two-year white sugar content and white sugar yield with an average of 15.02 and 7,280 kg ha⁻¹ in 2014 was more than 2013 with an average of 14.98 and 5,706 kg ha^{-1}, respectively. The abovementioned results over the two years are probably due to different temperature patterns (Figure 3.3). As mentioned earlier, in the second year (2014) from about day 50 to the end of the growing season, daily temperatures averaged 3.1°C higher than in the first year (2013). The temperature increase in 2014, as with a decrease in the water requirements of the plant and root yield, might have caused a decrease in the sugar content, α-amino-N, K, Na, and molasses sugar yield. In light of the aforementioned, the researchers believe that in the warmer year more root yield and lower sugar content are to be expected.

3.3.4 SUGAR CONTENT

The average comparison of both years in terms of irrigation treatments for sugar content is shown in Table 3.7. The results show that the maximum sugar content averaged 18.71% and was associated with PRD$_{65}$ treatment. The minimum sugar content at 16.82% was attained through FI treatment. The maximum sugar content among irrigation treatments for 2013 and 2014 was recorded in PRD$_{65}$ at 19.25% and 18.17%, respectively, and FI at 17.38% and 16.27% showed minimal sugar content. Topak et al. (2011) observed a general increase in sugar content in response to deficit irrigation treatments and reported sugar content ranging from 18.81% to 21.42%. Sahin et al. (2014) reported sugar content variations from 16.16% to 17.51% in irrigation level treatments. Kiymaz and Ertek (2015b) portrayed sugar content ranging from 15.29% to 17.43%. Topak et al. (2016) reported sugar content variations from 19.32% to 21.32%. Ghamarnia et al. (2012) reported sugar content ranging from 12.62% to 23.53% and stated that with increasing water deficit, the sugar content increased.

The results of the present study show that with an increase in water stress, sugar beet crops increase their sugar content when there is a water shortage. Research carried out by Barbieri (1982); Carter et al. (1980); Hoffmann et al. (2009); Tognetti et al. (2003); Winter (1980); and Topak et al. (2011) confirm this conclusion.

The average two-year sugar content difference between PRD and RDI at similar irrigation levels was nonsignificant (Table 3.7). As water stress intensified, the two-year average sugar content in PRD treatments increased from 17.52 (PRD$_{85}$) to 18.71 (PRD$_{65}$) and in RDI treatments improved from 17.15% (RDI$_{85}$) to 18.30% (RDI$_{65}$). Topak et al. (2016) stated that the highest root sugar content was 21.32% under the PRD$_{50}$ treatment as compared to the RDI$_{50}$ and RDI$_{75}$ treatments, which were 20.82% and 20.29%, respectively. Sahin et al. (2014) reported sugar content at 16.51% in the PRD technique and 16.23% under full irrigation technique for the experimental year of 2011, but no significant difference between PRD and full irrigation treatments was observed.

On the basis of the two-year average comparison test in Table 3.7, the sugar content in the f$_{75}$ fertilizer treatment, with an average of 17.87%, was more than the f$_{100}$ treatment (17.66%). Moreover, the f$_{75}$ fertilizer treatment showed more sugar content (18.35% and 17.39%) than the f$_{100}$ treatment (18.17% and 17.14%) for 2013 and 2014, respectively. Kiymaz and Ertek (2015a) found that in the highest nitrogen application sugar content was lower. They reported sugar content varies from 15.35% to 17.35%. Abdel-Motagally and Attia (2009) found similar results. Tsialtas and Maslaris (2013) reported sugar content ranged from 13.82% to 14.19% in a similar case.

The result of this study shows that the reduction of nitrogen fertilizer increases the sugar content between fertilizer treatments. Weeden (2000) stated that the decrease in sugar content occurred with the declining use of nitrogen fertilizer, due to more water being stored in the root. When nitrogen becomes deficient before harvesting, leaf initiation and expansion are slowed relative to photosynthesis, and photosynthesis produces sucrose, which accumulates in roots as storage rather than as new vegetative growth (Kaffka and Grantz, 2014).

3.3.5 IMPURITIES IN THE ROOTS (α-AMINO-N, K, AND Na)

A comparison of the average amount of impurities in the two-year study that is α-amino-N, K, and Na among the different irrigation treatments as per Table 3.7 shows that in RDI_{65} treatment an average of 2.90, 5.84, and 1.68 mmol 100 g^{-1} root, respectively, had the highest amount of impurities. Moreover, FI treatment with an average of 1.97, 4.66, and 1.04 mmol 100 g^{-1} root, respectively, had the lowest amount of impurities. Most impurities: α-amino-N, K, and Na in different irrigation treatments in 2013 with 4.19, 6.51, and 2.12 mmol 100 g^{-1} root, respectively, and in 2014 with 1.61, 5.17, and 1.24 mmol 100 g^{-1} root, respectively, are related to the RDI_{65} treatment. The least amount of impurities related to the FI treatment were also observed in 2013, with an average of 3.09, 5.52, and 1.52 mmol 100 g^{-1} root, respectively, and in 2014 with 0.85, 3.81, and 0.56 mmol 100 g^{-1} root, respectively. Tognetti et al. (2003) reported that the values of α-amino-N ranged from 2 to 2.5; K values ranged from 5.4 to 6; and Na values changed between 1.9 and 3 mmol 100 g^{-1} root. Topak et al. (2011) indicates a general increase in impurities in response to deficit irrigation treatments and reported the values of α-amino-N ranged from 2.11 to 4.31; K values ranged from 4.33 to 4.87; and Na values changed between 0.88 and 1.09 mmol 100 g^{-1} root. Kiymaz and Ertek (2015a) reported the values of α-amino-N ranged from 2.61 to 4.44; K values ranged from 4.19 to 5.10; and Na values changed between 1.21 and 2.32 mmol 100 g^{-1} root. Similar studies by Kiymaz and Ertek (2015b) report the values of α-amino-N as ranging from 3.97 to 6.17, K values ranging from 4.39 to 5.22, and Na values changing between 1.86 and 2.35 mmol 100 g^{-1} root. Tsialtas and Maslaris (2013) and Ghamarnia et al. (2012) found similar results. Overall, the current study's results are in accordance with the abovementioned study's findings.

Intense water stress causes nitrogenous compounds such as betaine and proline to be increasingly produced in leaves and from there transferred to the root, thus aiding in the absorption of impurities from the soil to the roots in order to adjust the osmotic pressure of the roots. Other researchers have confirmed these results (Chołuj et al., 2014; Gzik, 1996; Monreal et al., 2007).

On the basis of the results of Table 3.7, a two-year average comparison of the impurities α-amino-N, K, and Na among the different fertilizer treatments showed that the f_{100} treatment with an average of 2.40, 5.26, and 1.34 mmol 100 g^{-1} root held more impurities than the f_{75} treatment with an average of 2.32, 5.23, and 1.29 mmol 100 g^{-1} root, respectively. In the evaluation of the average rate of impurities such as α-amino-N, K, and Na in the fertilizer treatments carried out in 2013, the f_{100} treatment with 3.61, 5.98, and 1.78 mmol 100 g^{-1} root held more impurities than the f_{75} treatment with an average of 3.55, 5.94, and 1.75 mmol 100 g^{-1} root, respectively. In 2014, the f_{100} treatment with 1.18, 4.53, and 0.90 mmol 100 g^{-1} root held more impurities than the f_{75} treatment with an average of 1.09, 4.51, and 0.82 mmol 100 g^{-1} root, respectively. Abdel-Motagally and Attia (2009) reported that the values of α-amino-N ranged from 3.11 to 4.75; K values ranged from 4.84 to 5.41; Na values changed between 1.10 and 1.89 mmol 100 g^{-1} root. Kiymaz and Ertek (2015a) reported that the values of α-amino-N ranged from 3.01 to 4.44; K values ranged from 4.44 to 4.94; Na values changed between 1.36 and 1.98 mmol 100 g^{-1} root for nitrogen fertilizer plots. Tsialtas and Maslaris (2013) reported the values of α-amino-N as ranging from 1.49 to 2.04; K values varied from 8.23 to 8.53; and Na values changed between 3.45 and 4.74 mmol 100 g^{-1} root.

The results of the study show that higher nitrogen fertilizer application can increase nitrogen uptake by the root and causes an increase in the amount of impurities in the root of the sugar beet. The results of a seven-year study in Greece by Maslaris et al. (2010) showed that with the increased use of nitrogen fertilizer from 0 to 240 kg ha^{-1}, the concentration of α-amino-N, K, and Na in the roots increased by 54%, 5.9%, and 29.6%. Abdel-Motagally and Attia (2009) stated increasing nitrogen fertilizer causes an increase in impurities such as α-amino-N, K, and Na in the root of sugar beet. Tsialtas and Maslaris (2013) in a study carried out on the Eastern Thessaly Plain in Greece reported that an increasing nitrogen fertilizer application increased N assimilation significantly (as assessed by petiole NO_3-N and root α-amino-N), but selective absorption (the preferential uptake of K over Na in roots) decreased with increasing nitrogen fertilizer application and was negatively correlated.

The average two-year difference of impurities such as α-amino-N, K, and Na between PRD and RDI at similar irrigation levels was nonsignificant (except 65% irrigation treatment for Na). Generally, PRD irrigation treatments held less impurities of α-amino-N, K, and Na than the RDI treatments. This is probably due to the fact that PRD irrigation treatments import less water stress than RDI treatment to sugar beet. This leads to a reduction in the impurities such as K and Na absorption by the roots, in addition to a reduced production of nitrogenous compounds by leaves and their transference to the roots. According to the studies made by Stoll et al. (2000) in PRD irrigation, a net flow of water at night from the roots of the moist soil around the roots of the dry soil occurs. This in itself causes reduction in dry stress to the root and reduces the absorption of impurities in order to adjust root osmotic pressure.

3.3.6 Molasses Sugar

On the basis of the results of the comparison of the two-year average in Table 3.7, the lowest and highest molasses sugar among the different irrigation treatments belonged to FI and RDI_{65} treatments with an average of 1.83%–2.54%, respectively. Most molasses sugar in different irrigation treatments in 2013 and 2014 is related to RDI_{65} treatment with an average of 3.04% and 2.04%, respectively. In addition, the least molasses sugar in different irrigation treatments in 2013 and 2014 was related to FI treatment with an average of 2.39% and 1.27%, respectively. Ghamarnia et al. (2012) stated that molasses sugar varied from 3.76% to 5.03% in their research, but generally no significant difference has been found between molasses sugar content for different irrigation treatments.

With increasing water stress, the root of the sugar beet absorbs more root impurities such as K and Na and causes an increase of molasses sugar. The average two-year molasses sugar difference between PRD and RDI in similar irrigation levels was nonsignificant. With an increasing water stress in PRD treatments, molasses sugar increases from 1.96% to 2.35% and RDI treatments 2.05% to 2.54%, respectively.

The average two-year molasses sugar difference between fertilizer treatments was nonsignificant. Molasses sugar in f_{100} and f_{75} treatments was 2.18% and 2.14%, respectively. The sugar molasses in a comparison between fertilizer treatments in 2013 and 2014 showed that the f_{100} treatment with an average of 2.69% and 1.66% and the f_{75} treatments with 2.66% and 1.62%, respectively, had a higher increase. Abdel-Motagally and Attia (2009) reported molasses sugar ranging from 2.80% to 3.07% in N-fertilizer treatments. They stated higher rates of N-fertilizer (285 kg N ha^{-1}) cause an increase in soluble non-sugar in root juice (impurities), and they interfere with sugar extraction. This was reflected by raising the percentage of sugar losses to molasses and consequently reducing sugar recovery.

3.3.7 White Sugar Content

The results of the two-year average comparison show that white sugar content among different irrigation treatments held the highest and lowest white sugar content in the PRD_{65} and FI treatments with an average of 15.76% and 14.39%, respectively (Table 3.7). Among the various irrigation treatments carried out in 2013 and 2014, the highest white sugar content was found to be in the PRD_{65} treatment with an average of 15.80% and 15.72%, respectively. Furthermore, the lowest white sugar content in different irrigation treatments in 2013 and 2014 is related to FI treatment with an average of 14.38% and 14.40%, respectively. Topak et al. (2011) reported white sugar content ranging from 15.95% to 18.68% in experimental conditions. Sahin et al. (2014) stated that white sugar content varied from 14.17% to 15.36%. Kiymaz and Ertek (2015a) reported white sugar content ranging from 13.48% to 13.94% in different irrigation treatments. The sugar content values in this study were similar to those obtained by Sahin et al. (2014).

Although increasing water stress causes an increased absorption of impurities in the root of the sugar beet, an increase in the amount of sugar content causes an increase in white sugar content. The results obtained by Topak et al. (2011) closely paralleled the results obtained by the current study.

The two-year average comparison of white sugar content in PRD and RDI treatments at similar irrigation levels shows that white sugar content in PRD treatments (85% and 65% water requirement with 14.96% and 15.76%, respectively) is more than that observed in RDI treatments (with 14.50% and 15.17%, respectively). As already noted, less water stress on the sugar beet in PRD as RDI treatments causes a reduction in impurities such as K and Na absorption by the roots; this in itself results in an increase in white sugar content.

The average two-year white sugar content difference between fertilizer treatments was significant (P < 0.01). The white sugar content of the f_{75} treatment with an average of 15.13% was more than the f_{100} treatment at 14.88%. White sugar content in comparison with the fertilizer treatments in 2013 and 2014 showed the f_{75} treatments with an average of 15.09%–15.17% and the f_{100} treatments with 14.88% and 14.88%, respectively. Kiymaz and Ertek (2015a) reported white sugar content ranged from 13.35% to 14.49% in different N-fertilizer treatments.

The decreased use of nitrogen fertilizer with an increased sugar content and decreased accumulation of impurities cause an increase in white sugar content. Abdel-Motagally and Attia (2009) in their study indicate that by increasing nitrogen fertilizer, a decrease in white sugar content results; they further reported that the values of white sugar content varied from 11.25% to 12.56% in nitrogen fertilizer plots. Studies by Campbell (2002); Maslaris et al. (2010) also show that the increased use of nitrogen fertilizer causes white sugar content to decrease.

3.3.8   White Sugar Yield

The results of the two-year average comparison for white sugar yield in different irrigation treatments show that the highest and lowest white sugar yields are related to the FI and RDI_{65} treatment with an average of 7,803 and 4,635 kg ha^{-1}, respectively (Table 3.7). The maximum white sugar yield for 2013 and 2014, in different irrigation treatments, is related to the FI treatment with an average of 6,850 and 8,755 kg ha^{-1}, respectively. What is more, the minimum white sugar yield in different irrigation treatments in 2013 and 2014 belonged to the RDI_{65} treatment with an average of 3,947 and 5,323 kg ha^{-1}, respectively. Esmaeili and Yasari (2011) in a study carried out in the east of Iran reported that when using a basin irrigation method, the white sugar yield ranged from 5,560 to 6,740 kg ha^{-1}. Topak et al. (2011) reported white sugar yield varies from 5,250 to 12,750 kg ha^{-1} in the drip irrigation method. Sahin et al. (2014) stated that white sugar yield varies from 3,330 to 5,600 kg ha^{-1}. Stevens et al. (2015) reported white sugar yield ranges from 10,230 to 11,230 kg ha^{-1} when applying the sprinkler irrigation method. Kiymaz and Ertek (2015a) reported a white sugar yield ranged from 8,810 to 10,930 kg ha^{-1} in different irrigation treatments. Topak et al. (2016) reported white sugar yield ranging from 11,850 to 15,380 kg ha^{-1} in the drip irrigation method. Haghverdi et al. (2017) stated white sugar yield varies from 2,210 to 12,000 kg ha^{-1} in rainfed and sprinkler irrigation methods, respectively. Generally, the white sugar yield values in the present study occur within the same range as the abovementioned studies.

The results of the present research show that an increase in water stress causes a decrease in white sugar yield over different irrigation treatments. Although a decreased use of water for irrigation causes an increase in white sugar content, decreased root yield causes a decrease in white sugar yield. Sahin et al. (2014) indicated that lower water causes lower root and white sugar yield. In addition, some studies show that an increase in water stress causes a decrease in white sugar yield (Bondok, 1996; Ghamarnia et al., 2012; Topak et al., 2011, 2016).

The two-year average comparison on white sugar yield on PRD and RDI treatments with similar irrigation levels shows that white sugar yield on PRD treatments (85%, 75%, and 65% of water requirement with 7,644, 6,734, and 5,446 kg ha^{-1}, respectively) is more than RDI treatments (with 7,064, 6,125, and 4,635 kg ha^{-1}, respectively). Thus, white sugar yield of PRD$_{85}$, PRD$_{75}$, and PRD$_{65}$ treatments increased 8.2%, 9.9%, and 17.5% as compared to RDI$_{85}$, RDI$_{75}$, and RDI$_{65}$ treatments, respectively. Adu et al. (2018) stated crop yields between FI, RDI, and PRD vary significantly if crops are more frequently irrigated. It is concluded that RDI and PRDI result in yields lower than those of FI, but yields of RDI and PRDI are comparable. Topak et al. (2016) reported that 12,460 and 11,220 kg ha^{-1} white sugar yield is obtained for PRD$_{50}$ and RDI$_{50}$ treatments via a drip irrigation method and improved 11% white sugar yield of PRD$_{50}$ when compared to RDI$_{50}$ treatment.

As already noted, less water stress on PRD as RDI treatments causes a reduction in impurities absorbed by the roots, and it causes increased white sugar yield. PRD irrigation can wet the lower parts of the root zone (Wang et al., 2013), which in itself results in the root being better able to use soil nutrients and thus to be more influential in increasing root and white sugar yields. Sepaskhah and Ahmadi (2012) state that crop yields under PRD irrigation are better than yields under the RDI technique when the same amount of irrigation water is applied. Some studies have indicated that PRD irrigation may improve product quality (Du et al., 2008; Shahnazari et al., 2007; Topak et al., 2016; Wang et al., 2013). The results obtained in the current study verify the results obtained by the abovementioned studies.

Based on the average two-year white sugar yield; the difference between fertilizer treatments is significant (P < 0.01). White sugar yield in the f$_{100}$ treatment with an average of 6,659 kg ha^{-1} is more than f$_{75}$ treatment with 6,327 kg ha^{-1}. When comparing fertilizer treatments in 2013 and 2014, the white sugar yield showed that the f$_{100}$ treatment with an average of 5,847 and 7,470 kg ha^{-1}, respectively, was more than the f$_{75}$ treatments with 5,565 and 7,089 kg ha^{-1}, respectively. Abdel-Motagally and Attia (2009) reported white sugar yield varying from 6,260 to 7,100 kg ha^{-1} in nitrogen fertilizer treatments. Tsialtas and Maslaris (2013) reported white sugar yield varying from 10,580 to 11,320 kg ha^{-1}. Kiymaz and Ertek (2015a) reported white sugar yield ranging from 8,590 to 11,230 kg ha^{-1} in different N-fertilizer treatments.

As already noted, although the increased use of nitrogen fertilizer causes a decrease in sugar content, an increase in root yield causes an increase in white sugar yield. Moreover, El-Gizawy et al. (2014) and Kaffka and Grantz (2014) consider the fact that an increase in nitrogen fertilizer causes an increase in white sugar yield.

3.3.9 WATER PRODUCTIVITY

The combined analysis of variance effect of irrigation and nitrogen fertilizer treatments on the water productivity for biomass (WP (B)), root (WP (R)), and white sugar yield (WP (WSY)) of sugar beet for two years (2013 and 2014) is presented in Table 3.8. Also, a comparison of two-year averages of WP (B), WP (R), and WP (WSY) in different treatments of irrigation, N-fertilizer, and experimental years is presented in Figures 3.4 through 3.6, respectively.

According to Table 3.8, the two experimental years show differences in WP (B), WP (R), and WP (WSY) significantly at 5% probability level. The results show irrigation and fertilizer treatment effects are significant on WP (B), WP (R), and WP (WSY) at 1% probability level. The results show that the interaction of the irrigation and fertilizer treatments was nonsignificant on WP (B), WP (R), and WP (WSY). According to average comparison of a two-year test in Table 3.9, the WP (B), WP (R), and WP (WSY) in the second year (2014) with averages of 10.55, 5.46, and 0.82 kg m^{-3} were higher than the first year (2013) with 8.49, 4.35, and 0.65 kg m^{-3}, respectively. This occurred because the biomass, root, and white sugar yield was in the second year (2014) with averages of 94,409; 48,664; and 7,280 kg ha^{-1} higher than the first year (2013) with 74,946; 38,286; and 5,706 kg ha^{-1}, respectively.

The results of the two-year average comparison (Table 3.9 and Figure 3.4) of WP (B) among different irrigation treatments show the highest WP (B) allocated to the FI treatment with an average of 11.41 kg m^{-3}. Also, the lowest WP (B) belonged to the PRD$_{65}$ treatment with an average of 7.51 kg m^{-3}. The results in Figures 3.5 and 3.6 show that the maximum WP (R) and WP (WSY) allocated to

TABLE 3.8

Results of Combined Variance Analysis (Mean Square) the Effects of Irrigation and Nitrogen Fertilizer Treatments on Water Productivity for Biomass, Root and White Sugar Yield of Sugar Beet at Karafs Region (2013–2014)

S.O.V	df	WP (B)	WP (R)	WP (WSY)
Year	1	89.34[a]	25.5[a]	0.59[a]
Irrigation	6	28.43[b]	4.85[b]	0.085[b]
Year × Irrigation	6	0.23 ns	0.02 ns	0.0005 ns
error	24	0.40	0.07	0.0015
Fertilizer	1	9.66[b]	2.32[b]	0.032[b]
Year × Fertilizer	1	0.04 ns	0.07 ns	0.0006 ns
error	4	0.015	0.03	0.0005
Irrigation × Fertilizer	6	0.10 ns	0.04 ns	0.0008 ns
Year × Irrigation × Fertilizer	6	0.03 ns	0.013 ns	0.0006 ns
Residual error	24	0.08	0.03	0.00076
Generalized error	83	3.53	0.77	0.0152
Coefficients variance (C.V)	—	2.74	3.53	3.76

Note: ns, nonsignificant.

[a] Significant at the 5% of probability level (P < 0.05).

[b] Significant at the 1% of probability level (P < 0.01).

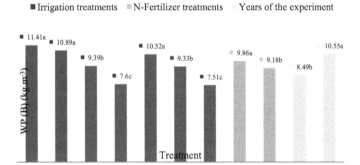

FIGURE 3.4   Comparison of two-year averages of biomass water productivity (WP (B)) index in different treatments of irrigation, N-fertilizer, and experimental years.

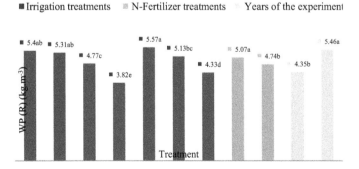

FIGURE 3.5   Comparison of two-year averages of root water productivity (WP (R)) index in different treatments of irrigation, N-fertilizer, and experimental years.

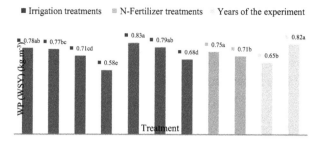

FIGURE 3.6   Comparison of two-year averages of water productivity of white sugar yield (WP (WSY)) index in different treatments of irrigation, N-fertilizer, and experimental years.

TABLE 3.9

Mean Comparison of the Effects of Irrigation and Nitrogen Fertilizer Treatments on Water Productivity for Biomass, Root and White Sugar Yield of Sugar Beet at Karafs Region (2013–2014)

Year	Factor	Treatment	WP (B)	WP (R)	WP (WSY)
				(kg m⁻³)	
Avg. 2013–2014	Irrigation	FI	11.41 a	5.40 ab	0.78 ab
		RDI_{85}	10.89 a	5.31 ab	0.77 bc
		RDI_{75}	9.39 b	4.77 c	0.71 cd
		RDI_{65}	7.60 c	3.82 e	0.58 e
		PRD_{85}	10.52 a	5.57 a	0.83 a
		PRD_{75}	9.33 b	5.13 bc	0.79 ab
		PRD_{65}	7.51 c	4.33 d	0.68 d
	Fertilizer	f_{100}	9.86 a	5.07 a	0.75 a
		f_{75}	9.18 b	4.74 b	0.71 b
	Year	Y_{2013}	8.49 b	4.35 b	0.65 b
		Y_{2014}	10.55 a	5.46 a	0.82 a
2013	Irrigation	FI	10.55 a	4.87 ab	0.70 ab
		RDI_{85}	9.73 ab	4.70 ab	0.68 abc
		RDI_{75}	8.39 b	4.24 bc	0.63 bc
		RDI_{65}	6.36 c	3.24 d	0.49 d
		PRD_{85}	9.46 ab	5.01 a	0.74 a
		PRD_{75}	8.45 b	4.63 ab	0.71 ab
		PRD_{65}	6.47 c	3.78 cd	0.60 c
	Fertilizer	f_{100}	8.81 a	4.49 a	0.67 a
		f_{75}	8.17 b	4.21 b	0.63 b
2014	Irrigation	FI	12.27 a	5.92 ab	0.853 ab
		RDI_{85}	12.05 a	5.93 ab	0.858 ab
		RDI_{75}	10.39 b	5.30 bc	0.795 b
		RDI_{65}	8.84 cd	4.41 d	0.67 c
		PRD_{85}	11.57 ab	6.13 a	0.92 a
		PRD_{75}	10.20 bc	5.63 ab	0.86 ab
		PRD_{65}	8.45 d	4.86 cd	0.767 bc
	Fertilizer	f_{100}	10.91 a	5.65 a	0.84 a
		f_{75}	10.19 b	5.26 b	0.80 b

Note: Means within the same column and factors, followed by the same letter using Tukey's Studentized Range (HSD) Test are not significantly different.

the PRD_{85} treatment with an average of 5.57 and 0.83 kg m^{-3}, respectively. Also, the minimum of these traits belonged to the RDI_{65} treatment with an average of 3.82 and 0.58 kg m^{-3}, respectively. The result shows that in general with increased water stress—because of biomass, root, and white sugar yield loss—WP (B), WP (R), and WP (WSY) decrease.

According to results depicted in Table 3.9, the average two-year WP (B) difference between PRD and RDI at 85%, 75%, and 65% of the water requirements was nonsignificant. Also, according to results shown in Table 3.9, WP (R) in all PRD treatments when compared to RDI treatments is higher at similar irrigation levels. The average two-year WP (R) difference between PRD and RDI at 85% and 75% of the water requirements was nonsignificant; however, with the intensification of tension up to 65% of water requirements, this difference became significant. The average WP (R) under the PRD_{65} and RDI_{65} treatment was 4.33 and 3.82 kg m^{-3} respectively, thus the WP (R) of PRD_{65} treatment increased 13.4% as compared to the RDI_{65} treatment. As Table 3.5 shows, the higher root yield in PRD treatments than RDI caused an increase in the WP (R) in PRD treatments compared to RDI (Figure 3.5).

The two-year average comparison of WP (WSY) on PRD and RDI treatments with similar irrigation levels shows that WP (WSY) on PRD treatments (85%, 75%, and 65% of water requirement with 0.83, 0.79, and 0.68 kg m^{-3}, respectively) is higher than RDI treatments (with 0.77, 0.71, and 0.58 kg m^{-3}, respectively). Thus, WP (WSY) of PRD_{85}, PRD_{75}, and PRD_{65} treatments increased 7.8%, 11.3%, and 17.2% as compared to RDI_{85}, RDI_{75}, and RDI_{65} treatments, respectively. As Table 3.7 shows, the higher white sugar yield in PRD treatments than RDI caused an increase in the WP (WSY) in PRD treatments compared to RDI (Figure 3.6).

The average two-year comparison of WP (B), WP (R), and WP (WSY) differences between N-fertilizer treatments was significant (P < 0.01). WP (B), WP (R), and WP (WSY) of f_{100} treatment with average 9.86, 5.07, and 0.75 kg m^{-3} were higher than f_{75} treatment with 9.18, 4.74, and 0.71 kg m^{-3}, respectively. The result shows that a decrease in the use of N-fertilizer—because of biomass, root, and white sugar yield loss—also results in WP (B), WP (R), and WP (WSY) decreasing.

3.4 CONCLUSION

The overall results of the trial-year comparison show that in warm years, the increase in the water requirement and crop growth rate cause an increase in the root yield and decreases sugar content. The current study's findings definitely show that regulating the amount of irrigation water and the utilization of a nitrogen fertilizer are an essential element for attaining the maximum quantity and quality of sugar beet. By increasing the amount of irrigation water, root and white sugar yields increase; however, the sugar content and white sugar content are reduced. Increasing water stress causes more impurities to be absorbed by the root of the sugar beet due to osmotic adjustments of the crop to deal with water stress conditions. All quantitative and qualitative characteristics except impurities and molasses sugar were affected by the amount of nitrogen fertilizer utilized. With the increasing use of nitrogen fertilizer, the root and white sugar yield increase, but the sugar content is reduced. It can be thus stated that the use of the partial root drying technique, as compared with the regulated deficit irrigation technique, increases white sugar content and reduces impurities thus causing an improvement in white sugar yield. Results of this study showed that root yield, white sugar yield, WP (R), and WP (WSY) in all PRD treatments when compared to RDI treatments are higher at similar irrigation levels. For example, the WP (R) of PRD_{65} treatment increased 13.4% as compared to the RDI_{65} treatment. WP (WSY) of PRD_{85}, PRD_{75}, and PRD_{65} treatments increased 7.8%, 11.3%, and 17.2% as compared to RDI_{85}, RDI_{75}, and RDI_{65} treatments, respectively. Also, PRD irrigation treatments held less impurities of α-amino-N, K, and Na than the RDI treatments. This is probably due to the fact that PRD irrigation treatments import less water stress than RDI treatment to sugar beet. This leads to a reduction in the impurities such as K and Na absorption by the roots, in addition to a reduced production of nitrogenous compounds by leaves and their transference to the roots. Therefore, the PRD irrigation compared with RDI irrigation causes reduction in dry stress to the root and reduces the absorption of impurities. Due to no significant differences between the partial root drying at 85% of crop water requirement with full irrigation treatments in terms of root and white sugar yield, and the saving of 15% of irrigation water, under such experimental conditions, the applying of the partial root drying at 85% of crop water requirement treatment is recommended. In terms of nitrogen fertilizer treatment, root and white sugar yields, a 100% of nitrogen fertilizer requirement treatment is better than 75% of the recommended nitrogen fertilizer requirement treatment. Thus, under such experimental conditions, using partial root drying at 85% of the crop water requirements and 100% of the nitrogen fertilizer requirement treatments is recommended.

LIST OF ABBREVIATIONS

α-amino-N	alpha-amino-nitrogen
FI	full irrigation
f_{100}	N-fertilizer level at 100% of N-fertilizer requirement
f_{75}	N-fertilizer level at 75% of N-fertilizer requirement
PRD	partial root drying
PRD_{85}	partial root drying at 85% of the crop water requirement
PRD_{75}	partial root drying at 75% of the crop water requirement
PRD_{65}	partial root drying at 65% of the crop water requirement
RDI	regulated deficit irrigation
RDI_{85}	regulated deficit irrigation at 85% of the crop water requirement
RDI_{75}	regulated deficit irrigation at 75% of the crop water requirement
RDI_{65}	regulated deficit irrigation at 65% of the crop water requirement
WP (R)	water productivity of root
WP (R)	water productivity of biomass
WP (WSY)	white sugar yield

REFERENCES

Abboud, S., Dbara, S., Abidi, W., Braham, M., 2019. Differential agro-physiological responses induced by partial root-zone drying irrigation in olive cultivars grown in semi-arid conditions. *Environ. Exp. Bot.* https://doi.org/10.1016/j.envexpbot.2019.103863

Abdel-Motagally, F.M.F., Attia, K.K., 2009. Response of sugar beet plants to nitrogen and potassium fertilization in sandy calcareous soil. *Int. J. Agric. Biol.* 11, 695–700.

Adu, M.O., Yawson, D.O., Armah, F.A., Asare, P.A., Frimpong, K.A., 2018. Meta-analysis of crop yields of full, deficit, and partial root-zone drying irrigation. *Agric. Water Manag.* 197, 79–90. https://doi.org/10.1016/j.agwat.2017.11.019

Aganchich, B., Wahbi, S., Loreto, F., Centritto, M., 2009. Partial root zone drying: Regulation of photosynthetic limitations and antioxidant enzymatic activities in young olive (*Olea europaea*) saplings. *Tree Physiol.* 29, 685–696. https://doi.org/10.1093/treephys/tpp012

Ahmadi, S.H., Agharezaee, M., Kamgar-Haghighi, A.A., Sepaskhah, A.R., 2014. Effects of dynamic and static deficit and partial root zone drying irrigation strategies on yield, tuber sizes distribution, and water productivity of two field grown potato cultivars. *Agric. Water Manag.* 134, 126–136. https://doi.org/10.1016/j.agwat.2013.11.015

Ahmadi, S.H., Andersen, M.N., Plauborg, F., Poulsen, R.T., Jensen, C.R., Sepaskhah, A.R., Hansen, S., 2010. Effects of irrigation strategies and soils on field-grown potatoes: Gas exchange and xylem [ABA]. *Agric. Water Manag.* 97, 1486–1494. https://doi.org/10.1016/j.agwat.2010.05.002

Albaji, M., Behzad, M., Nasab, S.B., Naseri, A.A., Shahnazari, A., Meskarbashee, M., Judy, F., Jovzi, M., 2011a. Investigation on the effects of conventional irrigation (CI), regulated deficit irrigation (RDI) and partial root zone drying (PRD) on yield and yield components of sunflower (*Helianthus annuus* L.). *Res. Crop.* 12, 142–154.

Albaji, M., Nasab, S.B., Behzad, M., Naseri, A., Shahnazari, A., Meskarbashee, M., Judy, F., Jovzi, M., Shokoohfar, A.R., 2011b. Water productivity and water use efficiency of sunflower under conventional and limited irrigation. *J. Food Agric. Environ.* 9, 202–209.

Allen, R.G., Pereira, L.S., Raes, D., Smith, M., 1998. Crop evapotranspiration: Guidelines for computing crop water requirements. *FAO Irrigation and Drainage Paper 56.* Rome, Italy.

Baigy, M.J., Sahebi, F.G., Pourkhiz, I., Asgari, A., Ejlali, F., 2012. Effect of deficit-irrigation management on components and yield of sugar beet. *Int. J. Agron. Plant Prod.* 3, 781–787.

Barbieri, G., 1982. Effect of irrigation and harvesting dates on the yield of spring-sown sugar-beet. *Agric. Water Manag.* 5, 345–357. https://doi.org/10.1016/0378-3774(82)90012-9

Bogale, A., Nagle, M., Latif, S., Aguila, M., Müller, J., 2016. Regulated deficit irrigation and partial root-zone drying irrigation impact bioactive compounds and antioxidant activity in two select tomato cultivars. *Sci. Hortic.* (Amsterdam). https://doi.org/10.1016/j.scienta.2016.10.029

Bondok, M.A., 1996. The role of boron in regulating growth, yield and hormonal balance in sugar beet, Beta vulgaris var. vulgaris. *Ann. Agric. Sci. Ain-Shams Univ.(Egypt)* 41, 15–33.

Campbell, L.G., 2002. Sugar beet quality improvement. *J. Crop Prod.* 5, 395–413. https://doi.org/10.1300/J144v05n01_16

Carter, J.N., Jensen, M.E., Traveller, D.J., 1980. Effect of mid-to late-season water stress on sugarbeet growth and yield1. *Agron. J.* 72, 806. https://doi.org/10.2134/agronj1980.00021962007200050028x

Chołuj, D., Wiśniewska, A., Szafrański, K.M., Cebula, J., Gozdowski, D., Podlaski, S., 2014. Assessment of the physiological responses to drought in different sugar beet genotypes in connection with their genetic distance. *J. Plant Physiol.* https://doi.org/10.1016/j.jplph.2014.04.016

Cooke, D.A., Scott, R.K. (Eds.), 1993. *The Sugar Beet Crop.* Springer Netherlands, Dordrecht. https://doi.org/10.1007/978-94-009-0373-9

Davies, W.J., Wilkinson, S., Loveys, B., 2002. Stomatal control by chemical signalling and the exploitation of this mechanism to increase water use efficiency in agriculture. *New Phytol.* 153, 449–460. https://doi.org/10.1046/j.0028-646X.2001.00345.x

Dbara, S., Haworth, M., Emiliani, G., Ben Mimoun, M., Gómez-Cadenas, A., Centritto, M., 2016. Partial root-zone drying of olive (*Olea europaea* var. "Chetoui") induces reduced yield under field conditions. *PLoS One* 11, 1–20. https://doi.org/10.1371/journal.pone.0157089

de Souza, C.R., Maroco, J.P., dos Santos, T.P., Rodrigues, M.L., Lopes, C., Pereira, J.S., Chaves, M.M., 2005. Control of stomatal aperture and carbon uptake by deficit irrigation in two grapevine cultivars. *Agric. Ecosyst. Environ.* 106, 261–274. https://doi.org/10.1016/j.agee.2004.10.014

Dry, P.R., Loveys, B.R., Düring, H., 2000. Partial drying of the rootzone of grape. II. Changes in the pattern of root development. *Vitis* 39, 9–12.

Du, T., Kang, S., Zhang, J., Li, F., 2008. Water use and yield responses of cotton to alternate partial root-zone drip irrigation in the arid area of north-west China. *Irrig. Sci.* 26, 147–159. https://doi.org/10.1007/s00271-007-0081-0

El-Gizawy, E., Shalaby, G., Mahmoud, E., 2014. Effects of tea plant compost and mineral nitrogen levels on yield and quality of sugar beet crop. *Commun. Soil Sci. Plant Anal.* 45, 1181–1194. https://doi.org/10.1080/00103624.2013.874028

English, M., Raja, S.N., 1996. Perspectives on deficit irrigation. *Agric. Water Manag.* 32, 1–14. https://doi.org/10.1016/S0378-3774(96)01255-3

Esmaeili, M.A., Yasari, E., 2011. Evaluation of the effects of water stress and different levels of nitrogen on sugar beet (*Beta Vulgaris*). *Int. J. Biol.* 3, 89–93. https://doi.org/10.5539/ijb.v3n2p89

Fabeiro, C., Martín de Santa Olalla, F., López, R., Domínguez, A., 2003. Production and quality of the sugar beet (*Beta vulgaris* L.) cultivated under controlled deficit irrigation conditions in a semi-arid climate. *Agric. Water Manag.* 62, 215–227. https://doi.org/10.1016/S0378-3774(03)00097-0

FAO, 2002. Deficit irrigation practices, natural resources management and environment bepartment. *FAO Water Reports 22,* Rome, Italy.

FAOSTAT, 2013. http://faostat.fao.org/site/567/default.aspx#ancor.

Fernández, J.E., Perez-Martin, A., Torres-Ruiz, J.M., Cuevas, M.V., Rodriguez-Dominguez, C.M., Elsayed-Farag, S., Morales-Sillero, A., García, J.M., Hernandez-Santana, V., Diaz-Espejo, A., 2013. A regulated deficit irrigation strategy for hedgerow olive orchards with high plant density. *Plant Soil* 372, 279–295. https://doi.org/10.1007/s11104-013-1704-2

Ghamarnia, H., Arji, I., Sepehri, S., Norozpour, S., Khodaei, E., 2012. Evaluation and comparison of drip and conventional irrigation methods on sugar beets in a semiarid region. *J. Irrig. Drain. Eng.* 138, 90–97. https://doi.org/10.1061/(ASCE)IR.1943-4774.0000362

Gzik, A., 1996. Accumulation of proline and pattern of α-amino acids in sugar beet plants in response to osmotic, water and salt stress. *Environ. Exp. Bot.* 36, 29–38. https://doi.org/10.1016/0098-8472(95)00046-1

Haghverdi, A., Yonts, C.D., Reichert, D.L., Irmak, S., 2017. Impact of irrigation, surface residue cover and plant population on sugarbeet growth and yield, irrigation water use efficiency and soil water dynamics. *Agric. Water Manag.* 180, 1–12. https://doi.org/10.1016/j.agwat.2016.10.018

Hamedan Meteorological Office, 2013. Meteorology Report of Karafs Plain. *Hamedaninan Meteorological Office,* 10 p. (In Persian).

Hassanli, A.M., Ahmadirad, S., Beecham, S., 2010. Evaluation of the influence of irrigation methods and water quality on sugar beet yield and water use efficiency. *Agric. Water Manag.* https://doi.org/10.1016/j.agwat.2009.10.010

Hernandez-Santana, V., Fernández, J.E., Cuevas, M.V., Perez-Martin, A., Diaz-Espejo, A., 2017. Photosynthetic limitations by water deficit: Effect on fruit and olive oil yield, leaf area and trunk diameter and its potential use to control vegetative growth of super-high density olive orchards. *Agric. Water Manag.* 184, 9–18. https://doi.org/10.1016/j.agwat.2016.12.016

Hoffmann, C.M., Huijbregts, T., van Swaaij, N., Jansen, R., 2009. Impact of different environments in Europe on yield and quality of sugar beet genotypes. *Eur. J. Agron.* 30, 17–26. https://doi.org/10.1016/j.eja.2008.06.004

Hu, T., Kang, S., Zhang, F., Zhang, J., 2006. Alternate application of osmotic and nitrogen stresses to partial root system: Effects on root growth and nitrogen use efficiency. *J. Plant Nutr.* 29, 2079–2092. https://doi.org/10.1080/01904160600972167

Jovanovic, Z., Stikic, R., Vucelic-Radovic, B., Paukovic, M., Brocic, Z., Matovic, G., Rovcanin, S., Mojevic, M., 2010. Partial root-zone drying increases WUE, N and antioxidant content in field potatoes. *Eur. J. Agron.* https://doi.org/10.1016/j.eja.2010.04.003

Kaffka, S.R., Grantz, D.A., 2014. Sugar crops, in: *Encyclopedia of Agriculture and Food Systems.* Elsevier, pp. 240–260. https://doi.org/10.1016/B978-0-444-52512-3.00150-9

Kang, S., Liang, Z., Pan, Y., Shi, P., Zhang, J., 2000. Alternate furrow irrigation for maize production in an arid area. *Agric. Water Manag.* 45, 267–274. http://dx.doi.org/10.1016/S0378-3774(00)00072-X

Kang, S., Zhang, J., 2004. Controlled alternate partial root-zone irrigation:Its physiological consequences and impact on water use efficiency. *J. Exp. Bot.* 55, 2437–2446. https://doi.org/10.1093/jxb/erh249

Kernchen, W., 1997. *Instruction for Installation and Operation*, Betalyzer, Dr Wolfgang Kernchen, GmBH, Germany. pp.14.

Kiymaz, S., Ertek, A., 2015a. Yield and quality of sugar beet (*Beta vulgaris* L.) at different water and nitrogen levels under the climatic conditions of Kırsehir, Turkey. *Agric. Water Manag.* 158, 156–165. https://doi.org/10.1016/j.agwat.2015.05.004

Kiymaz, S., Ertek, A., 2015b. Water use and yield of sugar beet (Beta vulgaris L.) under drip irrigation at different water regimes. *Agric. Water Manag.* https://doi.org/10.1016/j.agwat.2015.05.005

Kuhlmann, H., Barraclough, P.B., Weir, A.H., 1989. Utilization of mineral nitrogen in the subsoil by winter wheat. *Zeitschrift für Pflanzenernährung und Bodenkd.* 152, 291–295. https://doi.org/10.1002/jpln.19891520305

Kusakabe, A., Contreras-Barragan, B.A., Simpson, C.R., Enciso, J.M., Nelson, S.D., Melgar, J.C., 2016. Application of partial rootzone drying to improve irrigation water use efficiency in grapefruit trees. *Agric. Water Manag.* 178, 66–75. https://doi.org/10.1016/j.agwat.2016.09.012

Leib, B.G., Caspari, H.W., Redulla, C.A., Andrews, P.K., Jabro, J.J., 2006. Partial rootzone drying and deficit irrigation of "Fuji" apples in a semi-arid climate. *Irrig. Sci.* 24, 85–99. https://doi.org/10.1007/s00271-005-0013-9

Li, F., Liang, J., Kang, S., Zhang, J., 2007. Benefits of alternate partial root-zone irrigation on growth, water and nitrogen use efficiencies modified by fertilization and soil water status in maize. *Plant Soil* 295, 279–291. https://doi.org/10.1007/s11104-007-9283-8

Li, F., Wei, C., Zhang, F., Zhang, J., Nong, M., Kang, S., 2010. Water-use efficiency and physiological responses of maize under partial root-zone irrigation. *Agric. Water Manag.* 97, 1156–1164. https://doi.org/10.1016/j.agwat.2010.01.024

Loomis, R.S., Nevins, D.J., 1963. Interrupted nitrogen nutrition effects on growth, sucrose accumulation and foliar development of the sugar beet plant. *J. Am. Soc. Sugar Beet Technol.* 12, 309–322.

Mahmoodi, R., Maralian, H., Aghabarati, A., 2008. Effects of limited irrigation on root yield and quality of sugar beet (*Beta vulgaris* L.). *African J. Biotechnol.* 7, 4475–4478.

Martin, R.J., 1986. Growth of sugar beet crops in Canterbury. *New Zeal. J. Agric. Res.* 29, 391–400. http://dx.doi.org/10.1080/00288233.1986.10423491

Maslaris, N., Tsialtas, I.T., Ouzounidis, T., 2010. Soil factors affecting yield, quality, and response to nitrogen of sugar beets grown on light-textured soils in Northern Greece. *Commun. Soil Sci. Plant Anal.* 41, 1551–1564. https://doi.org/10.1080/00103624.2010.485236

Mehrabi, F., Sepaskhah, A.R., 2019. Partial root zone drying irrigation, planting methods and nitrogen fertilization influence on physiologic and agronomic parameters of winter wheat. *Agric. Water Manag.* 223, 105688. https://doi.org/10.1016/j.agwat.2019.105688

Mingo, D.M., Theobald, J.C., Bacon, M.A., Davies, W.J., Dodd, I.C., 2004. Biomass allocation in tomato (*Lycopersicon esculentum*) plants grown under partial rootzone drying: Enhancement of root growth. *Funct. Plant Biol.* 31, 971–978. http://dx.doi.org/10.1071/FP04020

Monreal, J.A., Jiménez, E.T., Remesal, E., Morillo-Velarde, R., García-Mauriño, S., Echevarría, C., 2007. Proline content of sugar beet storage roots: Response to water deficit and nitrogen fertilization at field conditions. *Environ. Exp. Bot.* 60, 257–267. https://doi.org/10.1016/j.envexpbot.2006.11.002

Mossad, A., Scalisi, A., Lo Bianco, R., 2018. Growth and water relations of field-grown "Valencia" orange trees under long-term partial rootzone drying. *Irrig. Sci.* 36, 9–24. https://doi.org/10.1007/s00271-017-0562-8

Nelson, J.M., 1978. Influence of planting date, nitrogen rate, and harvest date on yield and sucrose concentration of fall-planted sugarbeets in central Arizona. *J. Am. Soc. Sugar Beet Technol.* 20, 25–32.

Padilla-Díaz, C.M., Rodriguez-Dominguez, C.M., Hernandez-Santana, V., Perez-Martin, A., Fernández, J.E., 2016. Scheduling regulated deficit irrigation in a hedgerow olive orchard from leaf turgor pressure related measurements. *Agric. Water Manag.* 164, 28–37. https://doi.org/10.1016/j.agwat.2015.08.002

Posadas, A., Rojas, G., Malaga, M., Mares, V., Quiroz, R.A., 2008. *Partial Root-Zone Drying: An Alternative Irrigation Management to Improve the Water Use Efficiency of Potato Crops.* International Potato Center, Lima, Peru.

Reinefeld, E., Emmerich, A., Baumgarten, G., Winner, C., Beiss, U., 1974. Zur voraussage des melassezuckers aus rubenanalysen. *Zucker* 27, 2–15.

Sahin, U., Ors, S., Kiziloglu, F.M., Kuslu, Y., 2014. Evaluation of water use and yield responses of drip-irrigated sugar beet with different irrigation techniques. *Chil. J. Agric. Res.* https://doi.org/10.4067/S0718-58392014000300008

Santos, T.P. dos, Lopes, C.M., Rodrigues, M.L., Souza, C.R. de, Maroco, J.P., Pereira, J.S., Silva, J.R., Chaves, M.M., 2003. Partial rootzone drying: Effects on growth and fruit quality of field-grown grapevines (*Vitis vinifera*). *Funct. Plant Biol.* 30, 663–671. https://doi.org/10.1071/FP02180

SAS Institute, 2006. *The SAS Systems for Windows Release 9.1.* SAS Institute Inc., Cary, NC.

Savić, S., Stikić, R., Radović, B.V., Bogičević, B., Jovanović, Z., Šukalović, V.H.-T., 2008. Comparative effects of regulated deficit irrigation (RDI) and partial root-zone drying (PRD) on growth and cell wall peroxidase activity in tomato fruits. *Sci. Hortic. (Amsterdam).* 117, 15–20. https://doi.org/10.1016/j.scienta.2008.03.009

Sepaskhah, A.R., Ahmadi, S.H., 2012. A review on partial root-zone drying irrigation. *Int. J. Plant Prod.* 4, 241–258. https://doi.org/10.22069/ijpp.2012.708

Sepaskhah, A.R., Hosseini, S.N., 2008. Effects of alternate furrow irrigation and nitrogen application rates on yield and water-and nitrogen-use efficiency of winter wheat (*Triticum aestivum* L.). *Plant Prod. Sci.* 11, 250–259. https://doi.org/10.1626/pps.11.250

Sepaskhah, A.R., Kamgar-Haghighi, A.A., 1997. Water use and yields of sugarbeet grown under every-other-furrow irrigation with different irrigation intervals. *Agric. Water Manag.* 34, 71–79. https://doi.org/10.1016/S0378-3774(96)01290-5

Sezen, S.M., Yazar, A., Tekin, S., 2011. Effects of partial root zone drying and deficit irrigation on yield and oil quality of sunflower in a Mediterranean environment. *Irrig. Drain.* 60, 499–508. https://doi.org/10.1002/ird.607

Shahnazari, A., Liu, F., Andersen, M.N., Jacobsen, S.-E., Jensen, C.R., 2007. Effects of partial root-zone drying on yield, tuber size and water use efficiency in potato under field conditions. *F. Crop. Res.* 100, 117–124. http://dx.doi.org/10.1016/j.fcr.2006.05.010

Steduto, P., Hsiao, T.C., Fereres, E., Raes, D., 2012. Crop yield response to water, *FAO Irrigation and Drainage Paper 66.* Rome, Italy.

Stevens, W.B., Evans, R.G., Iversen, W.M., Jabro, J.D., Sainju, U.M., Allen, B.L., 2015. Strip tillage and high-efficiency irrigation applied to a sugarbeet–barley rotation. *Agron. J.* 107, 1250–1258. https://doi.org/10.2134/agronj14.0525

Stoll, M., Loveys, B., Dry, P., 2000. Hormonal changes induced by partial rootzone drying of irrigated grapevine. *J. Exp. Bot.* 51, 1627–1634. https://doi.org/10.1093/jexbot/51.350.1627

Tahi, H., Wahbi, S., EL Modafar, C., Aganchich, A., Serraj, R., 2008. Changes in antioxidant activities and phenol content in tomato plants subjected to partial root drying and regulated deficit irrigation. *Plant Biosyst.— An Int. J. Deal. with all Asp. Plant Biol.* 142, 550–562. https://doi.org/10.1080/11263500802410900

Tavakkoli, A.R., Oweis, T.Y., 2004. The role of supplemental irrigation and nitrogen in producing bread wheat in the highlands of Iran. *Agric. Water Manag.* 65, 225–236.

Tognetti, R., Palladino, M., Minnocci, A., Delfine, S., Alvino, A., 2003. The response of sugar beet to drip and low-pressure sprinkler irrigation in southern Italy. *Agric. Water Manag.* 60, 135–155. https://doi.org/10.1016/S0378-3774(02)00167-1

Topak, R., Acar, B., Uyanöz, R., Ceyhan, E., 2016. Performance of partial root-zone drip irrigation for sugar beet production in a semi-arid area. *Agric. Water Manag.* 176. https://doi.org/10.1016/j.agwat.2016.06.004

Topak, R., Süheri, S., Acar, B., 2011. Effect of different drip irrigation regimes on sugar beet (*Beta vulgaris* L.) yield, quality and water use efficiency in Middle Anatolian, Turkey. *Irrig. Sci.* 29, 79–89. https://doi.org/10.1007/s00271-010-0219-3

Tsialtas, J.T., Maslaris, N., 2013. Nitrogen effects on yield, quality and K/Na selectivity of sugar beets grown on clays under semi-arid, irrigated conditions. *Int. J. Plant Prod.* 7, 355–372.

Uçan, K., Gençoglan, C., 2004. The effect of water deficit on yield and yield components of sugar beet. *Turkish J. Agric. For.* 28, 163–172.

Waller, P., Yitayew, M., 2016. *Irrigation and Drainage Engineering.* Springer International Publishing. https://doi.org/10.1007/978-3-319-05699-9

Wang, H., Liu, F., Andersen, M.N., Jensen, C.R., 2009. Comparative effects of partial root-zone drying and deficit irrigation on nitrogen uptake in potatoes (*Solanum tuberosum* L.). *Irrig. Sci.* 27, 443–448. https://doi.org/10.1007/s00271-009-0159-y

Wang, Y., Liu, F., Jensen, L.S., de Neergaard, A., Jensen, C.R., 2013. Alternate partial root-zone irrigation improves fertilizer-N use efficiency in tomatoes. *Irrig. Sci.* https://doi.org/10.1007/s00271-012-0335-3

Weeden, B.R., 2000. Potential of Sugar Beet on the Atherton Tableland: A Report for the Rural Industries Research and Development Corporation. RIRDC.

Winter, S.R., 1980. Suitability of sugarbeets for limited irrigation in a semi-arid climate1. *Agron. J.* 72, 118–123. https://doi.org/10.2134/agronj1980.00021962007200010024x

Xie, K., Wang, X.-X., Zhang, R., Gong, X., Zhang, S., Mares, V., Gavilán, C., Posadas, A., Quiroz, R., 2012. Partial root-zone drying irrigation and water utilization efficiency by the potato crop in semi-arid regions in China. *Sci. Hortic. (Amsterdam).* 134, 20–25. https://doi.org/10.1016/j.scienta.2011.11.034

Yactayo, W., Ramírez, D.A., Gutiérrez, R., Mares, V., Posadas, A., Quiroz, R., 2013. Effect of partial root-zone drying irrigation timing on potato tuber yield and water use efficiency. *Agric. Water Manag.* 123, 65–70. https://doi.org/10.1016/j.agwat.2013.03.009

Yuan, J., Xu, M., Duan, W., Fan, P., Li, S., 2013. Effects of whole-root and half-root water stress on gas exchange and chlorophyll fluorescence parameters in apple trees. *J. Am. Soc. Hortic. Sci.* 138, 395–402. https://doi.org/10.21273/JASHS.138.5.395

Zegbe, J.A., Serna-Pérez, A., 2011. Partial rootzone drying maintains fruit quality of "Golden Delicious" apples at harvest and postharvest. *Sci. Hortic. (Amsterdam).* 127, 455–459. https://doi.org/10.1016/j.scienta.2010.10.022

Zhang, H., 2003. Improving water productivity through deficit irrigation: Examples from Syria, the North China Plain and Oregon, USA. In: *Water Productivity in Agriculture: Limits and Opportunities for Improvement,* J.W. Kijne, R. Barker and D. Molden (eds.), CABI Publishing, pp. 301–309.

4 Solute Leaching Modeling under Different Irrigation Regimes, Soil Moisture Conditions, and Organic Fertilizer Application

Hossein Bagheri, Hamid Zare Abyaneh, and Azizallah Izady

CONTENTS

4.1 INTRODUCTION

Efficient solute leaching is a great challenge in saline soils. The leaching process creates less osmotic potential difference between soil and crops and leads to optimized crop cultivation pattern, which results in less crop water requirement during growth season. Therefore, it is imperative to understand the leaching phenomenon through agricultural soils, especially for semiarid regions.

The downward transport of exceeding solute from root zone to lower soil layer is dependent on various factors, such as water flow rates of irrigation systems, soil hydraulic conditions, and the type of fertilizer (Knappenberger et al., 2014; Merdun et al., 2008; Tiemeyer et al., 2017). The water flow pattern through soil macropores and micropores is mainly controlled by flow intensity of irrigation systems and the soil moisture content at unsaturated conditions. Sommerfeldt et al. (1982) pointed out that an air-dried soil had more solute leaching compared to nearly saturated soil. Merdun et al. (2008) declared that solute transport was very dependent on the soil moisture levels. The high water flow rate increased the mobility of soil solute in macrochannels, and it decreased the time required for solutes to reach below the soil layers (Shang et al., 2008).

The organic fertilizers are another factors to change the behavior of solute leaching through soil. Vermicompost is one of the organic fertilizers that has been widely used for management and improvement of soil chemical-structural conditions. Because of high levels of nitrates, phosphates, calcium, potassium, and a considerable fraction of organic carbon in vermicompost, it is considered an excellent organic fertilizer for soil nutrition and crop yield (Lim et al., 2015). Despite the high nutrients in vermicompost, it initially makes soil salty by adding various solutes due to its raw material and production procedure. The organic matter fraction in vermicompost increases hydraulic conductivity, porosity, soil particle aggregation, and changes flow-field tortuosity and pore-size distribution (Abu, 2013; Lekfeldt et al., 2017; Nada et al., 2011; Schjønning et al., 2007). These physical hydraulic factors influence water movement and solute adsorption-desorption in soils (Bejat et al., 2000; Bradford et al., 2002). Understanding the experimental effects of the aforementioned factors on solute transport in soil is necessary for the analysis of solute leaching modeling.

4.1.1   WATER MOVEMENT MODELING

Modeling of water flow in saturated-unsaturated conditions is needed for solute transport modeling. Water movement through soil was mainly modeled by the Richards equation (Richards, 1931); its one-dimensional form in vertical direction is given as:

$$\frac{\partial \theta(h)}{\partial t} = \frac{\partial}{\partial z}\left[K(h)\left(\frac{\partial h}{\partial z}+1\right)\right] \tag{4.1}$$

where θ [L³/L³] is the soil water content, h [L] is the pressure head, K [L/T] is the unsaturated hydraulic conductivity, t [T] is the time, and z [L] is the soil depth positive downward. To solve Equation (4.1), van Genuchten (1980) and Mualem (1976) equations were used as follows:

$$\theta(h) = \theta_r + \frac{\theta_s - \theta_r}{\left(1+(\alpha h)^n\right)^m}; m = 1 - 1/n \tag{4.2}$$

$$K(h) = K_s S_e^l \left[1-\left(1-S_e^{\frac{1}{m}}\right)^m\right]^2, \quad S_e = \frac{\theta - \theta_r}{\theta_s - \theta_r} = \frac{1}{\left(1+(\alpha h)^n\right)^m}, \quad m = 1 - 1/n \tag{4.3}$$

where $K(\theta)$ represents the unsaturated hydraulic conductivity (L/T); K_s is the saturated hydraulic conductivity (L/T); θ_s is the saturated moisture (L³/L³); θ_r is the residual moisture (L³/L³); α, n, and m are experimental coefficients; S_e is the relative saturation; and l was assumed to be 0.5 (Schaap and Van Genuchten, 2006).

4.1.2   SOLUTE LEACHING MODELING

The movement of solute in soil is dependent on advection (mass flow), dispersion, diffusion, and soil-water-solute reaction processes. These phenomena were considered in the advection-dispersion equation (ADE). The one-dimensional form of this equation is as follows:

$$\frac{\partial \theta c}{\partial t} + \rho\left(\frac{\partial S}{\partial t}\right) = (\lambda \vartheta + D_0\tau)\theta\frac{\partial^2 c}{\partial z^2} - \vartheta\theta\frac{\partial c}{\partial z} \tag{4.4}$$

where c is the flux average or resident concentration of solute (mg/L), ρ is the soil bulk density (gr/cm^3), θ is the volumetric water content (cm^3/cm^3), t is the time (min), λ is the dispersivity (cm), z is the distance (cm), ϑ is the average pore-water velocity (cm.min^{-1}), S is the solute concentration on solid phase (mg/Kg), D_0 is the molecular diffusion coefficient in the water (L^2/T), and τ is the tortuosity factor that was driven by Moldrup et al. (1997) with equation of $\tau = 0.66(\theta/\theta_s)^{8/3}$. The inverse of this factor is the amount of flow path tortuosity.

Various models were presented to model solute adsorbed to solid phase (S) such as a chemical nonequilibrium model of two-site sorption. This model assumes that the sorption/desorption sites can be divided into two fractions: equilibrium (instantaneous) and kinetic sorption/desorption sites, as shown in Figure 4.1. The equations of the sorption model are as follows (Fetter and Thomas Boving, 2017):

$$S = S_1 + S_2 \tag{4.5}$$

$$S_1 = FK_D c \tag{4.6}$$

$$S_2 = (1-F)K_D c \tag{4.7}$$

$$\frac{\partial S_1}{\partial t} = FK_D \frac{\partial c}{\partial t} \tag{4.8}$$

$$\frac{\partial S_2}{\partial t} = \beta \left[(1-F)K_D c - S_2 \right] \tag{4.9}$$

$$\frac{\partial S}{\partial t} = \left[FK_D \frac{\partial c}{\partial t} \right] + \left[\beta \left[(1-F)K_D c - S_2 \right] \right] \tag{4.10}$$

where S_1 and S_2 are solute concentrations in Type-1 and -2 sites (M/M), F is the fraction of equilibrium exchange sites (−), K_D is the distribution coefficient of soluble salts (L^3/M), and β is the first-order rate constant for the sorption kinetic process (1/T).

Therefore, this study aimed to quantify and model the impacts of vermicompost, flow rate, initial soil moisture, and saturation status on solute leaching to better understand the displacement of solute in soil layers and manage more efficiently.

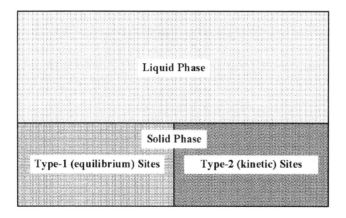

FIGURE 4.1 Schematic of the two-site sorption model.

4.2   MATERIALS AND METHODS

A sandy loam soil (unamended soil) was collected from a farmland located at Bu-Ali Sina University, Hamedan, Iran. This soil had 59.6% sand, 23.2% silt, and 17.2% clay, which were measured using a hydrometer method. The ratio of 1.45/100 g (vermicompost/soil) was used to enrich the soil. Vermicompost (VC) and unamended soil were uniformly mixed to achieve a homogeneous, porous consistency. Table 4.1 shows the characteristics of vermicompost that were determined according to the standard methods.

The main chemical and physical properties of unamended and amended soils were determined as follows. The soil porosity, bulk density, and saturated hydraulic conductivity of the natural and amended soils were measured by the saturation, cylinder, and constant head methods, respectively. The breakthrough curves of both soils were measured using pressure plates.

The soil organic matter, $EC_{1:5}$ and $pH_{1:5}$, were measured by the EC meter, pH meter and Walkley–Black method, respectively. The total dissolved solid (TDS) was calculated according to equation of $TDS(mg/L) = 524.86 \times EC(dS/m)$ in which the 524.86 is the instrument constant coefficient. The soluble-exchangeable sodium in 1:5 soil to 1 M ammonium acetate suspension was determined by flame photometry. Nitrate in 1:5 soil to 2 M KCl was extracted and measured using spectrophotometry (Carter and Gregorich, 2006).

4.2.1   PREPARATION OF SOIL LEACHING COLUMNS

The columns were constructed of a PVC pipe with an internal diameter of 5.95 cm and length of 10 cm. The end of each column was equipped with a nylon membrane filter with a 10-μm pore size and a stainless steel wire screen. The empty columns were filled with 0.35 Kg soil. Two initial saturated and air-dried conditions were applied within soil columns. Water was injected from the bottom of the column to liberate trapped air bubbles at initial saturated conditions (ISC). Next, the columns were placed in distilled water for 24 h to ensure that the soil samples were completely saturated. Under initial air-dried conditions (IAC), both soils were air-dried prior to the experiments.

4.2.2   SOIL LEACHING COLUMN EXPERIMENT

The experimental leaching conditions are summarized in Table 4.2. For leaching, three sets of flow rate were considered as representative of three irrigation regimes—basin irrigation method (2.7, 4.3, 3.8, and 5.1 mL/min), sprinkler system (2.5 mL/min), and drip system (0.4 mL/min). Water level of 3 cm was kept steady on the soil column surface at the highest flow rate (2.7, 4.3, 3.8, and 5.1 mL/min) and controlled by a Mariotte bottle. Ponded water was not considered for leaching under unsaturated flow rates of 2.5 and 0.4 mL/min, and the effluent of the Mariotte bottle was regulated by a control valve. Leaching of both soils was conducted in three replications for 24 h at the temperature of 28°C.

The distilled water with an EC of 5 μS/cm was applied to leach the soil solute. The effluents from each column were collected, and their EC values were analyzed for each time interval. Finally, the EC values were plotted versus the time and pore volume to compare their leaching behavior under different conditions.

TABLE 4.1
Some Characteristics of Vermicompost

Nitrogen (%)	EC (dS/m)	pH	Organic Carbon (%)	Na (%)	Ca (%)	Fe (g/kg)	TDS (mg/L)
1.77	9.34	7.5	25.31	0.38	6.38	11.15	4902

TABLE 4.2

Experimental Conditions for Salt Leaching from the Unamended and Amended Soil at Various Hydraulic Conditions of Soil and Water Flow

Soil	Initial Soil Conditions	Soil Condition during Leaching	Water Depth (cm)	Discharge (mL/min)	Pore-Water Velocity (cm/hr)	Water Saturation (S_w)	Tortuosity Factor (τ)
Unamended	Saturated	Saturated	3	2.7	9.9	1.00	0.66
		Unsaturated	0	2.5	9.5	0.98	0.63
		Unsaturated	0	0.4	1.7	0.90	0.50
	Air-dried	Unsaturated	3	4.3	16.4	0.98	0.63
		Unsaturated	0	2.5	10.0	0.93	0.54
		Unsaturated	0	0.4	1.8	0.88	0.47
Vermicompost-amended	Saturated	Saturated	3	3.8	14.0	1.00	0.66
		Unsaturated	0	2.5	9.3	0.98	0.63
		Unsaturated	0	0.4	1.7	0.89	0.48
	Air-dried	Unsaturated	3	5.1	19.7	0.94	0.56
		Unsaturated	0	2.5	10.1	0.91	0.51
		Unsaturated	0	0.4	1.9	0.84	0.41

4.2.3 DETERMINING PARAMETERS OF SOLUTE LEACHING MODELING

The required coefficients of Equations (4.2) and (4.3) including θ_r, θ_s, α, and n were determined using RETC software (van Genuchten et al., 1991) for both unamended and amended soils. Dispersivity of soils (λ) was determined using nitrate tracer as a function of the aforementioned factors (Bagheri et al., 2019).

The unknown parameters of Equations (4.4) and (4.9) including D_0, K_D, β, and F were estimated based on curves of EC variations for both soils under various conditions using HYDRUS-1D software with inverse method.

4.3 RESULTS AND DISCUSSION

4.3.1 THE EFFECT OF INITIAL SOIL MOISTURE ON FLOW RATE

According to Table 4.2, the highest flow rate under initial air-dried conditions was more than the highest flow rate under initial saturated conditions for both soils. Edwards et al. (1992) stated that preferential water flow in air-dried soil was increased due to the hydrophobicity of the soil surface and cracks. In this regard, there are many reasons for lower flow rate under initial saturated conditions such as the settlement of the soil under saturation state (Zainal and Al-Ebadi, 2016), sedimentation of colloid and clay particles in macropores in 24-hour saturated soil and trapping water packets in soil micropores, dead and intra-aggregated pores during leaching in initially saturated soil (Mohanty et al., 2016). In the ISC, the large portion of water moves in macropore paths or mobile zones (Šimůnek et al., 2003), while the presence of matrix potential increases flow rate in air-dried soil and causes the water to move in micropores' paths as well.

4.3.2 THE EFFECT OF VERMICOMPOST ON SOIL PHYSICAL AND CHEMICAL PROPERTIES

Some properties of the unamended and amended soils are shown in Table 4.3. Vermicompost increased soil nitrate, organic carbon, exchangeable sodium, EC, and TDS to 119%, 12%, 45%, 109%, and 109%, respectively, and decreased soil pH to 2.5%. This variation reveals high nutrients and salt in vermicompost that exacerbate soil salinity problems. The previous studies confirmed this

TABLE 4.3

The Physical and Chemical Properties of Unamended and Vermicompost-Amended Soils

Parameter	Unamended Soil	Vermicompost-Amended Soil	Change Percentage
Bulk density (g/cm^3)	1.21	1.16	−4.1
Porosity (%)	57.9	58.7	1.4
Saturated hydraulic conductivity (cm/hr)	4.4	6.3	43
Organic carbon (%)	0.94	1.05	12
Nitrate (mg/L)	38.4	84.1	119
Exchangeable sodium (mg/L)	33.2	48.1	45
pH	8.1	7.9	−2.5
EC (dS/m)	1.27	2.66	109
TDS (mg/L)	666.6	1396.1	109

finding (e.g., Abbaspour and Golchin, 2011; Paradelo et al., 2011). Lower pH of vermicompost than pH of unamended soil (Tables 4.1 and 4.3) and releasing of organic acid as a result of organic matter decomposition decreased vermicompost-amended soil (Brady and Weil, 2001).

The lower bulk density, higher porosity, and saturated hydraulic conductivity of amended soil than unamended soil were other effects of vermicompost (Table 4.3). The organic matter fraction of vermicompost did not show a considerable reduction effect on soil aggregation due to the shortage of the time of the experiment; hence, it slightly diminished soil bulk density and amplified soil porosity. Therefore, significant increase of saturated hydraulic conductivity can be related to location or distribution of vermicompost particles through soil columns.

4.3.3 THE EFFECT OF VERMICOMPOST ON SOIL WATER RETENTION CURVE

The soil water retention curves of both soils and their hydraulic parameters are presented in Figure 4.2 and Table 4.4, respectively. Vermicompost increased soil water capacity of saturation (θ_s) and permanent wilting points to 1.4% and 12%, respectively (Figure 4.2 and Table 4.4). Two soil retention curves intersected each other at suction 16 and 450 cm. The moisture content of VC-amended soil was reduced

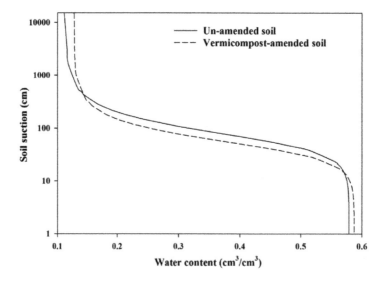

FIGURE 4.2 The soil water retention curves for unamended and vermicompost-amended soils.

TABLE 4.4
The Soil Hydraulic Coefficients Corresponding to Unamended and Amended Soils

Soil	θ_s (cm³/cm³)	θ_r (cm³/cm³)	α (1/cm)	n (–)
Unamended	0.579	0.114	0.016	2.417
Vermicompost-amended soil	0.587	0.128	0.023	2.508
Change percentage	1.4	12	44	3.8

with any suction within the mentioned domain. It shows an increase in macropores of vermicompost-amended soil, which are unable to retain soil water and drain rapidly before field capacity point. Increase in saturated hydraulic conductivity (Table 4.3) and α to 44% (Table 4.4) confirms more numbers of macropores in amended soil (Vervoort et al., 1999). Also, the increased micropores-adsorbent surfaces in amended soil retained more moisture content than those of unamended soil after 450-cm suction. The greater n for amended soil was another result of vermicompost application (Table 4.4).

4.3.4 THE EFFECT OF FLOW RATE, VERMICOMPOST, AND INITIAL SOIL MOISTURE ON SOIL LEACHING

The leaching curves of both soils are shown in Figure 4.3. This figure contains four subplots, which each shows EC variations in the effluent as function of flow rate, S_w, initial soil moisture,

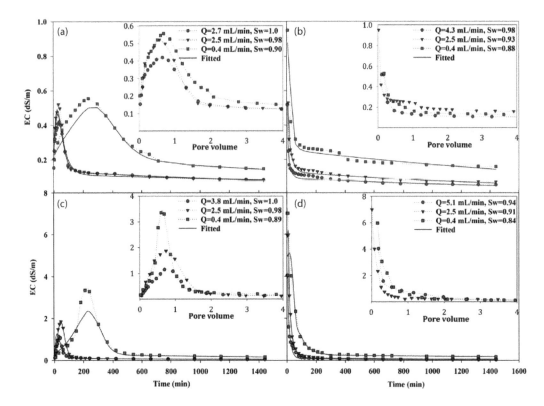

FIGURE 4.3 The curves of EC variations as a function of time and pore volume for unamended soil under initial saturated and air-dried conditions (a and b), and vermicompost-amended soil under initial saturated and air-dried conditions (c and d) with saturated-unsaturated flow rates.

and fertilizer application based on time (main subplots) and pore volumes (inserted subplots). The leaching behavior of soil solute under initial saturated conditions was different in comparison to initial air-dried conditions. EC values were maximum at the beginning of leaching under initial air-dried conditions due to high-releasing precipitated solute on surfaces of soil particles before great ion-exchange to soil microsites. EC values were continuously decreased until the end because of desorption of remained solutes from soil microsites (Figure 4.3b and d). Under initial saturated conditions, solute leaching was initially increased to maximum rate in both soils; then, it was gradually declined until the end. The behavior of solute leaching under initial saturated condition is dependent on the ionic exchange process, tendency to uniform distribution of ions at saturated media, and upward transfer of soluble salts during 24-hour saturating (Chaganti et al., 2015). Finally, these phenomena delayed solute leaching and decreased maximum EC value at the effluent.

According to Figure 4.3, more salts were extracted from the VC-amended soil in comparison to unamended soil under both initial soil moisture conditions due to high initial salt contents in the vermicompost and high hydraulic conductivity (Table 4.3). The flow rates of 4.3–5.1, 2.5, and 0.4 mL/min increased maximum EC values to 0.52, 0.53, and 0.95 dS/m in unamended soil and 4.1, 6, and 7 dS/m in amended soil under initial air-dried conditions, respectively (Figure 4.3b and d). In this condition, soil leaching with flow rate 0.4 mL/min (S_w = 0.84) had a lower EC value than other flow rates in VC-amended soil during leaching, while EC values for flow rate 0.4 mL/min (S_w = 0.88) were more than others in unamended soil. The lowest flow rate had less pore-water velocity, which increased time for ionic exchange processes from vermicompost particles to surfaces of soil particles, decreasing concentrations in effluent in VC-amended soil.

The EC values at the peak of leaching curves for flow rates of 2.7–3.8, 2.5, and 0.4 mL/min were 0.42, 0.52, and 0.55 dS/m in unamended soil, and 1.2, 1.9, and 3.4 dS/m in vermicompost-amended soil under initial saturated conditions, respectively (Figure 4.3a and c). The peak of curves was observed at lower one pore volume and was delayed by desaturation to more than 200 min, due to slower soil water velocity and increased residence and intact times of water in soil pores. These resulted in dissolution of salts at higher times and lower water volume to concentrate the EC of effluent.

The fitted solute leaching parameters consisting of F, D_0, K_D, and α are shown in Table 4.5. The values of RMSE ranged between 0.03 to 0.37 dS/m, which were increased by desaturation (lower flow rate). The lower accuracy of the two-site model was revealed in within the range of of peak of leaching curves due to a sharp change of solute leaching in effluent (Figure 4.3a and b). Overall, the accuracy of the model was acceptable for flow rates of 2.7–3.8, 4.3–5.1, and 2.5 mL/min.

The solute dispersivity ranged 0.11–0.15 cm in unamended soil and 0.14–0.22 cm in amended soil (Table 4.5). As stated earlier, the coefficient of longitudinal dispersivity was determined using a conservative tracer of nitrate under initial saturated conditions. The coefficients for all flow rates were also applied under air-dried conditions as an initial value in the modeling process. Determining dispersivity using injection of a tracer in air-dried soil is very complicated due to mixing the initial concentration of the tracer in soil with the influent tracer concentration, and the role of initial soil matric potential on movement of flow and tracer. The results showed that the model was not sensitive to the coefficient of the applied dispersivity for flow rates of 4.3–5.1 and 2.5 mL/min under air-dried conditions, while λ was increased for flow rate of 0.4 mL/min. Vermicompost increased dispersivity by changing the scale and form of soil pores and expanding the range of pore sizes (Bagheri et al., 2019). Dispersivity was increased for both soils by desaturation due to great tortuosity of flow paths (inverse amounts of τ in Table 4.2) and the reduction of cross section of pore-flow field (Tangkoonboribun et al., 2006).

As shown in Table 4.5, the fractions of instantaneous sites were found less than 0.5 (50%). It shows that the dominant adsorption/desorption process is kinetic during solute leaching. Higher fractions were noticed in VC-amended soil because of higher solute concentration and more

TABLE 4.5

The Fitted Parameters of Leaching Curves as a Function of Initial Soil Moisture and Flow Rate in Both Unamended and Amended Soils

Soil	Initial Conditions	Flow Rate (cm³/min)	λ (cm)	F (–)	D_0 (cm²/s)	K_D (L/Kg)	β (1/min)	RMSE (dS/m)
Unamended	Air-dried	4.3	0.11	0.014	0.054	10.3	0.00053	0.13
		2.5	0.12	0.029	0.052	10.3	0.00062	0.10
		0.4	0.15	0.0001	0.036	5.82	0.00054	0.37
	Saturated	2.7	0.11	0.007	0.0045	5.56	0.00044	0.04
		2.5	0.12	0.074	0.0068	0.54	0.00022	0.07
		0.4	0.13	0.13	0.0006	2.02	0.00025	0.20
Vermicompost-amended	Air-dried	5.1	0.14	0.19	0.11	2.76	0.00069	0.11
		2.5	0.15	0.15	0.067	1.82	0.00044	0.06
		0.4	0.22	0.1	0.034	2.14	0.00027	0.06
	Saturated	3.8	0.14	0.05	0.018	4.35	0.00033	0.10
		2.5	0.15	0.033	0.0015	1.52	0.00042	0.03
		0.4	0.19	0.27	3×10^{-7}	1.6	0.00043	0.26

cation-exchangeable soil surfaces. Flow rate reduction decreased the coefficient of F under air-dried conditions and increased under initial saturated conditions.

The diffusion coefficients of solute in solution (D_0) are shown in Table 4.5. Desaturation decreased diffusion coefficients of solute in both soils under both initial moisture conditions because dissolving soil salts in lower influent water volume caused more ionic strength of soil solution, great ionic interaction, higher viscosity, the thinner hydration layer of cations, and variety of ions (Gao et al., 2007; Tsaih and Chen, 1999). The coefficient of D_0 was decreased by changing initial soil moisture from air-dried to saturated conditions because of differences in ionic strength and ions variety after high dilution of solute with wetting front in air-dried conditions in comparison to uniform distribution of ions at saturated state (Table 4.5). Vermicompost increased the solute diffusion coefficient at a higher flow rate, while it decreased at a lower water saturation. Because lower pore-water velocity maintained realized solute from vermicompost at exchangeable soil sites, that decreased the tendency of ions to diffuse through the soil water solution.

The distribution coefficients of solute (K_D) are presented in Table 4.5. This coefficient means adsorption rate of solute to adsorbent soil surfaces to its desorption rate in equilibrium state, or the amount of solute at the solid phase to the presence of solute in the liquid phase. The flow rate reduction decreased values of K_D by streaming water in soil micropores, which increased intact time in soil column and released more solute from soil surface to liquid phase at a higher desorption rate. Solute leaching under initial air-dried conditions had higher K_D than initial saturated conditions due to great initial desorption by wetting front (Table 4.5, Figure 4.3b and d). Vermicompost application decreased values of solute distribution coefficient, which shows more tendency of adsorbent soil sites to the solute desorption process during leaching due to high release of solute.

The first-order rate constant (β) for the sorption/desorption kinetic process was in the range of 0.00022–0.00063 min^{-1} in unamended soil and 0.00027–0.00069 min^{-1} in amended soil (Table 4.5). An increase of β constant was observed as a result of changes in initial soil moisture conditions to air-dried state and vermicompost application at some flow rates. This shows more tendency of kinetic desorption to reach equilibrium with a liquid phase, or more desorption rate in VC-amended soil and initial air-dried conditions. Although desaturation changed values of β in both soils under both initial conditions, a clear trend was not noted as a result of desaturation or changing flow rate (Table 4.5).

4.4 CONCLUSION

The experimental and numerical modeling was conducted to investigate the effect of irrigation regimes, soil moisture conditions, and vermicompost application on solute leaching curves and their transport coefficients. The experimental results showed desaturation increased maximum EC values in effluent and delayed peak of leaching curves as a function of time. More solute was leached from vermicompost-amended soil in comparison to unamended soil due to higher salt in vermicompost. The leaching curves for initial air-dried conditions were different in comparison to initial saturated condition because of ionic exchange process and upward water flow during 24-hour saturating.

The modeling results revealed higher accuracy of a two-site model for higher flow rate. The fractions of instantaneous desorption sites (F) were less than 50% of total desorption sits, which shows dominance of kinetic desorption during the leaching. Higher instantaneous fraction and lower distribution coefficient (K_D) were found in soil amended with vermicompost. Change in initial soil moisture conditions to air-dried increased the first-order rate constant (β) and diffusion coefficient (D_0). The direct correlation was found between flow rate-water saturation and the values of diffusion coefficient. This coefficient was decreased by desaturation due to the significant ionic interaction and strength, higher viscosity, and the thinner hydration ion layer. Dispersivity was increased by desaturation and vermicompost application because of changes of pore-scale and pore-flow field.

The findings in this study disclosed the role of soil and water management to improve soil chemical conditions by exceeding solute leaching. Regarding the results, the reduction flow rate, initial soil moisture conditions, and organic fertilizer are effective and crucial facts for farmers and decision makers to transport solute from root zone more effectively with less water usage.

REFERENCES

Abbaspour, A., Golchin, A., 2011. Immobilization of heavy metals in a contaminated soil in Iran using di-ammonium phosphate, vermicompost and zeolite. *Environ. Earth Sci.* 63, 935–943. https://doi.org/10.1007/s12665-010-0762-5

Abu, S.T., 2013. Evaluating long-term impact of land use on selected soil physical quality indicators. *Soil Res.* 51, 471–476. https://doi.org/10.1071/SR12360

Bagheri, H., Zare Abyaneh, H., Izady, A., Brusseau, M.L., 2019. Modeling the transport of nitrate and natural multi-sized colloids in natural soil and soil amended with vermicompost. *Geoderma* 354, 113889. https://doi.org/10.1016/j.geoderma.2019.113889

Bejat, L., Perfect, E., Quisenberry, V.L., Coyne, M.S., Haszler, G.R., 2000. Solute transport as related to soil structure in unsaturated intact soil blocks. *Soil Sci. Soc. Am. J.* 64, 818–826. https://doi.org/10.2136/sssaj2000.643818x

Bradford, S.A., Yates, S.R., Bettahar, M., Simunek, J., 2002. Physical factors affecting the transport and fate of colloids in saturated porous media. *Water Resour. Res.* 38, 63-1–63-12. https://doi.org/10.1029/2002wr001340

Brady, N.C., Weil, R.R., 2001. *The Nature and Properties of Soils*, 13th ed. Prentice Hall, Upper Saddle River, NJ.

Carter, M.R., Gregorich, E., 2006. *Soil Sampling and Methods of Analysis, Measurement.* https://doi.org/10.1017/S0014479708006546

Chaganti, V.N., Crohn, D.M., Šimůnek, J., 2015. Leaching and reclamation of a biochar and compost amended saline-sodic soil with moderate SAR reclaimed water. *Agric. Water Manag.* 158, 255–265. https://doi.org/10.1016/j.agwat.2015.05.016

Edwards, W.M., Shipitalo, M.J., Dick, W.A., Owens, L.B., 1992. Rainfall intensity affects transport of water and chemicals through macropores in no-till soil. *Soil Sci. Soc. Am. J.* 56, 52–58. https://doi.org/10.2136/sssaj1992.03615995005600010008x

Fetter, C.W., Thomas Boving, D.K., 2017. *Contaminant Hydrogeology*, 3rd ed. Waveland Press, Long Grove, IL.

Gao, G.-H., Shi, H.-B., Yu, Y.-X., 2007. Mutual diffusion coefficients of concentrated 1:1 electrolyte from the modified mean spherical approximation. *Fluid Phase Equilib.* 256, 105–111. https://doi.org/10.1016/J.FLUID.2006.11.017

Knappenberger, T., Flury, M., Mattson, E.D., Harsh, J.B., 2014. Does water content or flow rate control colloid transport in unsaturated porous media? *Environ. Sci. Technol.* 48, 3791–3799. https://doi.org/10.1021/es404705d

Lekfeldt, J.D.S., Kjaergaard, C., Magid, J., 2017. Long-term effects of organic waste fertilizers on soil structure, tracer transport, and leaching of colloids. *J. Environ. Qual.* 46, 862–870. https://doi.org/10.2134/jeq2016.11.0457

Lim, S.L., Wu, T.Y., Lim, P.N., Shak, K.P.Y., 2015. The use of vermicompost in organic farming: Overview, effects on soil and economics. *J. Sci. Food Agric.* 95, 1143–1156. https://doi.org/10.1002/jsfa.6849

Merdun, H., Meral, R., Riza Demirkiran, A., 2008. Effect of the initial soil moisture content on the spatial distribution of the water retention. *Eurasian Soil Sci.* 41, 1098–1106. https://doi.org/10.1134/S1064229308100128

Mohanty, S.K., Saiers, J.E., Ryan, J.N., 2016. Colloid mobilization in a fractured soil: Effect of pore-water exchange between preferential flow paths and soil matrix. *Environ. Sci. Technol.* 50, 2310–2317. https://doi.org/10.1021/acs.est.5b04767

Moldrup, P., Olesen, T., Rolston, D.E., Yamaguchi, T., 1997. Modeling diffusion and reaction in soils: VII. Predicting gas and ion diffusivity in undisturbed and sieved soils. *Soil Sci.* 162, 632–640. https://doi.org/10.1097/00010694-199709000-00004

Mualem, Y., 1976. A new model for predicting the hydraulic conductivity of unsaturated porous media. *Water Resour. Res.* 12, 513–522. https://doi.org/10.1029/WR012i003p00513

Nada, W.M., van Rensburg, L., Claassens, S., Blumenstein, O., 2011. Effect of vermicompost on soil and plant properties of coal spoil in the Lusatian region (Eastern Germany). *Commun. Soil Sci. Plant Anal.* 42, 1945–1957. https://doi.org/10.1080/00103624.2011.591469

Paradelo, R., Moldes, A.B., Barral, M.T., 2011. Carbon and nitrogen mineralization in a vineyard soil amended with grape marc vermicompost. *Waste Manag. Res.* 29, 1177–1184. https://doi.org/10.1177/0734242X10380117

Richards, L.A., 1931. Capillary conduction of liquids through porous mediums. *J. Appl. Phys.* 1, 318–333. https://doi.org/10.1063/1.1745010

Schaap, M.G., Van Genuchten, M.T., 2006. A modified Mualem-van Genuchten formulation for improved description of the hydraulic conductivity near saturation. *Vadose Zo. J.* 5, 27–34. https://doi.org/10.2136/vzj2005.0005

Schjønning, P., Munkholm, L.J., Elmholt, S., Olesen, J.E., 2007. Organic matter and soil tilth in arable farming: Management makes a difference within 5–6 years. *Agric. Ecosyst. Environ.* 122, 157–172. https://doi.org/10.1016/j.agee.2006.12.029

Shang, J., Flury, M., Chen, G., Zhuang, J., 2008. Impact of flow rate, water content, and capillary forces on in situ colloid mobilization during infiltration in unsaturated sediments. *Water Resour. Res.* 44. https://doi.org/10.1029/2007WR006516

Šimůnek, J., Jarvis, N.J., Van Genuchten, M.T., Gärdenäs, A., 2003. Review and comparison of models for describing non-equilibrium and preferential flow and transport in the vadose zone. *J. Hydrol.* 272, 14–35. https://doi.org/10.1016/S0022-1694(02)00252-4

Sommerfeldt, T.G., Chang, C., Carefoot, J.M., 1982. A laboratory study on the effects of soil moisture content, texture, and timing of leaching on N loss from two southern Alberta soils. *Can. J. Soil Sci.* 62, 407–413. https://doi.org/10.4141/cjss82-044

Tangkoonboribun, R., Rauysoongnern, S., Rambo, P.V., Tumsan, B., 2006. Effect of organic and clay material amendment on physical properties of degraded sandy soil for sugarcane production. *Sugar Tech.* 8, 44–48. https://doi.org/10.1007/BF02943740

Tiemeyer, B., Pfaffner, N., Frank, S., Kaiser, K., Fiedler, S., 2017. Pore water velocity and ionic strength effects on DOC release from peat-sand mixtures: Results from laboratory and field experiments. *Geoderma* 296, 86–97. https://doi.org/10.1016/j.geoderma.2017.02.024

Tsaih, M.L., Chen, R.H., 1999. Effects of ionic strength and pH on the diffusion coefficients and conformation of chitosans molecule in solution. *J. Appl. Polym. Sci.* 73, 2041–2050. https://doi.org/10.1002/(SICI)1097-4628(19990906)73:10<2041::AID-APP22>3.0.CO;2-T

van Genuchten, M.T., 1980. Closed-form equation for predicting the hydraulic conductivity of unsaturated soils. *Soil Sci. Soc. Am. J.* 44, 892–898. https://doi.org/10.2136/sssaj1980.03615995004400050002x

van Genuchten, M.T., Leij, F.J., Yates, S.R., 1991. The RETC code for quantifying the hydraulic functions of unsaturated soils. Technical Report EPA/600/2-91/065, US Environmental Protection Agency, Oklahoma.

Vervoort, R.W., Radcliffe, D.E., West, L.T., 1999. Soil structure development and preferential solute flow. *Water Resour. Res.* 35, 913–928. https://doi.org/10.1029/98WR02289

Zainal, A. E., A Al-Ebadi, L.H., 2016. The effect of varying degree of saturation on settlement rate of. *Appl. Res. J.* 2, 27–42.

5 Performance Evaluation of the BUDGET Model in Simulating Different Strategies of Irrigation

*Mona Golabi, Mohammad Albaji, and
Saeed Boroomand Nasab*

CONTENTS

5.1 INTRODUCTION

In several countries, water resources' current availability and future water supplies are at risk as a result of population increase and urbanization. Rainfall, a major source of fresh water, is likely to be severely affected by changes in global climate that are expected to account for about 20% of the global increase in water scarcity. Agriculture is responsible for 70% of all water use globally, and water use efficiency in this sector is very low, not exceeding 45% (Zhang and Yang, 2004).

The arid climate in Khuzestan Province in the southwest of Iran makes water resources the main factor for sustainable agricultural development. Hence water shortage is one of the major challenges in the arid region of Iran. This challenge is likely to intensify with population growth (Ehsani, 2005). For instance, the population in Iran has increased with a factor of 6.8 during the last 90 years, from under 10 million in 1922 to 75 million in 2010. With the current population growth rate, Iran's population will reach 100 million by the year 2025, which may outweigh the growth of food production. The annual per capita utilizable fresh water in Iran has decreased from 13,000 m^3 in 1922 to 1,733 m^3 in 2010. Countries with annual per capita water availability of less than 1,700 m^3 are denoted as water stressed, and less than 1,000 m^3 as water scarce (Falkenmark et al., 1989). Taking into account the increase in population up to 100 million by the year 2025, Iran will need 170 billion m^3 of water per year to be above the water stress zone and 100 billion m^3 of water per year to avoid being a water-scarce country. However, the total annual renewable water resources in Iran are assessed at 130 billion m³, of which 95 billion m³ of surface water and 25 billion m³ of ground water are utilizable (Ehsani, 2005).

There are several strategies to face with water deficit such as low irrigation, using drain water, management of water resources and crops. Recently, water productivity is considered a main factor to management of water use. In addition, modeling techniques applied to agriculture can be useful to define research priorities and understand the basic interactions of the soil-plant-atmosphere system. Using a model to estimate the importance and the effect of certain parameters, a researcher can notice which factors can be most useful. Also, by using modeling, the complex relationship between parameters is considered.

One of the models that can be used in agriculture is the BUDGET model, which evaluates the effect of water balance components on yield, soil moisture content, and water productivity under different irrigation methods and intervals. BUDGET calculates the water storage in a soil profile as affected by the infiltration of rain and irrigation water and withdrawal of water by crop evapotranspiration and deep percolation for a given period (Raes, 2001).

Numerous models have been developed and used by researchers for simulation of water balance in the cropped field, which is mentioned in the following.

Kipkorir et al. (2002) performed seasonal water production functions and yield response factors for maize and onion in Perkerra, Kenya, by using the BUDGET model. Water production functions (wpf) giving the relation between crop yield and water application under furrow irrigation on a clay loam soil in the semiarid region in Kenya (Perkerra) were derived for maize and onion. The seasonal yield response factors, K_y, giving the relationship between evapotranspiration deficit and yield depression for maize and onion for the area were computed as 1.21 and 1.28, respectively. Analytical analysis using the derived wpf for maize and existing conditions in an irrigation system located in the area confirmed that if rainfall is significant, deficit irrigation will be more attractive, and at a certain point, it is profitable to cultivate all available area.

Application of crop water productivity models for better utilization of water resources in Uzbekistan was done by Nasyrov (2008). This research focused on the arid ecosystems located in the southern part of Uzbekistan. With the help of the soil water balance model BUDGET, an assessment was made of the region's potential productivity for different kinds of rainfall regimes. BUDGET is a robust simulation tool and only requires a minimum of input data, which is readily available or can be easily collected. A local meteorological station in Tim delivered most of the required climatic data, while most of the crop- and soil-related parameters were found in literature or were estimated by common sense. Once a renewed dataset is available, the BUDGET model can be calibrated and validated, and more specific results on the biomass production in the region can be expected.

Malekpour and Babazadeh (2011) performed simulation of yield decline as a result of water stress using the BUDGET model. To evaluate the model, the simulated yields for winter wheat (grown in Sharif Abad district) under various levels of water stress were compared with observed yields. The result showed simulated crop yields agreed well with observed yields for this location using a multiplicative approach (in comparison with minimal and seasonal approaches). The determination coefficient (R^2) between observed and simulated yields ranged from 0.81 to 0.92 with very high modeling efficiencies. The root of mean square error (RMSE) values was relatively small and ranged between 6 and 14. A sensitivity analysis showed that the model is robust and that good estimates can be obtained by using indicative values for the required crop and soil parameters.

Ali et al. (2011) estimated evapotranspiration using a simulation model. In this study, the BUDGET model was used to evaluate its performance to simulate water balance in a wheat field. The BUDGET model is composed of a set of validated subroutines describing the various processes involved in water extraction by plant roots and soil water movement in the absence of a water table. The model was run to simulate evapotranspiration values with the actual observed weather, crop, and soil data for three years (2002–2005), obtained from the experimental station of the Bangladesh

Institute of Nuclear Agriculture (BINA). The input data of the model are separated into four stages, and the value of Kc and root depth are different for each stage. Evaluation of model performance is done with both graphical display and statistical criteria. The simulated values fall close to the 1:1 line, indicating better performance. The statistical parameters such as standard deviation (SD), standard error (SE), and coefficient of variation (CV) of simulated and actual evapotranspiration values are found to be 21.07 and 29.23; 4.49 and 6.23; and 38.03 and 50.75, respectively. Both the standard error and coefficient of variation for simulated values are found lower than the observed values, indicating stability of the model output. The coefficient of determination value ($R^2 = 0.83$) is high for this model, which indicates good simulation performance. The relative error (RE) is 23.28% and model efficiency (EF) is 78.95%, which means that the simulation of actual evapotranspiration is satisfactory. The value of index of agreement (IA) is 0.918, which indicates a very good performance of the model. The overall statistical parameters of simulation period are at a satisfactory level. Therefore, the BUDGET model is able to predict actual evapotranspiration for any level of soil moisture with reasonable accuracy. The model can be used in planning, management, and operation of an irrigation project for judicious use of water with the limited inputs, and it is especially suitable for countries where modeling of crop yield is needed under water stress conditions.

Kenjabaev et al. (2013) performed evaluation of the BUDGET model in simulating cotton and wheat yields and soil moisture in the Fergana Valley. The objective of this study was to adapt and test the ability of the soil water balance model BUDGET (ver. 6.2) to simulate cotton as well as wheat yield and soil water content under current agronomic practices in the Fergana Valley. Crop yield and soil moisture content data, collected and measured from sites in 2010 and 2011, were compared with model simulations. Results showed that the BUDGET model can be used to predict cotton yield and soil water content with acceptable accuracy using the minimum approach. However, predicted wheat yield was high compared to the observed and reported yield. Overall, the relationship between the observed and predicted cotton and wheat yield for both sites combined produced R^2 of 0.91 and 0.15, RMSE of 0.24, and 1.64 t ha^{-1}, relative Nash-Sutcliffe efficiency (Erel) of 0.71 and 5.68 and index of agreement (d) of 0.48 and −0.54, respectively. Similarly, comparison of the observed and simulated soil moisture contents at the top 0–30 cm soil layer and soil water contents in the 90 cm profile yielded R^2 of 0.88 and 0.71–0.88, RMSE of 2.74% vol. and 21.4–28.7 mm, Erel of 0.87 and 0.53–0.81, respectively, and d around 1.0. Consequently, the BUDGET model can be a valuable tool for simulating both cotton yield and soil water content, particularly considering the fact that the model requires relatively minimal input data. Predicted soil water balance can be used to improve the current practice of irrigation water management, whereas simulated soil moisture content can be used to estimate capillary rise from groundwater in the UPFLOW model. However, performance of the model has to be evaluated under a wider range of agroclimatic and soil conditions in the future.

Misra et al. (2010) evaluated the amount of water use efficiency (WUE) in India on 20 durum wheat cultivars, under three water regimes (full irrigation, limited irrigation, and residual soil moisture) and during two seasons. WUE components were estimated using a soil water balance model (BUDGET), allowing comparison of environments in data-scarce situations. A highly significant correlation was noted between grain yield and grain Δ across water regimes. However, the associations between grain yield, Δ and m_a were found to depend highly on the water regime and environmental conditions. The association between grain yield and grain Δ was significant under full irrigation in season 1 and under residual soil moisture in season 2. Significant positive correlations were noted in both seasons between grain yield and leaf Δ under residual soil moisture and between grain yield and leaf ash content at anthesis under limited irrigation. A significant correlation was found across environments between grain and leaf Δ and T, the quantity of water transpired during the growth cycle, as estimated by the soil water balance model. T also significantly correlated to grain and leaf m_a. Variation in WUE across environments was driven more by runoff, drainage, and soil evaporation than by harvest index and transpiration. The associations between WUE and transpiration, runoff, and Δ were negative but not significant. WUE was significantly correlated with leaf and grain m_a at maturity. The study indicates that Δ and m_a can be used as indirect selection criteria for grain yield and suggests that m_a is a good predictor of transpiration, grain yield, and WUE across environments. The use of mechanistic models that allows differentiating between cultivars should permit in the near future to analyze the relationships between WUE, Δ, and m_a across cultivars and evaluate the possibility to use these traits as predictors of WUE in wheat-breeding programs.

However, far too little attention has been paid to application of the BUDGET model for simulation yield, soil moisture, and water productivity of sunflower in semiarid areas.

According to the importance of Khuzestan Province in agriculture, lack of water, and necessity of management, the present study was performed for simulation of sunflower yield, soil moisture content, and water productivity under different strategies of irrigation including conventional irrigation (CI), regulated deficit irrigation (RDI), and partial root zone drying (PRD) by using the soil water balance model BUDGET (ver. 6.2). Finally, the measurement and simulation data were compared.

5.2 MATERIALS AND METHODS

5.2.1 LOCATION AND DESCRIPTION OF STUDY SITES

The investigation was conducted in the research field of faculty of water sciences and engineering at Shahid Chamran University of Ahvaz. This area is located between 48° 39′ E longitude and 31° 18′ N latitude. The climatic condition of the study sites is characterized by data from the meteorological station "Ahvaz." The long-term (1966–2009) average annual temperature, precipitation, and evapotranspiration are 25.36°C, 230.3 mm, and 1,100 mm, respectively. Figure 5.1 shows the location of the study area in Iran.

The major irrigated broad-acre crops grown in this area are wheat, barley, and sunflower. But like other plains in Khuzstan Province, the important major crops are wheat and barley (Behzad et al., 2009; Naseri et al., 2009; Albaji et al., 2012b; Albaji and Alboshokeh. 2017). Also, like other plains in Khuzstan Province, the surface irrigation is the major irrigation system in Ahvaz plain (Landi et al., 2008; Albaji et al., 2008; Rezania et al., 2009; Boroomand Nasab et al., 2010; Albaji et al., 2012a; Albaji et al., 2014a; Albaji et al., 2014b; Albaji et al., 2016). The sunflower hybrid Hysun 33 was sown in agricultural year 2009–2010. Five irrigation treatments were applied. For all treatments, conventional irrigation was applied prior to V6–V8 stages (Plant with 6 and 8 leaves, respectively). Then, irrigation treatments were conducted as follows: CI—applied 100% of water requirements during the whole season; RDI70 and RDI50—applied 70% and 50% of water

FIGURE 5.1 The study area (black circle on map).

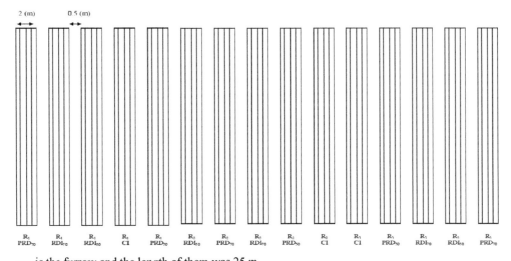

--- is the furrow and the length of them was 25 m

FIGURE 5.2   The field layout of the trial.

requirements, respectively; PRD70 and PRD50—both sides of plant row alternatively watered; applied 70% and 50% of water requirements, respectively. The layout of the experiments was a completely randomized block design with three iterations. Figure 5.2 presents the field layout of the trial. The furrow and the length of them were 25 m.

Before cultivation, field soil samples were collected, and the characteristics of soil were determined.

5.2.2   MODEL **BUDGET**

5.2.2.1   Model Description

BUDGET is composed of a set of validated subroutines describing the various processes involved in water extraction by plant roots and water movement in the soil profile in the absence of a water table. The model considers water storage in a soil profile affected by infiltration of rain and irrigation water including withdrawal of water by crop evapotranspiration and percolation for a given period (Raes, 2002).

To solve the one-dimensional vertical water flow and root water uptake, a finite difference technique is used. During periods of crop water stress, the resulting yield depression is estimated by means of yield response factors. By selecting appropriate time and depth criteria, irrigation schedules can be generated.

During periods of crop water stress, the resulting yield depression is estimated by means of yield response factors. By selecting appropriate time and depth criteria, irrigation schedules can be generated. Climatic data, crop parameters, and soil parameters are needed as input, while initial soil water and salt conditions and irrigation data are optional.

Simulations are performed in daily time steps. Finite difference technique is used to solve one-dimensional vertical water flow and root water uptake. Estimation of infiltration and percolation rates is based on exponential drainage function. Calculation of transpiration and separation of soil evaporation from evapotranspiration is based on the ground cover at maximum crop canopy, whereas on-site LAI measurements can be used to adjust ground canopy cover at specific growth stages. Relative yield decline, due to water stress during the growing stages, is based on yield response factor (K_y). Three approaches—such as seasonal, minimal, and multiplicative approaches—are considered in the BUDGET model to estimate expected crop yield. Further details of the subroutines, concepts, rationale, approaches, and procedures used to simulate the processes in the BUDGET model are given in its reference manual (Raes, 2002).

By calculating the water and salt content in a soil profile as affected by input and withdrawal of water and salt during the simulation period, the program is suitable:

- To assess crop water stress under different environmental conditions
- To estimate yield response to water
- To design irrigation schedules
- To study the building of salt in the root zone under averse irrigation conditions
- To evaluate irrigation strategies.

5.2.2.2 Model Input

The inputs of the model consist of climate, crop, soil, and irrigation management data. The climatic data option in BUDGET consists of daily, mean 10-day, or monthly ET0 (reference crop evapotranspiration) and rainfall observations. At run time, the 10-day and monthly data are processed to derive daily ET0 and rainfall data. By specifying and selecting a few appropriate crop parameters in a menu-driven environment, the program creates a complete set of parameters that can be displayed and updated if additional information is available. The soil profile may be composed of several soil layers, each with their specific characteristics. BUDGET contains a complete set of default characteristics that can be selected and adjusted for various types of soil layers. ETo was calculated using the "ETo Calculator" (Raes et al., 2009) based on the FAO Penman-Monteith equation (Allen et al., 1998). More details about the BUDGET model can be found in Raes (2002) and Raes et al. (2006).

5.2.2.3 Sensitivity Analysis

Before using a model, knowledge about its behavior and sensitivity to input parameters are necessary. Sensitivity analysis evaluates the effect of the input on output data. A sensitivity analysis was performed to identify the most influential factors on the model response. In sensitivity analysis, by using the input data, the model is run and output is considered as basic output. Then on the next step, one of the input models is changed and the model is performed. Finally, the result is compared with basic output and sensitivity coefficients that are calculated for each input. For calculation, a sensitivity coefficient formula is used formula (5.1).

$$S_c = \frac{\dfrac{\Delta W}{\overline{W}}}{\dfrac{\Delta P}{\overline{P}}} \tag{5.1}$$

where:
ΔW is the difference between output before and after change of input.
$\overline{W}$ is the average of output before and after change of input.
ΔP is the difference between the amounts of input data.
$\overline{P}$ is the average of input.

Compression between calculated coefficient and the information in Table 5.1, presented below, determine the degree of sensitivity of input data.

In current research, the investigated factors were Ritchie equation coefficients (f, c), depth of soil water extraction by evaporation and K_c for wet bare soil.

TABLE 5.1

The Amount of Variation of Sensitivity Coefficients

The Degree of Sensitivity	Amount of S_c
No sensitivity	$S_c = 0$
Low sensitivity	$0 \prec S_c \prec 0.3$
Moderate sensitivity	$0.3 \prec S_c \prec 1.5$
High sensitivity	$1.5 \prec S_c$

Source: Liu, H.F. et al., *J. Exp. Bot.*, 58, 3567–3580, 2007.

5.2.2.4 Model Calibration

One of the most important parts of application software is calibration. In this stage, some coefficients of software are determined locally. In the current paper, the data of one iteration were used for calibration. For calibration model, Ritchie equation coefficients (f, c) and K_c for wet bare soil locally were changed and in each step, the simulation data were compared with measurement data until simulation and measurement data presented acceptable statistical index.

5.2.2.5 Model Evaluation

5.2.2.5.1 Statistical Parameters Used to Compare the Model Output

For quantitative comparison of experimental and simulated analysis of residual errors, differences between measured and simulated values can be used to evaluate model performance. These are coefficient of determination (CD or R^2), normal root mean square error (NRMSE), Nash-Sutcliffe efficiency (NSE), and coefficient of residual mass (CRM). The mathematical expressions of these statistics are as follows:

$$1.\ R^2 = \frac{\left[\sum_{i=1}^{n} (O_i - O_{ave})(P_i - P_{ave}) \right]^2}{\sum_{i=1}^{n} (O_i - O_{ave})^2 \sum_{i=1}^{n} (P_i - P_{ave})^2}$$

$$2.\ NRMSE = \left[\frac{\sum_{i=1}^{n} (P_i - O_i)^2}{n} \right]^{1/2} \times \frac{100}{\overline{O}}$$

$$3.\ NSE = 1 - \frac{\sum_{i=1}^{n} (P_i - O_i)^2}{\sum_{i=1}^{n} (O_i - O_{ave})^2}$$

$$4.\ CRM = \frac{\sum_{i=1}^{n} O_i - \sum_{i=1}^{n} P_i}{\sum_{i=1}^{n} O_i}$$

where O and P are the observed and predicated values for the ith observation, and n is the number of observations. O_{ave} and P_{ave} are the average of observed and predicated values. If all observed and predicated values are equal, the amount of NRMSE and CRM will become zero and the amount of R^2 and NSE will become one (Moriasi et al., 2007).

5.3   RESULTS AND DISCUSSIONS

As mentioned before, soil samples from the field were collected, and physical and chemical properties of the soil were determined. The results are presented in Table 5.2.

According to Table 5.2 the average of bulk density is 1.55 (g/cm³) and the prevailing soil texture is loam.

Table 5.3 shows that according to the Wilcox diagram, the classification of water is C_3S_1. Sodium adsorption ratio was high and salinity was in the low level. This type of water is suitable for growing sunflower.

In order to adapt and test the ability of the soil water balance model BUDGET (ver. 6.2) to simulate sunflower yield, soil moisture content, and water productivity under different strategies, sensitive analysis was first performed for +50%. The result of this stage is shown in Figure 5.3.

TABLE 5.2
Physical and Chemical Properties of the Soil Within Experimental Field

Soil Depth (cm)	Sand (%)	Silt (%)	Clay (%)	Texture	Bulk Density (g/cm³)	EC *10³ (dS/m)	Field Capacity (%)	Permanent Wilting Point (%)
0–30	44	34	22	L	1.59	7.85	17.80	8.25
30–60	44	34	22	L	1.62	3.00	17.48	8.68
60–90	39	37	24	L	1.50	3.52	19.60	9.66
90–120	21	63	16	Si. L	1.49	18.70	8.90	8.90

TABLE 5.3
Average Properties of the Water During Study

EC (dS/m)	pH	TDS (mg/L)	SAR	Na⁺	Ca²⁺ (meq/L)	Mg²⁺	Classification
1.82	7.35	858	3.8	10.77	4.9	3.3	C_3S_1

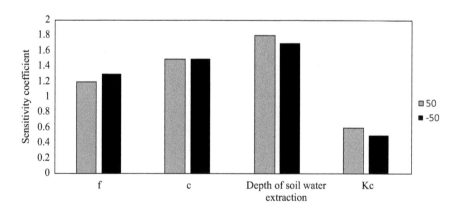

FIGURE 5.3   Sensitive analysis of BUDGET.

The results show that the model has moderate sensitivity for Ritchie equation coefficients (f, c) and K_c for wet bare soil ($0.3 < S_c < 1.5$). The amount of sensitivity for depth of soil water extraction by evaporation is high ($S_c > 1.5$). So should be attention to measurement or determination methods of these coefficients as input data.

After sensitive analysis, some coefficients of software were calibrated locally. As mentioned before, the data of one iteration were used for calibration. The main aim was determination of Ritchie equation coefficients (f, c) and K_c for wet bare soil locally. Results are presented in Table 5.4.

By using the result of calibration stage and input data, the model was performed. In order to conduct a comparison between measurement and simulation data, statistical indexes were used. The results are presented in Tables 5.5 through 5.7.

As mentioned before, results of comparison between simulated and observed yield are presented in Table 5.5. Results show that according to the R^2 values ranging from 0.87 to 0.95, a large fraction of the variation of observations is explained by the model. Observed and modeled sunflower

TABLE 5.4
Calibration Coefficients of Model

Fraction of Soil Surface Wetted for Alternated Furrows	Ritchie Equation's Coefficients		K_c for Wet Bare Soil	Depth of Soil Water Extraction by Evaporation
	f	c		
47	0.9	0.57	1.06	0.91

TABLE 5.5
Comparison Between Measured and Simulated Data of Yield

Parameter	Statistical Parameter	Conventional Irrigation (CI)	Regulated Deficit Irrigation		Partial Root Zone Drying	
			RDI_{70}	RDI_{50}	PRD_{70}	PRD_{50}
Yield	R^2	0.95	0.89	0.87	0.91	0.90
	NRMSE	11%	14%	15%	13%	14%
	NSE	0.83	0.79	0.78	0.81	0.80
	CRM	−0.006	−0.07	−0.07	−0.08	−0.07
Summation of ranking		4	13	16	9	11
Final ranking		1	4	5	2	3

TABLE 5.6
Comparison Between Measured and Simulated Data of Soil Moisture Content

Parameter	Statistical Parameter	Conventional Irrigation (CI)	Regulated Deficit Irrigation		Partial Root Zone Drying	
			RDI_{70}	RDI_{50}	PRD_{70}	PRD_{50}
Soil moisture content	R^2	0.76	0.72	0.71	0.74	0.74
	NRMSE	15%	17%	18%	15%	16%
	NSE	0.80	0.78	0.76	0.79	0.78
	CRM	−0.009	−0.06	−0.06	−0.05	−0.05
Summation of ranking		4	11	14	8	10
Final ranking		1	4	5	2	3

TABLE 5.7

Comparison Between Measured and Simulated Data Water Productivity

Parameter	Statistical Parameter	Conventional Irrigation (CI)	Regulated Deficit Irrigation		Partial Root Zone Drying	
			RDI_{70}	RDI_{50}	PRD_{70}	PRD_{50}
Water	R^2	0.88	0.82	0.80	0.85	0.83
productivity	NRMSE	8%	13%	15%	10%	10%
	NSE	0.85	0.81	0.75	0.83	0.84
	CRM	0.005	0.03	0.04	0.03	0.03
Summation of ranking		4	13	17	9	9
Final ranking		1	3	4	2	2

yield was correlated well giving the R^2. Conventional irrigation has the maximum amount of R^2 and NSE, and regulated deficit irrigation (RDI_{50}) has the minimum of them. Also, NRMSE and CRM of conventional irrigation are the minimum amount. Generally, the BUDGET model presents better simulation for conventional irrigation than other methods. The negative amount of CRM shows that the model estimates the amount of yield more than the actual value. Overestimation of yield in the current study is similar to Kenjabaev et al. (2013). Kenjabaev et al. (2013) showed that the BUDGET model estimated the amount of wheat yield more than the actual in Fergana Valley.

The analysis of coefficient of determination shows that the value of R^2 is between 0.71 to 0.76 and indicates that a suitable correlation exists between the simulated and observed. The magnitude of normal root mean square error (NRMSE) is also a useful parameter to evaluate model performance. If it tends to zero, the model will predict satisfactory results. According to Table 5.5, conventional irrigation has the least amount of NRMSE, and the BUDGET model estimated soil moisture content more than actual. Similar to yield, the BUDGET model simulated soil moisture content more than the actual amount. But a comparison between Tables 5.5 and 5.6 indicated that the accuracy of the model for prediction yield is more than the soil moisture content. It is related to the number of parameters that are used for estimation. The model applies more parameters for forecasting yield than soil moisture. Magodo (2007) used the BUDGET model for simulation yield and soil moisture of a maize farm in Zimbabwe; the amount of R^2 for yield was more than the soil moisture. Also, Kenjabaev et al. (2013) obtained a similar result.

The analysis of coefficient of determination shows that the value of R^2 is in general within the acceptable limit ($R^2 > 0.8$). The value of R^2 between 0.80 and 0.88 indicates that a good correlation exists between the simulated and observed water productivity. The magnitude of NRMSE is also a useful parameter to evaluate model performance. In an ideal condition, the values of 10%, 10%–20% are suitable; 20%–30% is moderate; more than 30% is inapplicable. According to Table 5.6, NRMSE is between 8% and 15% and shows ideal and suitable conditions. The NSE value is from negative infinity to 1. A higher value indicates a better agreement between the simulated and observed values. In this study, the value of NSE (0.75–0.85) shows a very good performance of the model. The above statistical parameters suggest that the overall performance of the BUDGET model in simulating actual water productivity from sunflower crop field is satisfactory. Conversely, for yield and soil moisture content, model was overestimated for simulation water productivity.

After evaluation of each irrigation strategy according to the statistical index, each treatment was ranked. The final ranking was obtained under consideration of the effect of all indexes. Results showed conventional irrigation in simulation was located in the first place, and forms of regulated deficit irrigation (70, 50) were in the next places. A similar result was obtained for simulation of soil moisture content and water productivity. Albaji et al. (2011) obtained the same result according to the statistical evaluation.

Based on the statistical indicators, it can be concluded that the simulation model BUDGET is able to predict yield, soil moisture, and water productivity with reasonable accuracy. Thus, the model can be used in planning, management, and operation of an irrigation project for judicious use of water.

5.4 CONCLUSIONS

This study was carried out to investigate the yield, soil moisture content, and water productivity using the BUDGET model to determine the irrigation method that gives the greatest production per unit irrigation water in the Ahvaz plain of Iran and to evaluate performance of the aforementioned software. Several parameters were used for the analysis of the field data and management; in order to manage the field process, it is necessary to use software. Software and models can consider the complex relationship between parameters and phenomena. One of these models is BUDGET, which is based on a water balance equation. The model was used in the current research, and the following conclusions can be drawn:

- The results show that the model has moderate sensitivity for Ritchie equation coefficients (f, c) and K_c for wet bare soil. The amount of sensitivity for depth of soil water extraction by evaporation is high.
- Some coefficients of the model were determined locally and used for the validation stage. Ritchie equation coefficients and K_c for wet bare soil and depth of soil water extraction by evaporation were obtained—0.9, 0.57, 1.06, and 0.91, respectively.
- As mentioned earlier, three irrigation methods were performed in the field. These methods consist of conventional irrigation (CI), partial root zone drying (PRD_{70}, and PRD_{50}), and regulated deficit irrigation (RDI_{70} and RDI_{50}). Overall, the result of CI was more acceptable than other methods. After CI, PRD and RDI were ranked second and third.
- The BUDGET model was overestimated for yield and soil moisture content and underestimated for water productivity.
- Based on the statistical indicators, the result of simulation of yield and water productivity was reasonably more accurate than soil moisture.
- According to the ranking of irrigation strategies, conventional irrigation and partial root zone drying (PRD_{70} and PRD_{50}) were in first and third places, respectively.
- The model will be useful to develop irrigation strategies and as a management tool.

Finally, under the local conditions in areas where water resources are more limited, according to the results of the BUDGET model simulation and statistical indicators, partial root zone drying (PRD_{70} and PRD_{50}) is an acceptable method for irrigation of sunflower in the same climate.

ACKNOWLEDGMENTS

We are grateful to the Research Council of Shahid Chamran University of Ahvaz for financial support (GN: SCU.WI98.281).

REFERENCES

Albaji, M., Boroomand Nasab, S., Kashkuli, H.A., Naseri, A.A., Sayyad, G., Jafari, S. 2008. Comparison of different irrigation methods based on the parametric evaluation approach in North Molasani Plain, Iran. *Journal of Agronomy* 7 (2), 187–191.
Albaji, M., Behzad, M., Borromand Nasab, S., Naseri, A.A., Shahnazari, A., Mesjarbashee, M., Judy, F, Jovzi, M. 2011. Investigation on the effects of conventional irrigation (CI), regulated deficit irrigation (RDI) and partial root zone drying (PRD) on yield and yield components of sunflower (*Helianthus annuus* L.). *Research on Crops* 12 (1), 142–154.

Albaji, M., Boroomand Nasab, S., Hemadi, J. 2012a. Comparison of Different Irrigation Methods Based on the Parametric Evaluation Approach in West North Ahvaz Plain. In: *Problems, Perspectives and Challenges of Agricultural Water Management*, M. Kumar, (ed.), InTech, Croatia, pp. 259–274.

Albaji, M., Papan, P., Hosseinzadeh, M., Barani, S. 2012b. Evaluation of land suitability for principal crops in the Hendijan region. *International Journal of Modern Agriculture* 1 (1), 24–32.

Albaji, M., Golabi, M., Boroomand Nasab, S., Jahanshahi. M. 2014a. Land suitability evaluation for surface, sprinkler and drip irrigation systems. *Transactions of the Royal Society of South Africa* 69 (2), 63–73.

Albaji, M., Golabi, M., Piroozfar, V.R., Egdernejad, A., Nazari Zadeh, F. 2014b. Evaluation of agricultural land resources for irrigation in the Ramhormoz Plain by using GIS. *Agriculturae Conspectus Scientificus* 79 (2), 93–102.

Albaji, M., Golabi, M., Hooshmand, A.R., Ahmadee, M. 2016. Investigation of surface, sprinkler and drip irrigation methods using GIS. *Jordan Journal of Agricultural Sciences* 12 (1), 211–222.

Albaji, M., Alboshokeh, A. 2017. Assessing agricultural land suitability in the Fakkeh region, Iran. *Outlook on Agriculture* 46 (1), 57–65.

Ali, M., H. Paul, H., Haque, M.R. 2011. Estimation of evapotranspiration using a simulation model. *Journal of the Bangladesh Agricultural University* 9(2), 257–266.

Allen, R.G., Pereira, L.S., Raes, D., Smith, M. 1998. Crop evapotranspiration: Guidelines for computing crop water requirements. *Irrigation and Drainage* Paper 56. FAO, Rome.

Behzad, M., Albaji, M., Papan, P., Boroomand Nasab, S. 2009. Evan region qualitative soil evaluation for wheat, barley, alfalfa and maize. *Journal of Food, Agriculture & Environment* 7 (2), 843–851.

Boroomand Nasab, S., Albaji, M., Naseri, A.A. 2010. Investigation of different irrigation systems based on the parametric evaluation approach in Boneh Basht plain, Iran. *African Journal of Agricultural Research* 5 (5), 372–379.

Ehsani, M. 2005. A vision on water resources situation, irrigation and agriculture production in Iran. *ICID 21st European Regional Conference*. Germany and Polland.

Falkenmark, M., Lundquist, J., Widstard, C. 1989. Macro-scale water scarcity requires micro-scale approaches: Aspect of vulnerability in semi-arid development. *Natural Resources Forum* 13, 258–267.

Kenjabaev, S., Forkutsa, I., Bach, M., Frede, H.G. 2013. Performance evaluation of the BUDGET model in simulating cotton and wheat yield and soil moisture in Fergana valley. *Center for International Development and Environmental Research, Forum of the International Conference-Natural Resource Use in Central Asia: Institutional Challenges and the Contribution of Capacity Building*. Giessen, Germany.

Kipkorir, E.C., Raes, D., Massawe, B. 2002. Seasonal water production functions and yield response factors for maize and onion in Perkerra, Kenya. *Agricultural Water Management* 56, 229–240.

Landi, A., Boroomand-Nasab, S., Behzad, M., Tondrow, M.R., Albaji, M., Jazaieri, A. 2008. Land suitability evaluation for surface, sprinkle and drip irrigation methods in Fakkeh Plain, Iran. *Journal of Applied Sciences* 8 (20), 3646–3653.

Liu, H.F., Genard, M., Guichard, S.S., Bertin, N. 2007. Model assisted analysis of tomato fruit growth in relation to carbon and water fluxes. *Journal of Experimental Botany* 58 (13), 3567–3580.

Malekpour, M., Babazadeh, H. 2011. Simulation of yield decline as a result of water stress using BUDGET model. *International Journal of Agricultural Science and Research* 2, 3.

Magodo, L. 2007. Determination of water productivity of maize varieties grown in Zimbabwe. *M. Sc Thesis*. University of Zimbabwe.

Misra, S.C., Shinde, S., Geerts, S., Rao, V.S., Monneveux, P. 2010. Can carbon isotope discrimination and ash content predict grain yield and water use efficiency in wheat? *Agricultural Water Management* 97, 57–65.

Moriasi, D.N., Arnold, J.G., Van Liew, M.W., Bingner, R.L., Harmel, R.D., Veith, T.L. 2007. Model evaluation guidelines for systematic quantification of accuracy in watershed simulations. *Transactions of the ASABE*, 50 (3), 885–900.

Naseri, A.A., Albaji, M., Boroomand Nasab, S., Landi, A., Papan, P., Bavi, A. 2009. Land suitability evaluation for principal crops in the Abbas Plain, Southwest Iran. *Journal of Food, Agriculture & Environment* 7 (1), 208–213.

Nasyrov, M.G. 2008. Application of crop water productivity models for better utilization of water resources in Uzbekistan. *Proceedings of the 1st Technical Meeting of Muslim Water Researchers Cooperation (MUWAREC)*. Malaysia.

Raes, D. 2001. *BUDGET: A Field Water Balance Model*. Institute for Land and Water Management, Leuven, Belgium.

Raes, D. 2002. *BUDGET—A Soil Water and Salt Balance Model. Reference Manual, Version 5.0. K.U. Leuven*, Institute for Land and Water Management, Leuven, Belgium. p. 79.

Raes, D., Geerts, S., Kipkorir, E., Wellens, J., Sahli, A. 2006. Simulation of yield decline as a result of water stress with a robust soil water balance model. *Agricultural Water Management* 81, 335–357.

Raes, D., Steduto, P., Hsiao, T.C., Fereres, E. 2009. AquaCrop-The FAO crop model to simulate yield response to water: II. Main algorithms and software description. *Agronomy Journal* 101, 438–447.

Rezania, A.R., Naseri, A.A., Albaji, M. 2009. Assessment of soil properties for irrigation methods in North Andimeshk Plain, Iran. *Journal of Food, Agriculture & Environment* 7 (3&4), 728–733.

Zhang, J., Yang, J. 2004. Crop yield and water use efficiency: A study case in rice. In: Bacon, M.A. (Ed.), *Water Use Efficiency in Plant Biology*. Blackwell, Victoria, Australia. pp. 198–227.

6 Investigation of Water Stress on *Kochia Scoparia* L. in Arid and Semiarid Areas

Mona Golabi, Reza Sadegh Mansouri,
Saeed Boroomand Nasab, and Maasomeh Salehi

CONTENTS

6.1 INTRODUCTION

The most important problem that threatens food security in Iran is the lack of water resources. Currently, agriculture consumes more than 92% of the country's water resources. The water crisis, which has been present in the country for many years, has caused the most damage to the agricultural sector. Given the importance of agriculture in terms of production and employment, it is necessary to use the least resources available in the country and continue to produce even under drought, rather than shutting down the agriculture production. On the other hand, the vast surface area of the country is naturally saline, and there is also a large amount of saline water resources not used properly. Although the salinity of water and soil resources is one of the main threats to agricultural development in the country, these resources can be viewed not only as a problem but also as an opportunity.

Due to the water crisis in Iran, using proper agricultural methods such as low-irrigation techniques, use of low-quality irrigation water (saline and brackish water), and use of dry and saline plants in the production of agricultural products are of particular importance. On the other hand, crop production in arid and semiarid lands inevitably requires leaching to control the salinity of the root zone, and increasing saline water consumption, in turn, leads to increasing salt intake. However, cultivation of salt-tolerant crops as well as halophytes usually reduces both plant water requirement and also the need for leaching (Khoorsandi et al., 2010).

The production of saline plants is one of the most sustainable ways of preserving desert ecosystems to produce food for the livestock and inhabitants of these areas (Kafi et al., 2010). Among saline plants, due to its high resistance to salinity and drought, *Kochia Scoparia* L. grows well in a variety of soils. In addition to growing well in nonsaline soils, it also grows well in saline soils where other crops are unable to grow and sustain. Numerous studies have shown that in saline lands, *Kochia Scoparia* L. produces significant biomass (Jami Al Ahmadi and Kafi, 2008; Steppuhn et al., 2005). Recent findings on salt tolerance of various Iranian ecotypes of *Kochia Scoparia* L. have been published, indicating that this plant has a very good potential for forage production (Kafi et al., 2010; Nabati et al., 2012b). In a study conducted by Riasi et al. (2008) on a number of saline plants, it was reported that Kochia had better nutritional and digestive value for ruminants than other halophytes studied.

Nabati et al. (2012a) investigated the effect of salinity on the quantitative and qualitative properties of Kochia forage in Mashhad. To that end, salinity at different levels (10, 0, 20, 30, 40, and 50 dS/m) was measured in three separate experiments from planting to the mentioned salinity levels, from seedling stage with gradual stress to desired levels, and gradual application from seedling stage to plant death (128 dS/m) using a completely randomized design with four replications in the natural environment in the pot. Results showed that plant height, number of branches, shoot fresh and dry weight, digestible dry matter yield, digestibility value, crude protein yield, and ash content reduced with increasing salinity in gradual application at planting and seedling stage and gradual stress application to the end of growth. On the other hand, with increasing salinity, dry matter digestibility, organic matter digestibility, crude protein as well as ash percentages increased in gradual application of stress at planting and seedling stage and also the gradual application of stress to the end of plant growth. In general, the gradual application of salinity increased Kochia's salinity tolerance to 128 dS/m.

Similarly, Sobhani and Majidian (2014) conducted a factorial experiment in randomized complete block design with three replications in Arak to investigate the effects of different levels of salinity and plant density on quantitative and qualitative yield of Kochia's forage and seeds. In this study, irrigation water salinity at three levels (including 4.1, 18, and 32 dS/m) and plant density at four levels (including 10, 20, 30 and 40 plants/m^2) were considered as experimental factors. The results showed that the highest fresh forage yield was related to irrigation with 4.1 dS/m with production of 33.91 t/ha, while the lowest one was for irrigation with 32 dS/m salinity. The results of this study showed that *Kochia Scoparia* L. with 18 dS/m salinity and 30 plants/m^2 for forage production and 20 plants/m^2 can be introduced for grain production in Arak and regions with similar climatic conditions.

Kafi et al. (2011) studied forage characteristics of different Kochia ecotypes with two irrigation salinity levels of 2.5 and 16.5 dS/m in Mashhad. The results showed that salinity stress levels and the ecotypes studied had no significant effect on studied properties except dry matter yield.

Salehi et al. (2009) also examined the effect of salinity on summer growth of Kochia, as a halophyte, and concluded that with increasing salinity above 7 dS/m, all growth parameters decreased. Therefore, at low salinity (7 dS/m) dry matter production and leaf area index improved. Due to the growth performance of these ecotypes under salinity, when the water is of poor quality, the plant has high potential for tolerating saline irrigation in summer.

Likewise, Nabati et al. (2012b) investigated the quantitative and qualitative edible properties of Kochia at different salinity and time levels in Mashhad. The results of using saline at the beginning of growth showed that Kochia was able to survive up to a salinity of 30 dS/m NaCl. The results also showed that increasing salinity at planting and at the stage where the seedling was in the pod decreased plant height, number of branches, as well as fresh and dry weight; however, the digestible dry matter, nutritional value, and crude protein yield increased.

Yamada et al. (2016) investigated the adsorption properties of K, Ca, Na, and P in *Kochia Scoparia* L. under salinity stress. The results showed that Na concentration in the leaves increased with salinity and tended to accumulate in the upper part of the plant, while other cations reduced. Kochia also accumulated large amounts of Na in its leaves and other upper parts to reduce transpiration, and it adapted to difficult conditions so as to keep up with food shortages.

Kochia is a common plant in grasslands, pastures, fields, arid lands, gardens, roads, and ponds (Whitson et al., 1991). It can also be found in areas with 152.4 mm of annual rainfall (Undersander et al., 1990). In addition to salinity resistance, as a four-carbon (C_4) plant, Kochia competes well for water absorption due to its deep roots and has a high water use efficiency (Madrid et al., 1996). Besides its rapid vegetative growth under conditions of salinity, drought and heat have made it a very valuable option for forage production in hot and dry regions (Jami Al Ahmadi and Kafi, 2008).

Another study in New Mexico showed that water use efficiency in Kochia is three times that of alfalfa (Foster, 1980). It is also reported that Kochia needs 830 mm of water for maximum biomass and seed production in Birjand and Mashhad by the end of the growing season (Ziaee et al., 2008), while in Golestan Province it requires 300 mm of water up to the forage harvesting stage in spring cultivation (Salehi et al., 2012). In summer cultivation, this amount reaches 228 mm by the time of forage harvest (Salehi, 2010).

Salehi et al. (2014) carried out an experiment with six water salinity levels (1.5, 7, 14, 21, 28, and 35 dS/m) and four levels of water use (50%, 75%, 100%, and 125% of water requirement) to evaluate and calculate Kochia's vegetation coefficient, productivity, and water sensitivity coefficient under salinity stress and irrigation regimes in spring 2009 in the northern Golestan Province. The results showed that salinity decreased water productivity and vegetation coefficient. The results also showed that up to 21 dS/m, water-deficit stress did not affect plant sensitivity, while at 28 and 35 dS/m salinity, Kochia's sensitivity coefficient increased in all water-deficit treatments.

Due to its high potential productivity and tolerance to salinity, Kochia has the potential to become a forage plant and biofuel using saline water in semiarid regions. Kochia's rapid vegetative growth under salinity, drought, and heat stress has made it a very valuable option for forage production in hot and dry regions (Jami Al Ahmadi and Kafi, 2008). Knipfel et al. (1989) reported that the nutritional quality of Koshia at 20% flowering was equal to alfalfa.

Given the scarcity of water resources in Khuzestan Province in recent years and the availability of saline water resources in arid and semiarid areas in the province, the implementation of horticulture projects in such areas seems required. Therefore, given the satisfactory results of Kochia's dry matter production in salinity and drought conditions, its forage quality, and high potential as a biofuel plant, this led to an experiment aimed at studying Kochia under water-deficit irrigation conditions. For cultivation and domestication of this plant, it is thus necessary to determine its best cultivation date, water productivity, sensitivity to drought, and salinity stresses and also its salinity tolerance threshold in the study area (Ahvaz). Since the excessive use of saline water resources along with the water used for leaching increases soil salinity and causes soil drainage problems, it is thus necessary to use low-irrigation management techniques without the need for leaching during Kochia's growing season in the south of Khuzestan Province (Ahvaz).

6.2 MATERIALS AND METHODS

This study was carried out in the experimental field of the Faculty of Water Engineering, Shahid Chamran University of Ahvaz in the 2017–2018 farming year. This field is located at 48°, 39 min, 68 s, east longitude and 31°, 18 min, 18 s, north latitude and has an area of 1.2 hectares. Figure 6.1 shows the location of the experimental field.

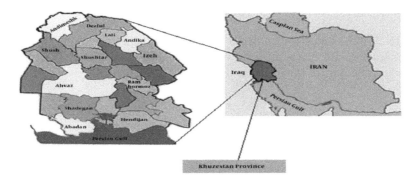

FIGURE 6.1   The location of study area.

Like other plains in Khuzstan Province, the important major crops in Ahvaz plain are wheat and barley (Albaji et al., 2012b; Albaji and Alboshokeh 2017; Behzad et al., 2009; Naseri et al., 2009). To conduct the experiment, Kochia was cultivated under different conditions to determine its planting date and the factors affecting its planting, germination, and growth. These factors included planting date, air temperature at planting and plant growth stages, bed soil salinity during planting, different light conditions for germination and growth of Kochia, as well as its direct and indirect planting.

To that end, Kochia was planted at different dates throughout the year (May, August, September, October, November, December, January, February, and March), under light conditions (artificial and natural light), at different temperature conditions (every four seasons), and in different substrates: natural substrate including light soil, heavy soil, the combination of light and heavy soil; the combination of heavy soil with animal manure; artificial substrate including peat moss and peat moss/perlite; and in soil with different salinity levels (up to 45 dS/m) using indirect and direct cultivation methods. Then, the most suitable date for the cultivation of Kochia under the specific conditions was determined experimentally in Ahvaz.

Also, like other plains in Khuzstan Province, the surface irrigation (by using Karoon River water) is the major irrigation system in Ahvaz plain (Albaji et al., 2008, 2012a, 2014a, 2014b, 2016; Boroomand Nasab et al., 2010; Landi et al., 2008; Rezania et al., 2009). In order to achieve these goals, Karoon River water, with an average electrical conductivity of 2.5 dS/m, and three levels of water deficit including 100%, 75%, and 50% of water requirement (I_1, I_2 and I_3, respectively) were used in three replications (R_1, R_2, and R_3) lysimetrically. The lysimeters used in this study were cylindrical with a radius of 0.3 and a height of 0.8 m. This design was done in the form of split plots with a randomized complete block design and is shown in Figure 6.2.

After determining the optimum cultivation date, Kochia was planted in early March 2018 and harvested in June 2018 as the first crop. Physical and chemical analyses of lysimeter soils were also performed before cultivation. Then, the required plant properties were measured during the growing season. Equation (6.1) was used to calculate the net depth of irrigation water.

$$SWD = (\theta_{FC} - \theta PWP).\rho_b.D_{rmax}.MAD \qquad (6.1)$$

In Equation (6.1), SWD stands for soil water deficit (in millimeters); θ_{FC} and θ_{PWP} are field capacity and permanent wilting point, respectively; ρb for apparent specific gravity (g/cm^3); D_{rmax} is the maximum of root development depth (in mm); and MAD is for the maximum allowable depletion moisture. Equation (6.2) was used to determine irrigation time.

$$f = \frac{SWD}{ET_{cmax}} \qquad (6.2)$$

In Equation (6.2), SWD stands for soil water deficit (in millimeters) and ET_{cmax} is the maximum crop evapotranspiration calculated using the evaporation data of the past 10 years, obtained from the Khuzestan Meteorological Organization (Equations 6.3 and 6.4).

R_1	I_2	I_1	I_3
R_2	I_1	I_3	I_2
R_3	I_3	I_2	I_1

FIGURE 6.2   Experimental design plan of the present study.

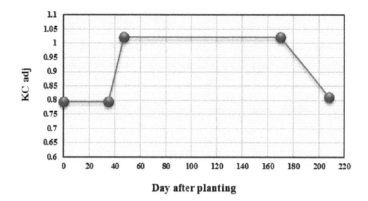

FIGURE 6.3   Diagram of coefficient of vegetation Kochia.

$$ET_{0max} = K_{pan} \bullet E_{panmax} \tag{6.3}$$

$$ET_{cmax} = K_{cmax} \bullet ET_{0max} \tag{6.4}$$

In the above equations, ET_{0max} is the maximum potential evapotranspiration, K_{pan} is the pan evaporation coefficient, E_{panmax} is the evaporation from evaporation pan, and K_{cmax} is the Kochia's maximum crop coefficient during the cultivation period.

Using Allen and Pruitt (1991) equation, Shokri et al. (2015) calculated pan evaporation coefficient from a 15-year data pan of the Ahvaz meteorological station. The evaporation coefficient was estimated to be 0.76, which was also used in the present study.

In order to determine crop coefficients in this study, data obtained from Salehi et al. (2012) and Jami Al Ahmadi and Kafi (2008) were used at different stages of Kochia growth in spring cultivation. Figure 6.3 shows the diagram of the crop coefficient of Kochia during the growth period. This curve was also used in the present study.

As an objective for the present study, Kochia's water use efficiency was obtained from the ratio of total fresh shoot weight (yield) to applied water (irrigation water). The coefficient of drought sensitivity was also calculated from Equation (6.5) (Doorenbos and Kassam, 1979).

$$K_y = \frac{1 - Y_r}{1 - ET_r} \tag{6.5}$$

where Y_r is the ratio of actual yield to maximum crop yield in the experiment, ET_r is the ratio of actual evapotranspiration to maximum evapotranspiration, and K_y is the coefficient of drought sensitivity of Kochia, respectively.

Finally, after harvesting Kochia, its yield and yield components were measured and analyzed.

6.3   RESULTS AND DISCUSSION

As stated above, Kochia was cultivated under different conditions to determine its most appropriate planting date and the factors affecting its planting, germination, and growth. According to the results and experiences obtained in different methods of planting Kochia under various time and different light and temperature conditions in different substrates, as well as its planting by transplanting and direct planting, the best date for its cultivation in the Ahvaz region was found to be its direct planting in mid-March (spring cultivation) and under natural soil conditions (it is best to mix the soil with a little animal manure) exposed to direct sunlight.

TABLE 6.1

Physical and Chemical Properties of Soil Lysimeters Before Cultivation

Depth (cm)	EC (dS/m)	pH	Na^+	K^+	Ca^{2+} (meq/L)	Mg^{2+}	Cl^-	ρ_b (gr/cm³)	θ_{FC}	θ_{PWP} %	Soil Texture
0–20	4.08	7.02	16.84	0.25	16	19	33	1.46	25.2	14	Loamy clay
20–40	5.2	7.37	20.16	0.25	16	19	43	1.5	25.2	14	Loamy clay

TABLE 6.2

Analysis of Variance of Yield and Yield Components of Kochia

		Mean Square					
Source	df	Total Fresh Weight	Leaf Fresh Weight	Stem Fresh Weight	Plant Height	Number of Lateral Branches	Stem Diameter
R	2	35.826	491.43	48.53	3.52	8.79	0.15
I	2	447.93[a]	13165.56[a]	10271.09[a]	1413.19[a]	134.55[a]	7.16[a]
Error	16	15.28	161.76	95.91	2.99	5.84	0.12
CV (%)		8.63	9.31	10.32	2.07	4.13	10.06

[a] was significant at the 1% levels.

Before cultivation, soil samples were collected from 0 to 20 and 20 to 40 cm of lysimeters then physically and chemically analyzed. The results are presented in Table 6.1.

As can be seen, the texture of the soil is loamy clay, which is of medium texture.

After harvesting Kochia, its yield components were measured and statistically analyzed. Table 6.2 shows the results of analysis of variance of yield and yield components.

The results of analysis of variance of measured indices in the experiment showed that the effects of water-deficit stress on shoot fresh weight, leaf and stem fresh weight, height, and number of lateral branches were separately significant at the 1% level of probability. Water-deficit stress at 1% probability level affected stem diameter. Water-deficit stress at 1% probability level affected stem diameter.

The results of mean comparison of yield and yield components under main effects of water-deficit stress are shown in Table 6.3.

Considering irrigation water depth, I_1 treatment with an average of 49.03 t/ha had the highest yield, while the lowest yield was in I_3 treatment with an average of 38.23 t/ha. The results showed that application of water-deficit stress up to 75% of plant water requirement had no effect on

TABLE 6.3

Results of Comparing Yield Averages and Yield Components Under Main Effects of Salinity and Water Deficit

Percent of Full Irrigation	Total Fresh Weight (ton/ha)	Leaf Fresh Weight (gram per bushes)	Stem Fresh Weight (gram per bushes)	Plant Height (cm)	Number of Lateral Branches	Stem Diameter (cm)
100%	49.03 a	167.47 a	125.88 a	95.71 a	62.33 a	4.24 a
75%	48.58 a	140.59 b	91.06 b	80.83 b	57.13 b	3.27 b
50%	38.23 b	101.59 c	67.75 c	74.58 c	56.08 b	2.72 c

[a–c] indicates significant difference.

Kochia's total shoot fresh weight, but the application of water deficit up to 50% of water requirement had significant effect on its total shoot fresh weight. Drought stress directly and indirectly affected photosynthesis and the accumulation of carbonate hydrates, and it ultimately reduced yield. Payero et al. considered one of the symptoms of water stress to be plant growth decline and stated that growth decline was due to decreased leaf area and plant height (Payero et al., 2006). Comparison of salinity treatments showed that Kochia's total shoot fresh weight loss due to increased salinity was more than that of moisture deficit.

In terms of the applied water depth (I), the highest stem fresh weight per plant belonged to I_1 treatment with a mean of 125.88, while the lowest to I_3 treatment with an average of 67.75 g/plant. The results showed that with increasing water-deficit stress, Kochia's stem fresh weight significantly decreased, so that in 50% water-deficit treatment, weight loss of 47.2 was observed.

In the present study, some plant properties were measured during the growing season, which are examined in the course of the change process.

The trend of changes in Kochia's height under water stress in different salinity treatments is shown in Figure 6.4.

The results of Figure 6.4 also showed that plant growth was rapid at the beginning of the growing season, and as the plants approached the end stages of growth period, the intensity of plant height decreased.

The trend of changes in the number of Kochia's lateral branches under water-deficit stress in different salinity treatments is shown in Figure 6.5.

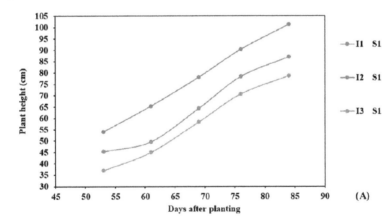

FIGURE 6.4   Changes in plant height at different water-deficit levels.

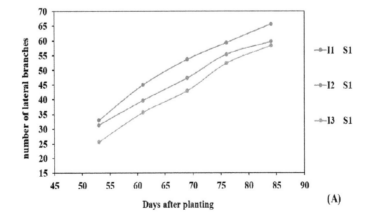

FIGURE 6.5   Changes in the number of lateral branches of the plant at different levels of water deficit.

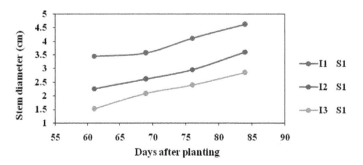

FIGURE 6.6   Changes in plant stem diameter at different water-deficit levels.

Under water-deficit stress with different salinity levels, Kochia's branching process was found to have a rapid growth from the beginning, while at the end of the growth the production rate decreased. It can be seen from Figure 6.5 that in all cases, the irrigation treatment (I_1) is superior. Investigating the impact of low irrigation on the forage yield of two indigenous Kochia ecotypes under 100%, 80%, 60%, and 40% irrigation regimes and measuring the trend changes in the number of branches during cultivation period, Soleimani et al. (2008) reported that the irrigation treatments had a significant effect on the number of branches, and in all cases during the cultivation period, the control irrigation treatment was superior in terms of lateral branch production.

The trend of stem diameter changes of Kochia under water-deficit stress is shown in Figure 6.6.

Contrary to the effects of salinity stress in the study, the effect of water stress on different salinity treatments on Kochia's stem diameter was significant. In all treatments, I_1 and I_3 treatments had the highest and the lowest increase in stem diameter, respectively. The results of Figure 6.6 also show that the increase in stem diameter has been developing throughout the study period. In general, it can be concluded that water stress had a greater effect on preventing stem diameter increase during the cultivation period than salinity stress.

6.4   CORRELATION BETWEEN PROPERTIES EVALUATED

All correlation coefficients between properties were evaluated in order to study and compare the correlation relationships between measured parameters in irrigation treatments (Table 6.4).

According to Table 6.4, it is observed that all properties showed positive and significant correlation with 1% probability level, indicating that increasing or decreasing each of these properties had

TABLE 6.4
Correlation Coefficients Between Different Traits

	Total Fresh Weight	Leaf Fresh Weight	Shoot Fresh Weight	Plant Height	Number of Lateral Branches	Stem Diameter
Total fresh weight	1					
Leaf fresh weight	0.756[a]	1				
Shoot fresh weight	0.602[a]	0.853[a]	1			
Plant height	0.636[a]	0.818[a]	0.908[a]	1		
Number of lateral branches	0.585[a]	0.709[a]	0.750[a]	0.812[a]	1	
Stem diameter	0.574[a]	0.818[a]	0.827[a]	0.898[a]	0.778[a]	1

[a] was significant at the 5% levels.

TABLE 6.5
Water Productivity for Kochia Plant

Treatments	The Volume of Water Used ($m^3 \cdot ha^{-1}$)	Total Performance ($ton. \, ha^{-1}$)	Water Productivity ($Kg. \, m^{-3}$)
I_1	5499.48	57.90	10.53
I_2	4396.50	56.32	12.81
I_3	3293.52	44.69	13.57

a direct effect on the forage yield of Kochia. In this regard, results of the present study confirmed those of Soleimani et al. (2008) investigating the effect of different water-deficit regimes on forage yield of two indigenous ecotypes of Kochia in saline irrigation conditions.

6.5 WATER PRODUCTIVITY

One of the indicators used in discussions about plant yield and water use is water productivity, which is determined by the ratio of crop yield to the amount of irrigation water consumed (Sepaskhah et al., 2006). Many engineering and technical research studies in recent years in the agricultural sector have been devoted to water productivity. In this study, water productivity of Kochia's yield under different irrigation treatments was determined. For this purpose, water yield per kg/ha was calculated by dividing the total yield values by the amount of irrigation water consumed. Table 6.5 shows the water consumption of Kochia.

It can be seen from Table 6.5 that in all cases, the irrigation treatment I_3 had the highest water productivity.

Overall, the results of this study and other research showed that the water productivity in Kochia forage production was high and acceptable even under severe stress conditions, with water deficit.

6.6 CONCLUSIONS

The lack of quality water resources, and soil and water salinity, is one of the main threats to agricultural development in Iran. In this study, evaluation of yield and yield components of Kochia under salinity stress showed that except for stem diameter, all components were significantly affected by salinity stress. Mean comparison of different properties showed that there were no significant differences in total shoot fresh weight between full irrigation and 75% irrigation treatments. Thus, it can be concluded that by reducing irrigation water by 25%, good yield can be achieved and more surface area can be cultivated. The results showed that with increasing salinity, salinity sensitivity coefficient (K_s) decreased. The results also showed that the K_s coefficients obtained in the combination of salinity stress with 50% water deficit were more than those of the combination of salinity stress with 75% water deficit. The mean of K_y coefficient obtained in this study indicates that Kochia has a low drought sensitivity coefficient, indicating its high potency and tolerance under drought and salinity conditions. Therefore, a low-irrigation method for Kochia up to a salinity of 20 dS/m is recommended. The results of this study showed that water productivity in Kochia forage production was high and acceptable even under severe stress conditions with high salinity and water deficit. It was also observed that salinity decreased water productivity. In turn, water productivity under low-irrigation conditions was found to be higher than that of full irrigation. Moreover, the findings revealed that Kochia can be a suitable forage plant in areas such as Ahvaz where irrigation water and soil are not of good quality, even without the need for leaching and soil remediation that cost the producers a great deal.

ACKNOWLEDGMENTS

We are grateful to the Research Council of Shahid Chamran University of Ahvaz for financial support (GN: SCU.WI98.281).

REFERENCES

Albaji, M., Boroomand Nasab, S., Kashkuli, H.A., Naseri, A.A., Sayyad, G., Jafari, S. 2008. Comparison of different irrigation methods based on the parametric evaluation approach in North Molasani Plain, Iran. *Journal of Agronomy* 7 (2), 187–191.

Albaji, M., Boroomand Nasab, S., Hemadi, J. 2012a. Comparison of different irrigation methods based on the parametric evaluation approach in West North Ahvaz Plain. In: *Problems, Perspectives and Challenges of Agricultural Water Management*. M. Kumar, (ed.), InTech, Croatia, pp. 259–274.

Albaji, M., Papan, P., Hosseinzadeh, M., Barani, S. 2012b. Evaluation of land suitability for principal crops in the Hendijan region. *International Journal of Modern Agriculture* 1 (1), 24–32.

Albaji, M., Golabi, M., Boroomand Nasab, S., Jahanshahi. M. 2014a. Land suitability evaluation for surface, sprinkler and drip irrigation systems. *Transactions of the Royal Society of South Africa* 69 (2), 63–73.

Albaji, M., Golabi, M., Piroozfar, V.R., Egdernejad, A., Nazari Zadeh, F. 2014b. Evaluation of agricultural land resources for irrigation in the Ramhormoz Plain by using GIS. *Agriculturae Conspectus Scientificus* 79 (2), 93–102.

Albaji, M., Golabi, M., Hooshmand, A.R., Ahmadee, M. 2016. Investigation of surface, sprinkler and drip irrigation methods using GIS. *Jordan Journal of Agricultural Sciences* 12 (1), 211–222.

Albaji, M., Alboshokeh, A. 2017. Assessing agricultural land suitability in the Fakkeh region, Iran. *Outlook on Agriculture* 46 (1), 57–65.

Allen, R.G., Pruitt, W.O. 1991. FAO-24 reference evapotranspiration factors. *Journal of Irrigation and Drainage Engineering* 117 (5), 758–773.

Behzad, M., Albaji, M., Papan, P., Boroomand Nasab, S. 2009. Evan region qualitative soil evaluation for wheat, barley, alfalfa and maize. *Journal of Food, Agriculture & Environment* 7 (2), 843–851.

Boroomand Nasab, S., Albaji, M., Naseri, A.A. 2010. Investigation of different irrigation systems based on the parametric evaluation approach in Boneh Basht plain, Iran. *African Journal of Agricultural Research* 5 (5), 372–379.

Doorenbos, J., Kassam, A.H. 1979. Yield response to water. *Irrigation and Drainage Paper*. 33, 257.

Foster, C. 1980. Kochia-poor man's alfalfa-shows potential as feed. *Rangelands Archives* 2 (1), 22–23.

Jami Al Ahmadi, M., Kafi, M. 2008. Kochia (*Kochia scoparia*): To be or not to be. *Crop and Forage Production Using Saline Waters*. Daya Publisher, New Delhi, India.

Kafi, M., Asadi, H., Ganjeali, A. 2010. Possible utilization of high-salinity waters and application of low amounts of water for production of the halophyte *Kochia scoparia* as alternative fodder in saline agro-ecosystems. *Agricultural Water Management* 97 (1), 139–147.

Kafi, M., Nabati, J., Khaninejad, S., Masomi, A., Zare Mehrjerdi, M. 2011. Evaluation of characteristics forage in different Kochia (*Kochia scoparia*) ecotypes in tow salinity levels irrigation. *Electronic Journal of Crop Production* 4 (1), 229–238. (In Persian).

Khoorsandi, F., Vaziri, J., Azizi zahan, A.A. 2010. Haloculture, sustainable use of saline soil and water resources in agriculture. *Iranian National Committee on Irrigation and Drainage (IRNCID)*, 1–320.

Knipfel, J.E., Kernan, J.A., Coxworth, E.C., Cohen, R.D.H. 1989. The effect of stage of maturity on the nutritive value of kochia. *Canadian Journal of Animal Science* 69 (4), 111–114.

Landi, A., Boroomand-Nasab, S., Behzad, M., Tondrow, M.R., Albaji, M., Jazaieri, A. 2008. Land suitability evaluation for surface, sprinkle and drip irrigation methods in Fakkeh Plain, Iran. *Journal of Applied Sciences* 8 (20), 3646–3653.

Madrid, J., Hernandez, F., Pulgar, M.A., Cid, J.M. 1996. Nutritive value of *Kochia scoparia* L. and ammoniated barley straw for goats. *Small Ruminant Research* 19 (3), 213–218.

Nabati, J., Kafi, M., Nezami, P., Rezavani Moghadam, A., Masoumi, A., Zare Mehrjerdi, M. 2012a. Investigating the effect of salinity stress in different growth stages on quantitative and qualitative features of kochia forage. *Electronic Journal of Crop Production* 5 (2), 111–128. (In Persian).

Nabati, J., Kafi, M., Nezami, A., Rezvani Moghaddam, P., Masoumi, A., Zare Mhregerdi, M. 2012b. Evaluation of quantitive and qualititive characteristic of forage kochia (*Kochia scoparia*) in different salinity levels and time. *Irrigation Journal of field crops Research* 12(4), 613–620 (In Persian).

Naseri, A.A., Albaji, M., Boroomand Nasab, S., Landi, A., Papan, P., Bavi, A. 2009. Land suitability evaluation for principal crops in the Abbas Plain, Southwest Iran. *Journal of Food, Agriculture & Environment* 7 (1), 208–213.

Payero, J.O., Melvin, S.R., Irmak, S., Tarkalson, D. 2006. Yield response of corn to deficit irrigation in a semi-arid climate. *Agricultural Water Management* 84(1–2), 101–112.

Rezania, A.R., Naseri, A.A., Albaji, M. 2009. Assessment of soil properties for irrigation methods in North Andimeshk Plain, Iran. *Journal of Food, Agriculture & Environment* 7 (3&4), 728–733.

Riasi, A., Mesgaran, M.D., Stern, M.D., Moreno, M.R. 2008. Chemical composition, in situ ruminal degradability and post-ruminal disappearance of dry matter and crude protein from the halophytic plants *Kochia scoparia*, atriplex dimorphostegia, suaeda arcuata and gamanthusgamacarpus. *Animal Feed Science and Technology* 141 (3–4), 209–219.

Salehi, M. 2010. The effect of salinity and water deficit stress on quantitative, qualitative and physiomorphological characteristics of kochia. *Ph.D. Thesis of Ferdowsi University*. Mashhad, Iran (In Persian).

Salehi, M., Kafi, M., Kiani, A. 2009. Growth analysis of kochia (*Kochia scoparia* (L.) schrad) irrigated with saline water in summer cropping. *Pakistan Journal of Botany* 41 (4), 1861–1870.

Salehi, M., Kafi, M., Kiani, A. 2012. Effect of salinity and water deficit stresses on biomass production of kochia (*Kochia scoparia*) and trend of soil salinity. *Seed and Plant production Journal* 2–27 (4), 417–433 (In Persian).

Salehi, M., Kafi, M., Kiani, A. 2014. Water efficiency, drought sensitivity and crop coefficient In the kochia plant under stress of salinity and limited irrigation in spring croppin. *Iranian Journal of Water Research* 7 (12), 89–98 (In Persian).

Sepaskhah, A.R., Tavakoli A.R., Mousavi, S.F. 2006. Principles and application of deficit irrigation. *Iranian National Committee on Irrigation and Drainage (IRNCID)*, 1–210.

Shokri, S., Hooshmand, A.R., Ghorbani, M. 2015. The estimation evaporation pan coefficient for calculating reference evapotranspiration in Ahvaz. *Journal of Irrigation Sciences and Engineering (JISE)* 40 (1), 1–12 (In Persian).

Sobhani, M.R, Majidian, M. 2014. Evaluation of different salinity stress and plant densities effects on quantitative and qualitative forage and grain yields of kochia in Arak region. *Journal of Plant Production Research* 21 (1), 91–110 (In Persian).

Soleimani, M.R., Kafi, M., Ziaee, M., Shabahang, J. 2008. Effect of limited irrigation with saline water on forage of two local populations of *Kochia scoparia* L. Schrad. *Journal of Agricultural Science and Technology* 22, 307–317 (In Persian).

Steppuhn, H., Van Genuchten, M.T., Grieve, C.M. 2005. Root-zone salinity: II. Indices for tolerance in agricultural crops. *Crop Science* 45 (1), 221–232.

Undersander, D.J., Oelke, E.A., Kaminski, A.R., Doll, J.D., Putnam, D.H., Combs, S.M., Hanson, C.V. 1990. *Alternative Field Crops Manual*. University of Wisconsin-Madison and Minnesota, St. Paul, Madison. 48.

Whitson, T.D., Burrill, L.C., Dewey, S.A., Cundey, D.W., Nelson, B.E., Lee, R.D., Parker, R. 1991. *Weeds of the West* Diane Pub Co, 1–630.

Yamada, S., Yamaguchi, T., Davaid Lopez Aguilar, R. 2016. Characteristics of Na, K, C, Mg and P absorption in kochia plant (*Kochia Scoparia* (L.) Schrad.) under sainity stress. *Sand Dune Research* 63 (1), 1–8.

Ziaee, S.M., Kafi, M., Khazaee, H.R., Shabahang, J.M., Soleimani, M.R. 2008. Effect of planting density and cutting frequency on forage and grain yields of kochia (*Kochia scoparia*) under saline water irrigation. *Iranian Journal of Agricultural Research* 6 (2), 335–342.

7 Coefficients of Infiltration Equations
Determination and Evaluation

Mona Golabi, Mohammad Albaji, and Saeid Eslamian

CONTENTS

7.1 INTRODUCTION

Infiltration is one of the major components of the hydrologic cycle. Water that falls as precipitation may run overland (eventually reaching streams, lakes, rivers, and oceans) causing erosion, flooding, and degradation of water quality; on the other hand, it may infiltrate through the soil surface, into the soil profile. This infiltration, which constitutes the sole source of water to sustain the growth of vegetation, is filtered by the soil, which removes many contaminants (through physical, chemical, and biological processes) and replenishes the groundwater supply to wells, springs, and streams (Rawls et al., 1993; Oram, 2005). Infiltration is critical because it supports life on the land. The ability to quantify infiltration is of great importance in watershed management. Prediction of flooding, erosion, and pollutant transport all depend on the rate of runoff, which is directly affected by the rate of infiltration. Quantification of infiltration is also necessary to determine the availability of water for crop growth and to estimate the amount of additional water needed for irrigation. Also, by understanding how infiltration rates are affected by surface conditions, measures can be taken to increase infiltration rates and reduce the erosion and flooding caused by overland flow. In order to develop improved hydrologic models, accurate methods for calculating infiltration are required (Shirmohammadi and Skaggs, 1984). Infiltration modeling approaches are often described as physically based, approximate, or empirical. The physically based approaches require solution of the Richards equation (Richards, 1931), which describes water flow in soil in terms of the hydraulic conductivity and the soil water pressure as functions of soil water content for specified boundary conditions. Solving this equation is extremely difficult for many flow problems requiring detailed data input and the use of numerical methods (Rawls et al., 1993).

In 1982, Skaggs and Khaleel stated that although numerical methods that allow the hydrologist to quantify the vertical percolation of water are critical for assessment of groundwater recharge and in the analysis of contaminant movement through soil, they are costly, data- and time-intensive computational procedures requiring numerous field measurements to be made and therefore are rarely used in practice. Since then, improvements in computer technology have greatly facilitated the use of numerical techniques. However, the quantity and the complexity of the measurements necessary to obtain much of the soil property data required for these numerical solutions impose a more severe limitation that has not diminished with time. Consequently, for many applications, equations that simplify the concepts involved in the infiltration process are advantageous (Rawls et al., 1993). Simplified approaches include empirical models such as Kostiakov, Horton, and Holtan, and approximate physically based models such as those of Green and Ampt, and Philip. Empirical models tend to be less restricted by assumptions of soil surface and soil profile conditions but more restricted by the conditions for which they were calibrated, since their parameters are determined based on actual field-measured infiltration data (Hillel, 1998; Skaggs and Khaleel, 1982). Equations that are physically based approximations use parameters that can be obtained from soil water properties and do not require measured infiltration data. It has been noted that different approximate equations for infiltration result in different predictions for infiltration rate, time of ponding, and time of runoff even when measurements from the same soil samples are used to derive parameter values. Also, different equations for infiltration require different parameters to be used. There are many factors that contribute to the infiltration rate including time from onset of rain or irrigation, initial water content of the soil, hydraulic conductivity, surface conditions, and profile depth and layering (Hillel, 1998). All the infiltration equations make use of some of these factors in characterizing infiltration. However, the more physically based equations rely more heavily on the soil hydraulic and physical properties occurring within the profile, such as saturated hydraulic conductivity, soil moisture gradient, and suction at the wetting front. Empirical models rely more on parameters that are determined by curve fitting or estimated by other means and thus may better reflect the effect of differences in surface conditions than the physical models, as long as parameters are calibrated separately for those different conditions. Additionally, sometimes approximate physically based models are used as empirical models with parameters determined in a similar manner. The assumptions, form, and intent of each equation need to be considered in deciding which equation to use for a particular application. Infiltration is the entrance of water from the soil surface into the top layer of the soil. Redistribution is the movement of water from point to point within the soil. These two processes cannot be separated because the rate of infiltration is strongly influenced by the rate of water movement within the soil below. After each infiltration event, soil water movement continues to redistribute the water below the surface of the soil (Rawls et al., 1993). Many of the same factors that control infiltration rate also have an important role in the redistribution of water below the soil surface during and after infiltration. Thus, a solid grasp of infiltration and the factors that affect it is important not only in the determination of surface runoff but also in understanding subsurface movement and storage of water within a watershed (Skaggs and Khaleel, 1982).

Factors that control infiltration rate include soil properties that are strongly affected by these three forces: hydraulic conductivity, diffusivity, and water holding capacity. These soil properties are related to the characteristics of soil texture, structure, composition, and degree of compaction, which influence soil matric forces and pore space. Additionally, antecedent water content, type of vegetative or other ground cover, slope, rainfall intensity, and movement and entrapment of soil air are important factors that also affect infiltration rates. The hydraulic conductivity is of critical importance to infiltration rate, since it expresses how easily water flows through soil and is a measure of the soil's resistance to flow. The unsaturated hydraulic conductivity is a function of pressure head (Serrano, 1997) and distribution of water in the soil matrix. The saturated hydraulic conductivity, the hydraulic conductivity at full saturation, is used as a parameter in many of the infiltration equations, since it is easier to determine than either the unsaturated hydraulic conductivity or the diffusivity.

The theory of infiltration of water to the soil has been described with mathematical equations. Infiltration equations have been divided into two categories (Larson, 1983):

7.1.1 BASIC EQUATIONS OF POROUS MEDIA FLOW

The base of this equation is the Richards equation (Richards, 1931), and it has been widely used in infiltration modeling. The Richards equation is represented below:

$$\frac{\partial \theta}{\partial t} = \frac{\partial}{\partial z}\left(D\frac{\theta}{z}\right) - \frac{\partial k}{\partial z} \qquad (7.1)$$

With boundary and initial conditions:

$$\theta = \theta_o, \quad Z \geq 0, \quad t = 0 \qquad (7.2)$$

$$\theta = \theta_s, \quad Z = 0, \quad t > 0 \qquad (7.3)$$

In these equations, θ_s is the soil moisture in saturation condition; θ_o is the initial moisture of the soil; θ is soil moisture in time t and at depth z; K is hydraulic conductivity; D is the diffusion coefficient; and z is distance from the soil surface that direct to down is positive. The numerical solution of the Richards equation for a given set of initial and boundary conditions allows the hydrologist to use the physical properties governing the movement of water and air through soils to precisely quantify the vertical percolation of water subject to a variety of conditions. These predictions are critical for assessment of groundwater recharge and in the analysis of contaminant movement through soil (Skaggs and Khaleel, 1982).

Additionally, the numerical solution of the Richards equation requires numerous measurements to be made to adequately describe variations in soil properties that occur both vertically in the soil profile and from point to point in the field (Skaggs and Khaleel, 1982); therefore, infiltration models with simplified data requirements are desirable for practical use. The rationale of simultaneous solutions of Darcy's law and the continuity equation would be highly desirable, but the required estimates of unsaturated hydraulic conductivities and diffusivities are difficult to obtain even in the laboratory. Valid estimates for field-scale applications are not available, and sequential treatment of successive soil horizons is extremely precarious in the anisotropic conditions characteristic of our watersheds (Holtan, England, and Shanholtz, 1967).

7.1.2 PHYSICAL AND EMPIRICAL EQUATIONS OR APPROXIMATE MODELS

Several equations that simplify the concepts involved in the infiltration process have been developed for field applications. Approximate models such as those of Philip, and Green and Ampt apply the physical principles governing infiltration for simplified boundary and initial conditions. They imply ponded surface conditions from time zero on (Hillel, 1998) and are based on assumptions of uniform movement of water from the surface down through deep homogenous soil with a well-defined wetting front, assumptions that are more valid for sandy soils than for clay soils (Haverkamp et al., 1987). These assumptions reduce the amount of physical soil data needed from that of numerical solutions but also limit their applicability under changing initial and boundary conditions (Haverkamp et al., 1987). Equations that are physically based approximations use parameters that can be obtained from soil water properties and do not require measured infiltration data. Thus, they should be able to produce estimates at lower cost than empirical equations. Other equations are partially or entirely empirical, and parameters must be obtained from measured infiltration data or roughly estimated by other means. Empirical equations such as those of Kostiakov and Horton are less restrictive as to mode of water application because they do not require the assumptions regarding soil surface and soil profile conditions that the physically based equations require (Hillel, 1998). Where soils are heterogeneous and factors such as macro pore flow and entrapped air complicate the infiltration process, empirical equations may potentially provide more accurate predictions, as long as they are used under similar conditions to those under which they were developed. This is because their initial parameters are determined based on actual field-measured infiltration data (Skaggs and Khaleel, 1982; Rawls et al., 1993). One characteristic of infiltration that all the equations predict is an initially rapid decrease in rate with time for ponded surfaces (Skaggs and Khaleel, 1982).

7.1.2.1 Kostiakov and Kostiakov–Lewis Equations

Kostiakov (1932) and, independently, Lewis (1937) proposed a simple empirical infiltration equation based on curve fitting from field data. It relates infiltration to time as a power function:

$$Z = kt^n \tag{7.4}$$

Here, Z is the cumulative depth of infiltration (cm), t is intake opportunity time (min), k is an empirical soil constant for infiltration, and n is an exponent. The equation describes the measured infiltration curve, and given the same soil and same initial water condition, it allows prediction of an infiltration curve using the same constants developed for those conditions.

Israelson and Hanson (1967) developed the modified Kostiakov equation and applied it for estimation of irrigation infiltration. Mbagwu (1993) recommended the modified Kostiakov equation for routine modeling of the infiltration process on soils with rapid water intake rates. The Kostiakov and modified Kostiakov equations tend to be the preferred models used for irrigation infiltration, probably because they are less restrictive as to the mode of water application than some other models. The SIRMOD model (Walker, 1998) simulates the hydraulics of surface irrigation (border, basin, and furrow) at the field level and employs the modified Kostiakov infiltration equation to represent infiltration characteristics.

Ghosh (1980, 1983) obtained better results with the Kostiakov equation than the Philip model for fields with wide spatial variability in the infiltration data. Clemmens (1983) found that the Kostiakov equation provided significantly better predictions than the theoretical equations of Philip and GA for border irrigation infiltration data. Naeth, Chanasyk, and Bailey (1991) found that the Kostiakov equation fit double-ring infiltrometer data very well for all three ecosystems that he studied. Naeth (1988) also found that the Kostiakov equation was sensitive to changes in infiltration capacity brought about through different grazing treatments. However, Gifford (1976) found that the Kostiakov equation did not fit infiltrometer data collected from semiarid rangelands in Australia and the United States. Gifford (1978) determined that the coefficients in the Kostiakov equation were more closely related to vegetation factors than to soil factors from infiltrometer data run with soils pre-wet to field capacity prior to the infiltration test.

Ghosh (1985) challenged the commonly accepted view that the value of the n term in the Kostiakov equation lies between zero and one, and proved mathematically that the value of n can be greater than one. Mbagwu (1990), however, found empirically that the value of n was consistently less than one. Fok (1986) showed that the K and α term of the Kostiakov equation do have physical meaning even though several authors have described it as purely empirical. Mbagwu (1994) found that the two soil properties with greatest influence over the K term are the effective porosity and bulk density. Bulk density, which correlated inversely with K, explained 43% of the variability; effective porosity, which is exponentially related to K, explained 78% of the variability in this parameter. Mbagwu (1994) found a critical effective porosity threshold of 15%–20%, below which the value for K was drastically reduced. He also found the saturated hydraulic conductivity to be linearly correlated with Kostiakov's K ($r = 0.9823$, $p \le 0.001$). Generally, the Kostiakov equation is usable for short time periods, but for the long run, this equation ignores basic infiltration. Long time should use other equations, such as the Kostikov–Lewis equation represented below:

$$Z = kt^n + f_o t \tag{7.5}$$

Here, Z is the depth of infiltration (cm), t is time (min), k and n are empirical coefficients, and f_o is the basic infiltration.

7.1.2.2 Green–Ampt Equation

Green and Ampt (GA) proposed in 1911 an approximate model that directly applies Darcy's law. The original equation was derived for infiltration from a ponded surface into a deep homogeneous soil with uniform initial water content. The GA model has been found to apply best to infiltration into uniform, initially dry, coarse-textured soils that exhibit a sharply defined wetting front. (Hillel and Gardner, 1970). This pattern is often called a piston displacement profile or plug flow. The transmission zone is a region of nearly constant water content above the wetting front, which lengthens as infiltration proceeds. The wetting front is characterized by a constant matric suction, regardless of time or position and is a plane of separation between the uniformly wetted infiltrated zone and the as-yet totally uninfiltrated zone (Hillel, 1998). These assumptions simplify the flow equation so that it can be solved analytically. Although measured infiltration data are not required to make predictions using the GA equation, Green and Ampt (1911) recommended that soil physical properties should be measured in the field so that undistrubed field conditions are reflected in the resulting values. The form of the GA equation that we used in our research has been represented below:

$$f = \frac{A}{F} + B \tag{7.6}$$

Here, f is the infiltration rate (cm/h), F infiltration depth (cm), A and B are the coefficient of the equation that is dependent on soil type.

7.1.2.3 Philip Equation

Philip (1957a) developed an infinite-series solution to solve the nonlinear partial differential Richards equation (Richards, 1931), which describes transient fluid flow in a porous medium for both vertical and horizontal infiltration. Philip's rapidly converging series solves the flow equation for a homogeneous deep soil with uniform initial water content under ponded conditions. For cumulative infiltration, the general form of the Philip infiltration model is expressed in powers of the square root of time, t, as:

$$z = St^{0.5} + At \tag{7.7}$$

where z is the cumulative infiltration, t is the opportunity of infiltration, S is the sorptivity coefficient, and A is the transmissivity coefficient.

Philip (1957b) defined sorptivity (S) as the measurable physical quantity that expresses the capacity of a porous medium for capillary uptake and release of a liquid.

White and Perroux (1987) referred to sorptivity as an integral property of the soil hydraulic diffusivity. S is constant, provided the water content at the inflow end is constant (Jury et al., 1991).

The form of Philip's truncated equation is very similar to that of Kostiakov. In fact, the modified Kostiakov equation with n equal to 0.5 is essentially the same equation. The Philip infiltration model is that the assumptions for which the equation is applicable are rarely found in the field on a large scale. Soil types vary both spatially and with depth, as do vegetation and surface conditions.

Although parameter values can be obtained by making point measurements in the field, variability limits the worth of test results for application to larger areas such as watersheds (Sullivan et al., 1996).

Whisler and Bouwer (1970) found that determining the values of the parameters S and A for the Philip equation from physical soil properties was very time consuming and yielded results that were not in agreement with the experimental curve. They were able to obtain close agreement with experimental values when they determined parameter values by curve fitting but lost the physical significance of the parameters by using this method.

7.1.2.4 Horton Equation

The Horton model of infiltration (Horton, 1939, 1940) is one of the best-known models in hydrology. The form of Horton's equation represent below.

$$f = f_c + (f_o + f_c)e^{-kt} \qquad (7.8)$$

where f is the infiltration rate in each time, f_o is the initial infiltration rate, f_c is the final infiltration rate (mm/h), t is the time (min), and k is the coefficient of equation.

Horton recognized that infiltration capacity (f) decreased with time until it approached a minimum constant rate (f_c). He attributed this decrease in infiltration primarily to factors operating at the soil surface rather than to flow processes within the soil (Xu, 2003). Beven (2004) discovered, upon making a study of Horton's archived scientific papers, that Horton's perceptual model of infiltration processes was far more sophisticated and complete than normally presented in hydrological texts. Furthermore, his understanding of the surface controls on infiltration continues to have relevance today (Beven, 2004).

When experimental value f_c is subtracted from experimental values for f and the natural log of the resulting values are plotted as a function of time, k can be determined from the slope of the line and f_o can be determined from the intercept. Other methods for finding parameters include a least squares method (Blake, 1968). Horton's equation has advantages over the Kostiakov equation. First, at t equals 0, the infiltration capacity is not infinite but takes on the finite value f_o. Also, as t approaches infinity, the infiltration capacity approaches a nonzero constant minimum value of f_c (Horton, 1940; Hillel, 1998). Horton's equation has been widely used because it generally provides a good fit to data. Although the Horton equation is empirical in that k, f_c, and f_o must be calculated from experimental data, rather than measured in the laboratory, it does reflect the laws and basic equations of soil physics (Chow et al., 1988).

However, the Horton equation is cumbersome in practice because it contains three constants that must be evaluated experimentally (Hillel, 1998). A further limitation is that it is applicable only when rainfall intensity exceeds f_c (Rawls et al., 1993). Horton's approach has also been criticized because he neglects the role of capillary potential gradients in the decline of infiltration capacity over time and attributes control almost entirely to surface conditions (Beven, 2004). Another criticism of the Horton model is that it assumes that hydraulic conductivity is independent of the soil water content (Novotny and Olem, 1994).

7.1.2.5 SCS Equation

Soil Conservation Society of America (USDA, 1974) represented SCS equation.

$$Z = kt^n + 0.6985 \qquad (7.9)$$

Parameters were previously defined.

Attempts to characterize infiltration for field applications usually involve expression of the infiltration rate or cumulative infiltration algebraically in terms of time and certain soil parameters. The principles governing soil water movement have been applied for simplified boundary and initial conditions in order to develop some of the approximate models, including Green–Ampt, Philip, and Smith-Parlange equations. The parameters for these physical models can be determined from soil water properties when they are available. Other models such as Kostiakov's equations are strictly empirical, and the parameters must be obtained from measured infiltration data or from more approximate estimation procedures. Still others, including the Horton equation, are intermediate and have some empirical characteristics while still reflecting physical laws of soil water movement. Although attributed to different physical phenomena, all of the approximate models show a rapid decrease in infiltration rate with time during the initial stage of an infiltration event under ponded conditions (Skaggs and Khaleel, 1982). Different equations that describe infiltration produce different predictions for infiltration rates. These equations use different parameters, and many were developed for different purposes. Each equation has some shortcomings.

The purely physically based equations, such as GA and Philip equations, are advantageous in not requiring measured infiltration data but are based on assumptions that can never be entirely valid. Specifically, they assume homogeneous soils, uniform initial water content, and piston flow, and they neglect the effect of entrapped air.

Both of these equations were originally developed for use under ponded conditions and for deep homogenous soils, but the GA equation was subsequently shown to be more versatile, as it can be applied validly under nonponded conditions and also with a variety of nonhomogeneous soil profiles. It has been applied with good results to soil profiles that become denser with depth (Childs and Bybordi, 1969), for profiles where hydraulic conductivity decreases (Bouwer, 1969) or increases with depth (Bouwer, 1976), and for soils with partially sealed surfaces (Hillel and Gardner, 1970). Bouwer (1969) also demonstrated that it could be used with nonuniform initial water contents. Morel-Seytoux and Khanji (1974) discovered that the equation could be used with slight modification when simultaneous movement of water and air is considered. The GA equation, although the earliest proposed, has proven to be the most versatile, and most widely used of all the infiltration equations.

The empirical equations, such as Kostiakov and Horton equations, provide infiltration rates based on measured field data and therefore provide more realistic estimates when measurements can be provided for the same or very similar conditions to the site for which the prediction is to be made. However, the equations have less value as predictive tools when the measured infiltration data on which the parameter values are based, is obtained from a site that differs significantly, from the site of application. Although the parameters depend on initial water content, rainfall application rate, and soil properties, their values cannot be determined by making such measurements and therefore cannot be easily adjusted to accommodate changes in initial conditions. Actual field measurements of infiltration are required to determine these parameters, making these models much less versatile.

For comparison infiltration equations, some researchers evaluated these equations such as comparison of Philip and Kostiakov equations by Fangmeier and Ramsey (1978); spatial and temporal variability of soil physical properties following tillage of Norfolk loam sand by Cassel (1983); evaluation infiltration of clay loam and clay soil in southeast Australia by Maheshwari and Jayawardane (1992); evaluation variation infiltration in furrow irrigation by Gates and Clyma (1984); and spatial and temporal variation of furrow infiltration by Childs et al. (1993).

7.2 MATERIALS AND METHODS

7.2.1 EXPERIMENT SITE

The study area has been located in the Shavoor Plain, one of the Khuzestan plains in southwest Iran. The area of the study region is 77,706 hectare and located 40 km north of Ahvaz on Ahvaz-Haft Tapeh road. The Shavoor Plain is located between 48° 15′ to 48° 40′ 40″ eastern longitude and 31° 37′ 30″ to 32° 3′ 30″ northern latitude.

The average annual temperature and rainfall are 25.6°C and 233.7 millimeters, respectively. According to the Amberzheh climoscope, the study area is medium hot desert. Thermal regime of soil is hypertermic, and the humidity regime is ustic and aquic.

TABLE 7.1
General Characteristics of Study Zones

Zone	Soil Class	Soil Texture	Bulk Density (g/cm³)	Initial Moisture (%)	Hydraulic Conductivity (m/day)	Impermeable Layer Depth (m)	Water Table Depth (m)
1	S_3A_3	Silt clay loam	1.50	14.75	2.20	2.50	0.92
2	S_3A_3	Silt clay loam	1.50	14.75	0.18	2.30	1.11
3	S_4A_3	Silt loam	1.43	10.28	0.43	1.80	0.45
4	S_4A_3	Silt clay loam	1.50	14.75	0.89	>3.00	>3.00

For this study, four stations of the Shavoor Plain were selected. General characteristics of these areas have been represented in Table 7.1.

Over much of the Shavoor Plain, the use of surface irrigation systems has been applied specifically for field crops to meet the water demand of both summer and winter crops. The major irrigated broad-acre crops grown in this area are wheat, barley, and maize, in addition to fruits, melons, watermelons, and vegetables such as tomatoes and cucumbers. But like other plains in Khuzstan Province, the important major crops are wheat and barley (Behzad et al., 2009; Naseri et al., 2009; Albaji et al., 2012b; Albaji and Alboshokeh, 2017).

As with other plains in Khuzstan Province, there are very few instances of sprinkle and drip irrigation on large-area farms in the Shavoor Plain, and surface irrigation is the major irrigation system in this plain (Landi et al., 2008; Rezania et al., 2009; Boroomand Nasab et al., 2010; Albaji et al., 2008, 2012a, 2014a, 2014b, 2016).

7.2.2 THE METHOD OF EXPERIMENT

Generally, infiltration is measured by some methods such as double rings, blocked furrow, recycling furrow infiltration, and two points. One of the current methods for the measurement of infiltration is use of double rings. This method is an ancient one; nevertheless, it is used more than other methods for determination coefficients of infiltration equations.

By using double rings in two-class soil of the Shavoor Plain (each class had two stations and three iterations), depths of infiltration in different times were measured. Figures 7.1 and 7.2 show double rings and requirement tools for measurement infiltration.

After that, by using SPSS13.0 and measurement data, coefficients of infiltration equations were obtained.

FIGURE 7.1   Requirement tools for infiltration test.

FIGURE 7.2   Performance test requirement by using double rings.

7.3   RESULTS AND DISCUSSION

By using data that were obtained from field experiment coefficients, Green and Ampt, Philip, Kostiakov, Kostiakov–Lewis, SCS, and Horton equations were calculated by SSPS13.0. To evaluate the model, the coefficient of determination (R^2) has been used. Formulas for calculating R^2 are given as follows:

$$R^2 = \frac{\left[\sum_{i=1}^{n} \left(O_i - O_{ave} \right) \left(P_i - P_{ave} \right) \right]}{\sum_{i=1}^{n} \left(O_i - O_{ave} \right)^2 \sum_{i=1}^{n} \left(P_i - P_{ave} \right)^2} \tag{7.10}$$

where P_i is the predicted data, O_i is the measured data, O_{ave} is the average of measurement data, P_{ave} is the average of predicted data, and n is the number of samples. The R^2 values indicate the degree to which data variations are explained by each model. If the model curve closely parallels the observation curve, the coefficient of determination (R^2) will be close to 1. The result for each station is represented in Tables 7.2 through 7.5.

Results show that the coefficient of determination of Kostiakov–Lewis in four stations had the most correlation, and the coefficient of determination of the Horton equation had the least correlation. In station 1, soil was S_3A_3, hydraulic conductivity was 2.2 (m/day), depth of water table was 0.92, and impermeable layer depth was 2.50 (m). In this zone, respectively, the Kostiakov–Lewis, Kostiakov, Philip, SCS, Green–Ampt, and Horton equations can be used for

TABLE 7.2
Infiltration Equation on Station 1

The Name of Equation	Formula	Coefficient of Determination
Green–Ampt	f = 12.742/F + 2.143	0.99622
Philip	$Z = 0.590t^{0.5} + 0.027t$	0.99951
Kostiakov	$Z = 0.441t^{0.648}$	0.99986
Kostiakov–Lewis	$Z = 0.0047t^{0.605} + 0.000099t$	0.99990
SCS	$Z = 0.278t^{0.726} + 0.6985$	0.99668
Horton	$f = 26.4 + 178.6e^{-5.4116t}$	0.89440

TABLE 7.3
Infiltration Equation on Station 2

The Name of Equation	Formula	Coefficient of Determination
Green–Ampt	$f = 29.392/F + 1.411$	0.91768
Philip	$Z = 0.914t^{0.5} + 0.015t$	0.99844
Kostiakov	$Z = 0.786t^{0.567}$	0.99937
Kostiakov–Lewis	$Z = 0.00818t^{0.550} + 0.00005t$	0.99962
SCS	$Z = 0.563t^{0.623} + 0.6985$	0.99750
Horton	$f = 26.3 + 263.7e^{-4.9699t}$	0.91768

TABLE 7.4
Infiltration Equation on Station 3

The Name of Equation	Formula	Coefficient of Determination
Green–Ampt	$f = 35.042/F + 5.828$	0.92135
Philip	$Z = 1.114t^{0.5} + 0.063t$	0.99373
Kostiakov	$Z = 0.793t^{0.676}$	0.99694
Kostiakov–Lewis	$Z = 0.00775t^{0.610} + 0.00046t$	0.99884
SCS	$Z = 0.639t^{0.712} + 0.6985$	0.99533
Horton	$f = 62.8 + 307.2e^{-6.173t}$	0.72987

TABLE 7.5
Infiltration Equation on Station 4

The Name of Equation	Formula	Coefficient of Determination
Green–Ampt	$f = 18.025/F + 2.527$	0.89279
Philip	$Z = 0.501t^{0.5} + 0.003t$	0.99954
Kostiakov	$Z = 0.484t^{0.520}$	0.99925
Kostiakov-Lewis	$Z = 0.0056t^{0.456} + 0.000069t$	0.99979
SCS	$Z = 0.241t^{0.635} + 0.6985$	0.99778
Horton	$f = 8.7 + 231.3e^{-8.4776t}$	0.87038

determination infiltration. In station 2, soil was S_3A_3, hydraulic conductivity is 0.18 (m/day), depth of water table was 1.11, and impermeable layer depth was 2.30 (m). In this zone, respectively, the Kostiakov–Lewis, Kostiakov, Philip, Green–Ampt, SCS, and Horton equations can be used for determination infiltration. In addition, in station 3, soil was S_4A_3, hydraulic conductivity was 0.43 (m/day), depth of water table was 0.45, and impermeable layer depth was 1.80 (m). In this zone, respectively, the Kostiakov–Lewis, Kostiakov, SCS, Philip, Green–Ampt, and Horton equations can be used for determination infiltration. Station 4 had similar soil class with station 3, with hydraulic conductivity equal to 0.89 (m/day) and depth of water table and impermeable layer depth more than 3 (m); the Kostiakov–Lewis, Philip, Kostiakov, SCS, Green–Ampt, and Horton equations can be used for determination infiltration. Differences in equation degrees in stations 3 and 4, which had similar soil classes, can be related to soil texture, hydraulic conductivity, and water table depth.

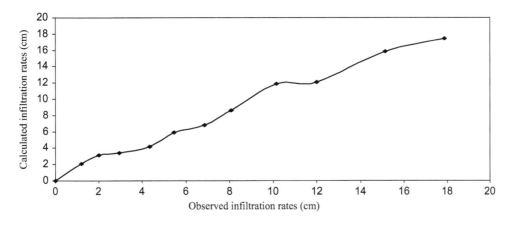

FIGURE 7.3   Calculated infiltration rates plotted against observed values (station 2).

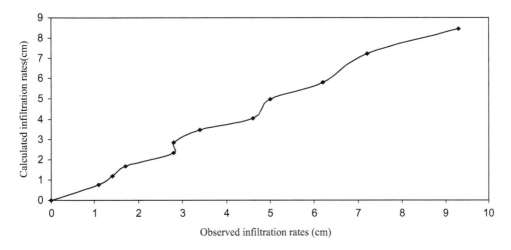

FIGURE 7.4   Calculated infiltration rates plotted against observed values (station 4).

After determining the best equation in each station, due to some limitations, stations 2 and 4 were selected for validation of the best model of the stations. In these stations, depth of infiltration was measured using double rings after two months.

Initially, depth of infiltration was calculated using the best equation (Kostiakov–Lewis equation) on each station in similar time of measurement. Then, measurement and calculated data were comprised. Figures 7.3 and 7.4 show predicted infiltration rates plotted against observed value.

Figures 7.3 and 7.4 show the Kostiakov–Lewis equation in these stations can predict infiltration very successful. Also, R^2 in these stations is more than 90%, which shows good agreement between calculated and observed values. Because characteristics of station 1 were similar to station 2 and, characteristics of station 4 were similar to station 3, the Kostiakov–Lewis equation can be used for predicted infiltration rates.

7.4   CONCLUSION

Generally, the Green–Ampt equation has been represented like $f = \frac{A}{F} + B$. In four study stations, A was more than 10 and B was more than 1. Also, A and B were dependent on soil hydraulic conductivity, depth of water on soil, soil matrix potential, and distance between soil surfaces to moisture front (Alizadeh, 1999). The Philip equation consists of a matrix and gravity head; S was more than 0.5, and A was more than 0.001. In the Kostikov equation, n and k were approximately more than 0.5. The considerable point in this equation k is equal to S in the Philip equation. This shows that k was dependent on physical characteristics of soil. Philip (1957a) and Haverkamp et al. (1977) showed that in initial time, the value of k is equal to S. Value of n in four stations is more than 0.5, and this was according to Hartley's (1992) research. Value of k in the Horton equation was dependent on soil type and was more than 4; and in soil, S_4A_3 was more than S_3A_3.

Generally, Kostiakov–Lewis, Kostaikov, and Philip in four stations have represented better results than others. But the Kostiakov equation gives suitable results for a short time and has a limitation. In this equation, if time tends to infinity, the rate of infiltration will tend to zero. Also, coefficients that were obtained can be used for similar situations (similar moisture) (Alizadeh, 1999). So, applications of the Kostikov and Kostiakov–Lewis equations are not recommended because of empirical basics. However, the Philip equation is recommended because flow in porous media was obtained, which gave a suitable result for the study area.

ACKNOWLEDGMENTS

We are grateful to the Research Council of Shahid Chamran University of Ahvaz for financial support (GN: SCU.WI98.281).

REFERENCES

Albaji, M., Boroomand Nasab, S., Kashkuli, H.A., Naseri, A.A., Sayyad, G., Jafari, S. 2008. Comparison of different irrigation methods based on the parametric evaluation approach in North Molasani Plain, Iran. *Journal of Agronomy* 7 (2): 187–191.

Albaji, M., Boroomand Nasab, S., Hemadi, J. 2012a. Comparison of different irrigation methods based on the parametric evaluation approach in West North Ahvaz Plain. In: *Problems, Perspectives and Challenges of Agricultural Water Management*, M. Kumar, (ed.), InTech, Croatia, pp. 259–274.

Albaji, M., Papan, P., Hosseinzadeh, M., Barani, S. 2012b. Evaluation of land suitability for principal crops in the Hendijan region. *International Journal of Modern Agriculture* 1 (1): 24–32.

Albaji, M., Golabi, M., Boroomand Nasab, S., Jahanshahi. M. 2014a. Land suitability evaluation for surface, sprinkler and drip irrigation systems. *Transactions of the Royal Society of South Africa* 69 (2): 63–73.

Albaji, M., Golabi, M., Piroozfar, V.R., Egdernejad, A., Nazari Zadeh, F. 2014b. Evaluation of agricultural land resources for irrigation in the Ramhormoz plain by using GIS. *Agriculturae Conspectus Scientificus* 79 (2): 93–102.

Albaji, M., Golabi, M., Hooshmand, A.R., Ahmadee, M. 2016. Investigation of surface, sprinkler and drip irrigation methods using GIS. *Jordan Journal of Agricultural Sciences* 12 (1): 211–222.

Albaji, M., Alboshokeh, A. 2017. Assessing agricultural land suitability in the Fakkeh region, Iran. *Outlook on Agriculture* 46 (1): 57–65.

Alizadeh, A. 1999. *Relationship of Water, Soil and Plant*, Emam Reza University, Ltd., Mashhad, Iran.

Behzad, M., Albaji, M., Papan, P., Boroomand Nasab, S. 2009. Evan region qualitative soil evaluation for wheat, barley, alfalfa and maize. *Journal of Food, Agriculture & Environment* 7 (2): 843–851.

Beven, K.J. 2004. *Rainfall-Runoff Modeling: The Primer*, John Wiley & Sons, Ltd., New York.

Blake, G.J. 1968. Infiltration at the Puketurua experimental basin. *Journal of Hydrology* 7(1): 38–46.

Boroomand Nasab, S., Albaji, M., Naseri, A.A. 2010. Investigation of different irrigation systems based on the parametric evaluation approach in Boneh Basht plain, Iran. *African Journal of Agricultural Research* 5 (5): 372–379.

Bouwer, H. 1969. Infiltration of water into nonuniform soil. *Journal of Irrigation and Drainage Division*, *ASCE* 95(IR4): 451–462.

Bouwer, H. 1976. Infiltration into increasingly permeable soils. *Journal of Irrigation and Drainage Division*, *ASCE* 102(IR1): 127–136.

Cassel, D.K. 1983. Spatialand temporal variability of soil physical properties following tillage of Norfolk loam sand. *Soil Science Society of America Journal* 47: 196–201.

Childs, E.C., Bybordi, M. 1969. The vertical movement of water in stratified porous material. 1. Infiltration. *Water Resources Research* 5 (2): 446–459.

Childs, J.L., Wallender, W., Hopmans, J.W. 1993. Spatial and temporal variation of furrow infiltration. *Journal of the Irrigation and Drainage Engineering* 119 (1): 74–90.

Chow, V.T., Maidment, D.R., Mays, L.W. 1988. *Applied Hydrology*. McGraw-Hill, New York.

Clemmens, A.J. 1983. Infiltration equations for border irrigation models. P266-274. In *Advances in Infiltration. Proceedings of the National Conference on Advances in Infiltration*. December 12–13, 1983. Chicago, IL. ASAE Pub. 11-83. St. Joseph, Mo.

Fangmeier, D.D., Ramsey, M.K. 1978. Intake characteristics of irrigation furrows. *Transactions of the ASAE*, 21 (4): 696–700.

Fok, Y.S. 1986. Derivation of Lewis-Kostiakov intake equation. *Journal of Irrigation and Drainage Engineering* 112: 164–171.

Gates, T.K., Clyma, W. 1984. Designing furrow irrigation systems for improved seasonal performance. *Transactions of the American Society of Agricultural Engineers* 27 (6): 1817–1824.

Gifford, G.F. 1976. Applicability of some infiltration formulae to rangeland infiltrometer data. *Journal of Hydrology* 28: 1–11.

Gifford, G.F. 1978. Use of infiltration coefficients as an aid in defining hydrologic impacts of range management schemes. *Journal of Range Management* 31: 115–117.

Ghosh, R.K. 1980. Modeling infiltration. *Soil Science* 130: 297–302.

Ghosh, R.K. 1983. A note on the infiltration equation. *Soil Science* 136: 333–338.

Ghosh, R.K. 1985. A note on Lewis-Kostiakov's infiltration equation. *Soil Science* 139: 193–196.

Green, W.H., Ampt, G. 1911. Studies of Soil Physics, Part 1. The flow of air and water through soils. *Journal of Agricultural Science* 4: 1–24.

Hartley, D.M. 1992. Interpretation of Kostiakov infiltration parameters for borders. *Journal of Irrigation and Drainage Engineering* 118 (1): 156–165.

Haverkamp, R., Vauclin, M., Touma, J., Wierenga, P.J., Vachaucl, G. 1977. A comparison of numerical simulation models for one-dimensional infiltration. *Soil Science Society of America Journal* 41: 285–293.

Haverkamp, R., Rendon, L., Vachaud, G. 1987. Infiltration equations and their applicability for predictive use. In Y.-S. Fok (ed.), *Infiltration Development and Application*, pp. 142–152, Honolulu, Hawaii.

Hillel, D. 1998. *Environmental Soil Physics*, Academic Press, San Diego, CA.

Hillel, D., Gardner, W.R. 1970. Transient infiltration into crust topped profiles. *Soil Science* 109: 69–76.

Holtan, H.N., England, C.B., Shanholtz, V.O. 1967. Concepts in hydrologic soil grouping. *Transactions of the ASAE*. Paper No. 65-739: 407–410.

Horton, R.E. 1939. Analysis of runoff plot experiments with varying infiltration capacity. *Transactions of the American Geophysicists*. Union, Part IV: 693–694.

Horton, R.E. 1940. An approach towards a physical interpretation of infiltration capacity. *Soil Science Society of America* 5: 399–417.

Israelson, O.W., Hansen, V.E. 1967. *Irrigation Principles and Practices*, pp. 198–200, John Wiley & Sons, New Delhi.

Jury, W.A., Gardener, W.R., Gardener, W.H. 1991. *Soil Physics*, 5th ed. Wiley, New York.

Kostiakov, A.N. 1932. On the dynamics of the coefficient of water-percolation in soils and on the necessity for studying it from a dynamic point of view for purposes of amelioration. In *Transactions Congress International Society for Soil Science*, 6th, Moscow, Part A, pp. 17–21.

Landi, A., Boroomand-Nasab, S., Behzad, M., Tondrow, M.R., Albaji, M., Jazaieri, A. 2008. Land suitability evaluation for surface, sprinkle and drip irrigation methods in Fakkeh Plain, Iran. *Journal of Applied Sciences* 8 (20): 3646–3653.

Larson, C.L. 1983. Advances in infiltration a summary. In *Advances in Infiltration Proceeding of the National Conference on Advances in Infiltration*, Michigan, ASAE, pp. 364–373.

Lewis, M.R. 1937. The rate of infiltration of water in irrigation practice. *Transactions of the American Geophysical* Union 18: 361–368.

Maheshwari, B.L., Jayawardane, N.S. 1992. Infiltration characteristics of some clayey soils measured during boarder irrigation. *Agricultural Water Management* 21: 265–279.

Mbagwu, J.S.C. 1990. Mulch and tillage effects on water transmission characteristics of an Ultisol and maize grain yield in SE, Nigeria. *Pedologie* 40: 155–168.

Mbagwu, J.S.C. 1993. Testing the goodness of fit of selected infiltration models on soils with different land use histories. International Centre for Theoretical Physics, Trieste, Italy.

Mbagwu, J.S.C. 1994. Soil physical properties influencing the fitting parameters in Philip and Kostiakov infiltration models. International Centre for Theoretical Physics. Trieste, Italy.

Morel-Seytoux, H.J., Khanji, J. 1974. Derivation of an equation for infiltration. *Water Resources Research* 10 (4): 795–800.

Naeth, M.A. 1988. The impact of grazing on litter and hydrology in mixed prairie and fescue grassland eco-systems of Alberta. Ph.D. Thesis. University of Alberta, Edmonton, Canada.

Naeth, M.A., Chanasyk, D.S., Bailey, A.W. 1991. Applicability of the Kostiakov equation to mixed prairie and fescue grasslands of Alberta. *Journal of Range Management* 44 (1): 18–21.

Naseri, A.A., Albaji, M., Boroomand Nasab, S., Landi, A., Papan, P., Bavi, A. 2009. Land suitability evaluation for principal crops in the Abbas Plain, Southwest Iran. *Journal of Food, Agriculture & Environment* 7 (1): 208–213.

Novotny, V., Olem, H. 1994. *Water Quality; Prevention, Identification, and Management of Diffuse Pollution.* Van Nostrand Reinhold, New York.

Oram, B. 2005. Hydrological Cycle. Watershed assessment, education, training, monitoring resources in Northeastern Pennsylvania. Wilkes University, Environmental Engineering and Earth Sciences Department, Wilkes-Barre, PA. http://www.water-research. net/watershed/hydrologicalcycle.htm. Accessed August 29, 2006.

Philip, J.R. 1957a. The theory of infiltration: 1. The infiltration equation and its solution. *Soil Science* 83: 345–357.

Philip, J.R. 1957b. The theory of infiltration: 4. Sorptivity and algebraic infiltration equations. *Soil Science* 84: 257–264.

Rawls, W.J., Ahuja, L.R., Brakensiek, D.L., Shirmohammadi, A. 1993. Infiltration and soil water movement. In Maidment, D.R. (Ed.), *Handbook of Hydrology*, McGraw-Hill. Inc, New York.

Rezania, A.R., Naseri, A.A., Albaji, M. 2009. Assessment of soil properties for irrigation methods in North Andimeshk Plain, Iran. *Journal of Food, Agriculture & Environment* 7 (3–4): 728–733.

Richards, L.A. 1931. Capillary conduction through porous mediums. *Physics* 1: 313–318.

Serrano, S.E. 1997. The hydrology of river pollution. *Hydrology for Engineers, Geologists and Environmental Professionals*, Hydroscience, Inc, Lexington, pp. 341–368.

Shirmohammadi, A., Skaggs, R.W. 1984. Effect of soil surface conditions on infiltration for shallow water table soils. *Transactions of the ASAE* 27 (6): 1780–1787.

Skaggs, R.W., Khaleel, R. 1982. Chapter 4: Infiltration. In *Hydrology of Small Watersheds*, ASAE, St. Joseph, MI.

Sullivan, M., Warwick, J.J., Tyler, S.W. 1996. Quantifying and delineating spatial variations of surface infiltration in a small watershed. *Journal of Hydrology*, 181 (1–4): 149–168.

USDA. 1974. Border Irrigation. Chapter 4, Section 15. *U. S. Soil Conservation Service National Engineering Handbook*. U.S. Govt. Printing Office, Washington, DC, 50 p.

Walker, W. 1998. SIRMOD – Surface Irrigation Modeling Software. Utah State University.

Whisler, F.D., Bouwer, H. 1970. Comparison of methods for calculating vertical drainage and infiltration for soils. *Journal of Hydrology* 10 (1): 1–19.

White, I., Perroux, K.M. 1987. Use of sorptivity to determine field hydraulic properties. *Soil Science Society of America Journal* 51: 1093–1101.

Xu, C.Y. 2003. Approximate infiltration models. Section 5.3. In *Hydrologic Models*, Uppsala University Department of Earth, Air and Water Sciences. Uppsala, Sweden.

8 Determination of the Optimized Equation of Leaching

Mona Golabi, Mohammad Albaji, and Saeid Eslamian

CONTENTS

8.1 INTRODUCTION

Decreasing soil salinity and increasing potential of production are the result of desalinization. The best method for desalinization is leaching soil by water with or without emendator material. This is done by putting water on soil for some time until the water infiltrates the soil and transforms the drain water to drainage or the bottom layer of soil. In a study project, evaluation of the possibility of reclamation of saline and alkaline soils and determination of water requirement for leaching with field-testing were recommended. With these, testing can be obtained through imperial and theoretical models, desalinization, and desodification curve.

Imperial models are types of mathematical equations that are obtained from measured and observation data. Therefore, these models do not have any mathematic or physics presuppositions but have some limitations (can be used for special location and problem) and advantages:

1. Application of imperial models can be used in approximate estimation.
2. Imperial models can be part of a numerical complex model.
3. Imperials models have some limitation but in practical applications, these models are better than theoretical models (Water Resources Management of Iran, 2006).

In a review, Reeve et al. (1955) and Reeve (1957) showed that a leaching curve generally had a special shape. They did some investigations with these situations: soil texture was silt clay loam and initial electrical conductivity on 0–30 cm depth of soil was 40 ds/m with continued leaching. Reeve et al. obtained the following inverse imperial equation:

$$\frac{D_W}{D_S} = \frac{1}{5\left(\dfrac{EC_f}{EC_i}\right)} + 0.15 = \frac{1}{5}\left(\frac{EC_i}{EC_f}\right) + 0.15 \qquad (8.1)$$

where:

D_W is the depth of application water (cm)

D_S is the soil depth (cm)

EC_i is the electrical conductivity before leaching

EC_f is the electrical conductivity after leaching.

According to investigation and field-testing in Hansa-Hariana in India, Leffelaar and Sharma (1977) represented that results of Reeve's model gave the depth of leaching water for light soil (sandy loam to silt loam) more than the requirement amount. They conducted an alternate experiment of leaching in soil with initial electrical conductivity of 30 ds/m and obtained the following inverse imperial equation:

$$\frac{EC_f - EC_e}{EC_i - EC_e} = \frac{0.062}{\dfrac{D_{lw}}{D_s}} + 0.034 \tag{8.2}$$

Hoffman (1980) used data obtained from the field in the USA and other countries, which are represented in the equation below:

$$\frac{D_W}{D_S} = K \frac{EC_i - EC_{eq}}{EC_f - EC_{eq}} \tag{8.3}$$

where:

D_W is the depth of application water (cm)

D_S is the soil depth (cm)

EC_i is the electrical conductivity before leaching

EC_f is the electrical conductivity after leaching

EC_{eq} is the equivalent electrical conductivity

K is the dimensionless imperial coefficient.

Pazira and Kawachi (1981), according to a study and some experiments done during several years in the central part of Khuzestan for silt clay to clay silt soil with electrical conductivity equivalent to 65–80 ds/m from surface to 150 cm of soil depth, represented the inverse and imperial equation as follows:

$$\frac{EC_f - EC_e}{EC_i - EC_e} = \frac{0.070}{\dfrac{D_{lw}}{D_s}} + 0.023 \tag{8.4}$$

Some variables have defined in equation (8.3) and D_{lw} is the net of irrigation water (cm).

Pazira et al. (1998) did field experiments on saline and sodic soil of southeast Khuzestan Province in Iran. The experiments were done on clay loam to silt clay with initial electrical conductivity 38.2–46.5 ds/m from surface to 1 m depth of soil. They obtained data for the alternate leaching equation:

$$\frac{EC_f - EC_e}{EC_i - EC_e} = 0.0764 \times \left(\frac{D_{lw}}{D_s}\right)^{-0.864} \tag{8.5}$$

The variables have been defined before.

Generally, infiltration equations can be classified as imperial models and mathematical methods. The imperial models of Reeve (1957), Leffelaar and Sharma (1977), Hoffman (1980), and Pazira and Kawachi (1981) are inverse equations; the imperial model of Pazira et al. (1998) is power; and the imperial models of Dieleman (1963) are semilogarithmic equations.

In this paper with field data, mathematical models have been obtained and finally, the best model for the study area has been suggested.

8.2 MATERIALS AND METHODS

8.2.1 EXPERIMENT SITE

For evaluation of leaching, mathematical modeling research was conducted in the Shavoor Plain, in Khuzestan Province. The area of the study region is 77,706 hectare and located 40 km north of Ahvaz in Ahvaz-Hafttapeh road. This area from the south terminates in the Tavana canal in the south to Elhaee village, from north to Shavoor village, from east to Tehran-Ahvaz railway, and from west to the Karkheh River. The field study was done in March 2007 at the Shavoor Plain, which is located between 48° 15′ to 48° 40′ 40″ eastern longitude and 31° 37′ 30″ to 32° 3′ 30″ northern latitude.

The average annual temperature and rainfall are 25.6°C and 233.7 millimeters, respectively. According to the Amberzheh climoscope, the study area is medium hot desert. Thermal regime of soil is hypertermic, and humidity regime is ustic and aquic.

Over much of the Shavoor Plain, the use of surface irrigation systems has been applied specifically for field crops to meet the water demand of both summer and winter crops.

The major irrigated broad-acre crops grown in this area are wheat, barley, and maize, in addition to fruits, melons, watermelons, and vegetables such as tomatoes and cucumbers. But like other plains in Khuzstan Province, the important major crops are wheat and barley (Behzad et al., 2009; Naseri et al., 2009; Albaji et al., 2012b; Albaji and Alboshokeh, 2017).

As with other plains in Khuzstan Province, there are very few instances of sprinkle and drip irrigation on large-area farms in the Shavoor Plain, and surface irrigation is the major irrigation system in this plain (Landi et al., 2008; Rezania et al., 2009; Boroomand Nasab et al., 2010; Albaji et al., 2008, 2012a, 2014a, 2014b, 2016).

8.2.2 THE METHOD OF EXPERIMENT

For this current study, two zones of the Shavoor Plain were selected. General characteristics of these areas are represented in Table 8.1.

Zone 1 is located near Seyed Ghazban village, and zone 2 is located on the north of Mazraeh village. To obtain information about chemical and physical characteristics of soil in the two zones, and water that was applied for leaching, chemical and physical parameters of soil and water before leaching were measured. The results are represented in Tables 8.2 through 8.6.

In each zone, eight plots with 1×1 m size were created. In four plots before leaching, 0.27 L sulfuric acid was mixed with soil. The whole water that was applied in all plots in each frequency was 100 cm and after, increases of 25 cm of water were provided to the soil samples.

In the first round, 25 cm or 250-liter water increased to each plot. Then, one plot was selected. After, the gravity water was removed, and 0 to 25 cm depth of soil was provided as samples; the experiment continued on the remaining plots. The 25 cm water was increased to the remaining plots; later, the gravity water was removed from the 25 to 50 cm of soil depth and the soil was sampled. This method was repeated for 75, 100, 125, and 150 cm of soil depth, and 75 and 100 cm of water. After collection of samples, EC_e, pH, CEC, ESP, $CaSO_4$, anions, and cations were measured

TABLE 8.1
General Characteristics of Study Zones

Zone	Class Before/ After Leaching	Soil Texture	Infiltration Rate (cm/h)	Hydraulic Conductivity (m/day)	Impermeable Layer Depth (m)	Water Table Depth (m)
1	S_4A_2/ S_3A_1	Silt loam & Silt clay loam	6.55	1.30	>3.00	2.85
2	S_4A_3/ S_2A_2	Silt clay loam	0.87	2.77	>3.00	1.14

TABLE 8.2

Soil Chemical Quality Before Leaching in Zone No. 1

Soil Depth (cm)	ECe (ds/m)	pH	Gypsum $\left(\dfrac{\text{meq}}{\text{100g soil}}\right)$	C.E.C $\left(\dfrac{\text{meq}}{\text{100g soil}}\right)$	EX.Na $\left(\dfrac{\text{cmol}}{\text{kg soil}}\right)$	SAR	ESP[a]	ESP[b]
0–25	56.84	7.73	25.3	14.4	3.01	19.05	20.90	21.16
25–50	29.68	7.77	18.2	15.3	2.33	13.65	15.23	15.88
50–75	10.18	7.70	9.1	15.4	1.06	5.13	6.88	5.93
75–100	6.87	7.56	8.2	15.6	0.96	4.55	6.15	5.17
100–150	4.57	7.63	7.8	15.7	0.43	1.09	2.74	0.35

[a] $\text{ESP} = \dfrac{\text{Ex.Na}}{\text{C.E.C}} \times 100$

[b] $\text{ESP} = \dfrac{100\left(-0.0126 + 0.01475\,\text{SAR}\right)}{1 + \left(-0.0126 + 0.01475\,\text{SAR}\right)}$

TABLE 8.3

Soil Chemical Quality Before Leaching in Zone No. 2

Soil Depth (cm)	ECe (ds/m)	pH	Gypsum $\left(\dfrac{\text{meq}}{\text{100g soil}}\right)$	C.E.C $\left(\dfrac{\text{meq}}{\text{100g soil}}\right)$	EX.Na $\left(\dfrac{\text{c mol}}{\text{kg soil}}\right)$	SAR	ESP[a]	ESP[b]
0–25	64.44	7.78	36.8	15.1	5.08	31.99	33.64	31.47
25–50	28.60	7.77	29.1	15.4	4.90	29.98	31.82	30.05
50–75	14.81	7.88	11.2	15.8	3.67	21.78	23.23	23.59
75–100	10.54	7.80	9.3	15.7	3.33	20.35	21.21	22.33
100–150	9.72	7.96	7.1	15.6	3.08	18.88	19.74	21.00

[a] $\text{ESP} = \dfrac{\text{Ex.Na}}{\text{C.E.C}} \times 100$

[b] $\text{ESP} = \dfrac{100\left(-0.0126 + 0.01475\,\text{SAR}\right)}{1 + \left(-0.0126 + 0.01475\,\text{SAR}\right)}$

TABLE 8.4

Soil Physical Characteristics (Definite Moisture in Soil Layers on Zone No. 1)

Soil Depth (cm)	Layer Depth (cm)	Percentage Weight Moisture θ_{mc}	θ_{mFC}	$\rho_d\left(\dfrac{\text{g}}{\text{cm}^3}\right)$	Deficit Moisture[a] (cm)	Totally
0–25	25	18.60	19.85	1.42	0.44	0.44
25–50	25	20.20	21.43	1.46	0.45	0.89
50–75	25	20.45	21.87	1.48	0.53	1.42
75–100	25	21.90	22.00	1.49	0.04	1.46
100–150	25	22.10	22.32	1.51	0.0	1.46

[a] $d = \dfrac{\left(\theta_{mc} - \theta_{mFC}\right) \times \rho_d \times D}{100}$

TABLE 8.5

Soil Physical Characteristics (Definite Moisture of Soil Layers in Zone No. 2)

Soil Depth (cm)	Layer Depth (cm)	Percentage Weight Moisture		$\rho_d\left(\dfrac{g}{cm^3}\right)$	Deficit Moisture[a](cm)	Totally
		θ_{mc}	θ_{mFC}			
0–25	25	14.75	21.61	1.50	2.57	2.57
25–50	25	21.12	21.93	1.51	0.31	2.88
50–75	25	21.27	23.02	1.54	0.67	3.55
75–100	25	21.52	22.71	1.53	0.46	4.01
100–150	25	20.13	22.63	1.53	1.78	5.79

a $d = \dfrac{\left(\theta_{mc} - \theta_{mFC}\right) \times \rho_d \times D}{100}$

TABLE 8.6

Characteristics of Water Irrigation Quality

Date Sampling			EC (µohms/cm)	pH	$Ca^{2+} + Mg^{2+}$ $\dfrac{meq}{l}$	Na^{1+} $\dfrac{meq}{l}$	SAR
Year	Month	Day					
2007	04	08	898	7.40	4.8	4.9	3.16

in a laboratory. In addition, after the fourth round of leaching from a depth of 0–5 cm, soil was sampled for measurement equivalently electrical conductivity (EC).

To obtain mathematical models, SPSS12.0 software was used. The ratio of net depth of irrigation to soil depth as independent variable (X), and the ratio difference between final electrical conductivity and equivalent electrical conductivity to the difference between initial electrical conductivity and equivalent electrical conductivity as dependent variable (Y) were used as input of SPSS. Then, 11 mathematical models were obtained. Similar methods were done for zone 2.

In addition, analysis of residual errors, differences between measured and simulated values, can be used to evaluate model performance. These are maximum error (ME), root mean square error (RMSE), coefficient of determination (CD or R^2), modeling efficiency (EF), and coefficient of residual mass (CRM). The mathematical expressions of these statistics are as follows (Homaee et al. 2002):

$$ME = Max\left|P_i - O_i\right|_{i=1}^{n} \tag{8.6}$$

$$RMSE = \left[\frac{\sum_{i=1}^{n}\left(P_i - O_i\right)^2}{n}\right]^{\frac{1}{2}} \times \frac{100}{\overline{\overline{O}}} \tag{8.7}$$

$$R^2 = \frac{\left[\sum_{i=1}^{n}\left(O_i - \overline{O}\right)\left(P_i - \overline{P}\right)\right]^2}{\sum_{i=1}^{n}\left(O_i - \overline{O}\right)^2 \sum_{i=1}^{n}\left(P_i - \overline{P}\right)^2} \tag{8.8}$$

$$EF = \frac{\sum_{i=1}^{n}\left(O_i - \overline{O}\right)^2 - \sum_{i=1}^{n}\left(P_i - O_i\right)^2}{\sum_{i=1}^{n}\left(O_i - \overline{O}\right)^2}$$ (8.9)

$$CRM = \frac{\sum_{i=1}^{n} O_i - \sum_{i=1}^{n} P_i}{\sum_{i=1}^{n} O_i}$$ (8.10)

where P_i is the predicted data, O_i is the measured data, and n is the number of samples. For each field statistical factors that mention above between the results of the model and laboratory experiment were calculated. Then, the degree of each statistical factor for selection fields was determined. Finally, the sum of degree and ranking for each field was determined.

8.3 RESULTS AND DISCUSSION

According to Tables 8.4 and 8.5, 100 cm of water that were given to soil as irrigation water and the total definite of moisture from the surface to 150 cm depth in zones 1 and 2, respectively, were 1.46 and 5.79 cm. So, depth of irrigation water that leached soil was 98.54 and 94.21 cm, respectively.

Results showed that in zone 1 with application acid before leaching, the maximum of electrical conductivity in depth of 0–25 cm was 56.84 ds/m; after leaching, it became 18.07 ds/m. In 150 cm depth of soil with use of 100 cm irrigation water, the average of electrical conductivity decreased from 32.63 to 18.99 ds/m.

In zone 1, without application acid before leaching, the maximum of electrical conductivity in depth of 0–25 cm was 56.84 ds/m; after leaching, it became 23.22 ds/m. Also, in 150 cm depth with use of 100 cm irrigation water, the average of electrical conductivity decreased from 32.63 to 26.29 ds/m.

In zone 2, with application acid before leaching, the maximum of electrical conductivity in depth of 0–25 cm was 64.44 ds/m; after leaching, it became 14.96 ds/m. In 150 cm depth with use of 100 cm irrigation water, the average of electrical conductivity decreased from 37.07 to 16.89 ds/m.

In addition, in zone 2—without application acid before leaching—the maximum of electrical conductivity in depth of 0–25 cm was 64.44 ds/m; after leaching, it became 13.95 ds/m. In 150 cm depth with use of 100 cm irrigation water, the average of electrical conductivity decreased from 37.07 to 16.36 ds/m.

Generally, in zone 1 with application acid electrical conductivity of 13.64 ds/m and without acid electrical conductivity, there was a decrease of 6.34 ds/m. In zone 2 with application acid electrical conductivity 20.18 ds/m and without acid EC, there was a 20.71 ds/m decrease.

In study areas (zones 1 and 2), equivalents of electrical conductivity of soil with application acid 3.85 and 7.61 ds/m and without application acid 6.13 and 6.34 were obtained. If equivalent electrical conductivity in this situation is compared with water electrical conductivity (898 microohms/cm), it would show that the salinity of soil cannot decrease more than this amount. This is because the equivalent electrical conductivity of soil is approximately 1.5–2 multiples of water electrical conductivity (Mohsenifar et al., 2006).

The main aim of the current study was to obtain desalinization mathematical models. Therefore, by using measurement data, 11 equations were obtained. These equations are represented in Tables 8.7 through 8.10. Equations represented consist of linear, logarithmic, inverse, quadratic, cubic, power, compound, S, logistic, growth, and exponential, and for choosing, the best equation was applied statistical indicators (Equations 8.7–8.10).

TABLE 8.7

Imperial Desalinization Models and Statistical Indicators in Zone No. 1 (With Application Acid)

Model	Desalinization Formula	ME	RMSE	R^2	EF	CRM	Total Ranking
Linear	$Y = -0.3285X + 1.0302$	0.950	5.284	0.548	−0.838	−0.143	4
Logarithmic	$Y = -0.4410 LnX + 0.5554$	1.736	7.572	0.798	−2.448	−0.143	3
Inverse	$Y = 0.2217/X + 0.2800$	3.915	9.865	0.720	−6.269	−0.263	6
Quadratic	$Y = 0.1755X^2 - 0.9760X + 1.3600$	1.242	3.403	0.758	−1.990	−0.130	2
Cubic	$Y = -0.1062X^3 + 0.7893X^2 - 1.8512^X + 1.6334$	1.224	0.003	0.830	−2.620	−0.111	1
Power	$Y = 0.4173X^{-0.8477}$	4.470	8.149	0.754	−8.742	−0.312	4
Compound	$Y = 1.1704 \times 0.4726^X$	1.058	7.096	0.730	−1.305	−0.034	5
S	$Y = e^{(0.3656/X - 1.2889)}$	2.021	4.755	0.502	−5.780	7.370	7
Logistic	$Y = 1/((1/1.446) + 0.1398 \times 4.3846X)$	1.986	6.585	0.432	−1.973	0.085	9
Growth	$Y = e^{(-0.7496X + 0.1574)}$	1.059	7.593	0.730	−1.302	0.035	8
Exponential	$Y = 1.1704e^{-0.7496X}$	1.059	3.450	0.730	−1.302	0.035	4

TABLE 8.8

Imperial Desalinization Models and Statistical Indicators in Zone No. 1 (Without Application Acid)

Model	Desalinization Formula	ME	RMSE	R^2	EF	CRM	Total Ranking
Linear	$Y = -0.3257X + 1.2536$	0.853	6.509	0.784	−0.837	−0.200	4
Logarithmic	$Y = -0.3829 LnX + 0.8007$	2.706	8.072	0.876	−3.400	−0.491	3
Inverse	$Y = 0.1685/X + 0.6069$	3.891	9.222	0.606	−6.003	−0.345	7
Quadratic	$Y = 0.1014X^2 - 0.6997X + 1.4441$	1.128	3.603	0.887	−1.041	−0.496	2
Cubic	$Y = -0.0529X^3 + 0.4070X^2 - 1.1355X + 1.5802$	1.008	0.025	0.913	−2.399	−0.498	1
Power	$Y = 0.7013X^{-0.5453}$	5.658	9.947	0.749	−9.742	−0.291	8
Compound	$Y = 1.4267 \times 0.5894^X$	1.080	8.006	0.871	−1.437	−0.129	5
S	$Y = e^{(0.2147/X - 0.5830)}$	1.317	4.839	0.415	−4.099	6.927	7
Logistic	$Y = 1/((1/1.429) + 0.0621 \times 4.0827^X)$	1.532	6.889	0.470	−1.965	0.073	9
Growth	$Y = e^{(-0.5286X + 0.3554)}$	1.081	6.056	0.871	−1.002	0.054	6
Exponential	$Y = 1.4267e^{-0.52866X}$	1.081	3.332	0.871	−1.002	0.054	3

TABLE 8.9

Imperial Desalinization Models and Statistical Indicators in Zone No. 2 (With Application Acid)

Model	Desalinization Formula	ME	RMSE	R^2	EF	CRM	Total Ranking
Linear	$Y = -0.2042X + 0.7910$	0.920	4.284	0.802	−0.543	−1.543	1
Logarithmic	$Y = -0.2119 LnX + 0.5111$	1.321	8.572	0.789	−1.567	−0.543	7
Inverse	$Y = 0.0717/X + 0.4437$	3.965	8.365	0.427	−6.653	−0.943	6
Quadratic	$Y = 0.0542X^2 - 0.3984X + 0.8852$	2.242	3.954	0.876	−1.760	−0.640	3
Cubic	$Y = -0.0021X^3 + 0.0659X^2 - 0.4145X + 0.8900$	1.824	0.653	0.876	−1.920	−0.181	1
Power	$Y = 0.4572X^{-0.4594}$	4.470	8.769	0.758	−6.723	−0.732	7
Compound	$Y = 0.8754 \times 0.6145^X$	1.858	7.546	0.931	−1.905	−0.144	5
S	$Y = e^{(0.1444/X - 0.9053)}$	2.021	4.233	0.353	−5.430	7.650	7
Logistic	$Y = 1/((1/0.846) + 0.0718 \times 4.3179^X)$	1.386	5.123	0.503	−2.534	0.231	4
Growth	$Y = e^{(-0.4869X - 0.1331)}$	0.959	7.328	0.931	−1.923	0.534	2
Exponential	$Y = 0.8754e^{-0.4869X}$	2.059	4.289	0.931	−0.642	1.095	8

TABLE 8.10

Imperial Desalinization Models and Statistical Indicators in Zone No. 2 (Without Application Acid)

Model	Desalinization Formula	ME	RMSE	R^2	EF	CRM	Total Ranking
Linear	$Y = -0.7872X + 0.1931$	1.350	3.994	0.838	-1.838	-0.983	2
Logarithmic	$Y = -0.2020 \ln X + 0.5219$	1.956	7.092	0.838	-2.768	-0.653	6
Inverse	$Y = 0.0707/X + 0.4529$	4.015	9.015	0.485	-5.289	-1.653	7
Quadratic	$Y = 0.0481X^2 - 0.3655X + 0.8709$	1.762	2.129	0.906	-1.543	-0.980	3
Cubic	$Y = -0.0123X^3 + 0.1174X^2 - 0.4606X + 0.8989$	1.544	0.993	0.910	-2.765	-0.811	1
Power	$Y = 0.4746X^{-0.4253}$	3.770	7.149	0.758	-8.054	-2.362	3
Compound	$Y = 0.8653 \times 0.6377^X$	1.548	7.566	0.928	-2.875	-0.734	3
S	$Y = e^{(0.1366/X - 0.8649)}$	1.921	4.565	0.369	-5.864	7.054	5
Logistic	$Y = 1/((1/0.837) + 0.0790 \times 4.0405^X)$	1.646	6.950	0.567	-3.063	1.805	5
Growth	$Y = e^{(-0.4498X - 0.1447)}$	0.059	5.973	0.928	-1.972	0.975	3
Exponential	$Y = 0.8653e^{-0.4498X}$	1.959	3.876	0.928	-2.122	0.633	4

ME indicates the worst-case performance of the model, and according to Table 8.7, linear equation has the least amount. The RMSE value shows how much the simulation overestimates or underestimates the measurements. This parameter for the cubic equation was the least and for inverse equation was the most.

The R^2 gives the ratio between the scatter of the simulation and measurement. R^2 of the cubic equation and logistic was the most and least amount, respectively. Comparison between the simulated data and the average measured values was determined with EF. This parameter for the power and linear equations was the least and most, respectively.

The CRM is a measure of the tendency of the model to overestimate or underestimate the measurements. The CRM of the power equation was the least and S equation was the most.

Generally, the cubic equation was the best model and the logistic equation was the worst for determination of leaching amount in zone 1 with application acid.

According to Table 8.8, the linear equation had the least amount of ME. The cubic equation had the least RMSE value.

The R^2 of the cubic equation had the most amounts. The power and cubic equations had amounts of EF and CRM, respectively.

Generally, the cubic equation was the best model, and the logistic equation was the worst for determination of leaching amounts in zone 1 without application acid.

In addition, in zone 2, the ME of the linear equation, RMSE of the cubic equation, EF of the power equation, and CRM of the linear equation were the least amounts. R^2 of the compound equation was the most. Generally, the cubic equation was the best equation in this zone.

Finally, in zone 2 without application acid, the ME of the growth equation, RMSE of the cubic equation, EF of the power equation, and CRM of the power equation were the least amounts. R^2 of the compound equation was the most. Generally, the cubic equation was the best equation in this zone.

8.4　CONCLUSION

For leaching salt of soil it is necessary to add water on soil, which is dependent on the situation with or without emendator material. If the amount of additional water is low, it will not dissolve salt. Also, if more water is applied than is required, cost will increase. In this investigation, results show that in zone 1 the application emendator material (sulfuric acid) caused more decrease on electrical conductivity, but in zone 2, similar results occurred with and without application acid.

In zone 1, leaching of salt with application acid was more effective than when water was used without acid. This was because the water table in zone 1 had a low level and soil texture was lighter than zone 2. Therefore, salt from the soil profile was easily removed. In addition, hydraulic conductivity was low in this zone. Therefore, emendator material had enough time to combine with the cations (especially calcium and magnesium) of the soil. In the sequel, exited salt of the soil was increased. In this area, concentrations of sulfate and sodium in saturation emulsion of soil after leaching (with acid) increased. Increasing these ions caused sodium replacement to calcium and a decrease in the exchange sodium percentage.

In zone 2 because the soil had a heavy texture and infiltration was low, the water had enough time to leach the soil. So, application of emendator material was not necessary. This case caused decreasing electrical conductivity and exchange sodium percentage with and without acid to become similar.

Statistical analyses of mathematic models indicate that in zone 1, with and without acid, the cubic equation was the best and the logistic equation was the worst equation. In addition, in zone 2, with and without acid, the cubic equation was the best; exponential and inverse equations were the worst model. So, generally, in the Shavoor Plain in the north of Khuzestan Province in southwest Iran, the cubic equation can be used for determining the leaching amount. By using all data of the two zones with and without acid, a cubic equation for the whole area was obtained (Equation 8.11).

$$Y = -0.0384X^3 + 0.3102X^2 - 0.9001X + 1.2209 \qquad (8.11)$$

Equation (8.11) indicates that coefficients of this equation are approximately average coefficients of cubic equations of Tables 8.7 through 8.10. Finally, Equation (8.11) with coefficient of determination equivalent 0.7 is suggested for determining leaching in this area.

ACKNOWLEDGMENTS

We are grateful to the Research Council of Shahid Chamran University of Ahvaz for financial support (GN: SCU.WI98.281).

REFERENCES

Albaji, M., Boroomand Nasab, S., Kashkuli, H.A., Naseri, A.A., Sayyad, G., Jafari, S. 2008. Comparison of different irrigation methods based on the parametric evaluation approach in North Molasani Plain, Iran. *Journal of Agronomy* 7 (2), 187–191.

Albaji, M., Boroomand Nasab, S., Hemadi, J. 2012a. Comparison of different irrigation methods based on the parametric evaluation approach in West North Ahvaz Plain. In: *Problems, Perspectives and Challenges of Agricultural Water Management*, K. Kumar, (ed.), InTech, Croatia, pp. 259–274.

Albaji, M., Papan, P., Hosseinzadeh, M., Barani, S. 2012b. Evaluation of land suitability for principal crops in the Hendijan region. *International Journal of Modern Agriculture* 1 (1), 24–32.

Albaji, M., Golabi, M., Boroomand Nasab, S., Jahanshahi. M. 2014a. Land suitability evaluation for surface, sprinkler and drip irrigation systems. *Transactions of the Royal Society of South Africa* 69 (2), 63–73.

Albaji, M., Golabi, M., Piroozfar, V.R., Egdernejad, A., Nazari Zadeh, F. 2014b. Evaluation of agricultural land resources for irrigation in the Ramhormoz Plain by using GIS. *Agriculturae Conspectus Scientificus* 79 (2), 93–102.

Albaji, M., Golabi, M., Hooshmand, A.R., Ahmadee, M. 2016. Investigation of surface, sprinkler and drip irrigation methods using GIS. *Jordan Journal of Agricultural Sciences* 12 (1), 211–222.

Albaji, M., Alboshokeh, A. 2017. Assessing agricultural land suitability in the Fakkeh region, Iran. *Outlook on Agriculture* 46 (1), 57–65.

Behzad, M., Albaji, M., Papan, P., Boroomand Nasab, S. 2009. Evan region qualitative soil evaluation for wheat, barley, alfalfa and maize. *Journal of Food, Agriculture & Environment* 7 (2), 843–851.

Boroomand Nasab, S., Albaji, M., Naseri, A.A. 2010. Investigation of different irrigation systems based on the parametric evaluation approach in Boneh Basht plain, Iran. *African Journal of Agricultural Research* 5 (5), 372–379.

Dieleman, P.J. 1963. *Reclamation of Salt-affected Soils in Iraq.* Veenman, Wageningen, 175 p.

Hoffman, G.J. 1980. Guideline for reclamation of salt-affected soils. In *Proceeding of International American Salinity and Water Management*, Technical Conference, Juarez, Mexico, pp. 49–64.

Homaee, M., Dirksen, C., Feddes, R.A. 2002. Simulation of root water uptake I. Non-uniform transient salinity using different macroscopic reduction function. *Agriculture Water Management* 57, 89–109.

Landi, A., Boroomand-Nasab, S., Behzad, M., Tondrow, M.R., Albaji, M., Jazaieri, A. 2008. Land suitability evaluation for surface, sprinkle and drip irrigation methods in Fakkeh Plain, Iran. *Journal of Applied Sciences* 8 (20), 3646–3653.

Leffelaar, P.A., Sharma, P. 1977. Leaching of a highly saline-sodic soil. *Journal of Hydrology* 32, 203–218.

Mohsenifar, K., Pazira, E., Najafi, P. 2006. Evaluation leaching models in two area of southeast Khuzestan province (Persian). In *Proceedings of the First Regional of Optimize Utilization of Water Resources of Karoon and Zayandehrood Watershed*, Shahrekor University, Shahrekor, pp. 2026–2037.

Naseri, A.A., Albaji, M., Boroomand Nasab, S., Landi, A., Papan, P., Bavi, A. 2009. Land suitability evaluation for principal crops in the Abbas Plain, Southwest Iran. *Journal of Food, Agriculture & Environment* 7 (1), 208–213.

Pazira, E., Kawachi, T. 1981. Studies on appropriate depth of leaching water, Iran. A case study. *Journal of Intergraded Agricultural Water Use and Freshing Reservoirs*, Kyoto University Japan, 6, 39–49.

Pazira, E., Keshavarz, A., Torii, K. 1998. Studies on appropriate depth of leaching water. In *International Workshop on Use of Saline and Brackish-Water for Irrigation*, Indonesia.

Reeve, R.C. 1957. The relation of salinity to irrigation and drainage requirements. Third Congress of International Commission on Irrigation and Drainage. *Transactions* 5(10), 175–187.

Reeve, R.C., Pillsbury, A.F., Wilcox, L.V. 1955. Reclamation of saline and high boron soil in the Coachella Valley of California. *Hilgardia* 24 (4), 69–91.

Rezania, A.R., Naseri, A.A., Albaji, M. 2009. Assessment of soil properties for irrigation methods in North Andimeshk Plain, Iran. *Journal of Food, Agriculture & Environment* 7 (3–4), 728–733.

Water Resources Management of Iran. 2006. *Guideline of Application Imperial and Theoretical Leaching Models on Salinity Soil*, pp. 1–115. Water Resources Management of Iran, Publication No. 359: 1–115.

Section III

Case Studies
Irrigation Method Selection for the Crossroads of
Central and Southeast Europe and South Asia

9 Multi-Criteria Decision-Making for Irrigation Management in Serbia

Boško Blagojević, Jovana Bezdan, Atila Bezdan, Milica Vranešević, and Radovan Savić

CONTENTS

9.1 INTRODUCTION

The climatic region of Vojvodina is characterized by the variability of meteorological conditions, especially precipitation, which vary both in quantity and in terms of distribution (Pejić et al., 2011a). The specificity of the agroclimatic conditions of Vojvodina, together with the strict requirements of agricultural production, makes this territory very vulnerable to the occurrence of drought or occurrence of excess water on the other hand (Bezdan, 2014). Numerous experts agree that in the climatic conditions of Vojvodina, drought is a common occurrence leading to a decrease in crop yields (Bošnjak, 2001; Dragović, 2001; Pejić et al., 2011a). The percentage of yield reduction ranges from a few to 50% (Dragović, 2012), and in the years of extreme drought, it can go up to 80%–100% (Dragović, 2001). Studies show that the frequency of droughts has increased over the last few decades (Stricevic et al., 2011; Rajic and Bezdan, 2012; Armenski et al., 2014; Dragincic et al., 2017 etc.) and that this natural hazard has become increasingly important for strategic development of the whole country (Stricevic and Djurovic, 2013). The importance of drought monitoring is especially necessary for the countries whose economic sustainability is strongly linked to agriculture (Lessel et al., 2016), which is the case in Serbia, where agriculture is one of the main activities, while Vojvodina province is the most important agricultural region of Serbia (Bezdan, 2019).

A large number of drought indices have been developed worldwide (see Heim, 2002; Quiring, 2009; Mishra and Singh, 2010; Zargar et al., 2011; Bachmair et al., 2016 etc.), and each has its advantages and disadvantages (Mishra and Singh, 2010). A wide range of drought indices are in use, and new indices are continually being introduced depending on the purpose of the conducted analyses (Fabian and Zelenhasic, 2016). When referring to the number of indices developed so far, Zargar et al. (2011) provide a list and descriptions of 74 indices out of nearly 150 available. It is assumed, considering the current literature, that the number of indices today is much greater (Bezdan, 2019). Nevertheless, there are possibilities for further improvements and development of new indices.

In order to perceive impact of drought on agricultural production in Vojvodina, as well as in Serbia, it is important to consider the state and prospects of irrigation in this area, as the main measure to combat agricultural drought (Bezdan, 2019). Irrigation in Vojvodina is not at a satisfactory level, nor is it in line with the requirements and existing potentials of agriculture and water management (Savić et al., 2013). According to the data of the public water management company "Vode Vojvodine," 936,000 hectares of agricultural land can be irrigated in Vojvodina. Currently, only a small percentage of these areas are irrigated, although there is a great potential for irrigation development and available water resources. Because of that, one of Serbian agricultural priorities is to expand the area of land under irrigation, especially in Vojvodina Province.

Decisions related to previously described problems—drought monitoring and defining spatial priorities for irrigation development—need to be made in the context of multiple, usually conflicting criteria. Multi-criteria decision analysis (MCDA) can be an appropriate approach for this type of decision-making problem and can be described "as an umbrella term to describe a collection of formal approaches which seek to take explicit account of multiple criteria in helping individuals or groups explore decisions that matter" (Belton and Stewart, 2002).

The structure of the chapter is as follows. MCDA methods are described in Section 9.2. Section 9.3 describes the study area. Examples of MCDA applications in irrigation management are presented in Sections 9.4 and 9.5. Finally, Section 9.6 closes the chapter with concluding remarks.

9.2 MCDA

9.2.1 GENERAL MCDA METHODOLOGY

The methodology of MCDA can be simplified in the following four basic steps: (1) structuring the decision problem, (2) assessing possible impacts of each alternative, (3) determining preferences (values) of decision makers, and (4) evaluating and comparing alternatives (Keeney, 1982;

Belton and Stewart, 2002; Nordström, 2010). Detailed description of MCDA phases can be found in Blagojevic et al. (2019b), while here is presented a short version.

The first step is to define the type of decision-making problems (individual or group) and to select decision makers (DMs). Then, in group decision-making, weights of DMs need to be defined. This can be done by a supra decision maker, which may be very difficult, or by using one of several existing methods for that purpose (see Blagojevic et al., 2016a). After that, the decision hierarchy must be structured; this means that the goal or overall objective (statement of what DM[s] want[s] to achieve by the decision), criteria, subcriteria (if any), and alternatives should be defined (Keeney, 1992). This is usually done by previously selected DMs (from above) with the help of a decision analyst (Blagojevic et al., 2019b). Selected criteria should be essential, controllable, complete, measurable, operational, decomposable, independent, concise, and understandable (Keeney, 1992; Kangas et al., 2015). It is also important to explore and include in analysis all relevant decision alternatives, which is not always an easy task. This phase (structuring decision problems) is essential for quality of the decision-making process because poor structure of a decision-making problem will lead to poor decision. After the decision problem is structured, it is necessary to obtain performance data of alternatives—which can be quantitative (measured, computed, simulated) or descriptive—with respect to all of the selected criteria.

The next step in MCDA methodology is to define the importance or weights of criteria. Although so-called "objective" methods can be used—such as the Criteria Importance Through Inter-criteria Correlation (CRITIC) (Diakoulaki et al., 1995)—the most common way to define the weights assigned to criteria is still to elicit preference values (subjective judgments) from decision makers (DMs). There are several weighting methods for deriving weights from preference statements (Danielson and Ekenberg, 2019), the most common being direct point allocation (DIRECT), simple multi-attribute rating technique (SMART) (Edwards, 1977; von Winterfeldt and Edwards, 1986), analytic hierarchy process (AHP) (Saaty, 1980), Trade-off (Keeney and Raiffa, 1976), and SWING (von Winterfeldt and Edwards, 1986). Weights of criteria must sum to 1. Finally, in the fourth step, alternatives will be evaluated and ranked using the selected MCDA method.

9.2.2 MCDA METHODS

According to Hajkowicz et al. (2000), all MCDA methods can be divided on continuous and discrete methods, based on the nature of the alternatives to be evaluated (Janssen, 1991). Continuous methods aim to identify an optimal solution from an infinitive number of alternatives, while discrete MCDA methods can be defined as decision support techniques that have a finite number of alternatives. In this section, we presented several discrete MCDA methods that have been most often used in natural resources management. Most of those that are not listed here are based on similar theories and assumptions as the methods presented in this section.

9.2.2.1 Simple Additive Weighting (SAW)

SAW (Hwang and Yoon, 1981) is a simple and frequently used method that can be directly applied for the given decision matrix $R = (r_{ij})$, where columns represent criteria and rows represent alternatives. Entries r_{ij} represent ratings (or performance) of alternatives with respect to all criteria.

$$
R = \begin{array}{c}
\begin{array}{cccc} C_1 & C_2 & \cdots & C_m \\ w_1 & w_2 & \cdots & w_m \end{array} \\
\begin{array}{c} A_1 \\ A_2 \\ \vdots \\ A_n \end{array}
\left[\begin{array}{cccc}
r_{11} & r_{12} & \cdots & r_{1m} \\
r_{21} & r_{22} & \cdots & r_{2m} \\
\vdots & \vdots & \ddots & \vdots \\
r_{n1} & r_{n2} & \cdots & r_{nm}
\end{array} \right]
\end{array}
\tag{9.1}
$$

Fully comparable values are obtained by normalizing ratings in each column. When those values are multiplied across rows with the corresponding weights of criteria, a total utility for every alternative is calculated by using Equation (9.2):

$$S_i = \sum_{j=1}^{n} w_j x_{ij}, \quad i = 1, \ldots, m, \tag{9.2}$$

where m is the number of alternatives and n is the number of criteria. S_i is the total utility of alternative i, x_{ij} is a normalized rating of alternative i for criterion j, and w_j is the weight of criterion j. The best alternative is the one with the highest value of S_i.

9.2.2.2 Compromise Programming (CP)

CP is a technique that ranks alternatives according to their closeness to an ideal alternative (Zeleny 1982). The best alternative is the one with the least distance from an ideal alternative in the set of efficient alternatives. The distance measure used in CP is the family of L_p–metrics given as

$$L_p(i) = \left[\sum_{j=1}^{m} w_j^{\ p} \left| \frac{r_j^* - r_{ij}}{r_j^* - r_j^{**}} \right|^p \right]^{1/p} \tag{9.3}$$

where $L_p(i)$ is the L_p-metric for alternative A_i; r_{ij} is the rating of alternative A_i for criterion C_j; and r_j^* and r_j^{**} are the best and the worst values, respectively, over the set of all alternatives for criterion C_j. A parameter p implicitly expresses the DM's attitude to balance criteria (p = 1), to accept decreasing marginal utility (p > 1), or to search for an absolutely dominant solution (p = ∞); the last is the well-known min-max Chebishev criterion. An alternative with minimum L_p–metric is considered the best.

9.2.2.3 TOPSIS

The Technique for Order Preference by Similarity to Ideal Solution (TOPSIS) method is based on order preference by similarity to an ideal alternative (Hwang and Yoon, 1981). The main idea is that most preferred alternative should not only have the shortest distance from the "ideal" alternative but also the longest distance from the "negative-ideal" alternative. The method evaluates a decision matrix (9.1) in several steps (Triantaphyllou and Lin, 1996), starting by normalizing columns of a decision matrix and then multiplying values in columns by the corresponding criterion's weights:

$$x_{ij} = \frac{r_{ij}}{\sqrt{\sum_{i=1}^{n} r_{ij}^2}}, \quad i = 1, \ldots, n, \quad j = 1, \ldots, m, \tag{9.4}$$

$$v_{ij} = w_j x_{ij}. \tag{9.5}$$

TOPSIS then identifies best and worst values in each column and creates two sets of these values across all columns named "ideal alternative" (v_j^+) and "negative-ideal alternative" (v_j^-), respectively. In the next step, using the relation (9.6) and (9.7), so-called separation measures for all alternatives are computed based on their Euclidean distances from ideal (D_i^+) and negative-ideal solutions (D_i^-) (across all criteria):

$$D_i^+ = \sqrt{\sum_{j=1}^{m}\left(v_{ij} - v_j^+\right)^2}, \quad i = 1,\ldots,n, \tag{9.6}$$

$$D_i^- = \sqrt{\sum_{j=1}^{m}\left(v_{ij} - v_j^-\right)^2}, \quad i = 1,\ldots,n \tag{9.7}$$

Finally, the relative closeness to ideal alternative (RC_i) is calculated for each alternative using Equation (9.8):

$$RC_i = \frac{D_i^-}{D_i^+ + D_i^-}, \quad i = 1,\ldots,n. \tag{9.8}$$

First-ranked alternative has minimal RC value.

9.2.2.4 Multi-attribute Utility Theory (MAUT)

In MAUT (Keeney and Raiffa, 1976), the preferences of DMs are represented by a sub-utility function for each criterion. This subfunction(s) must be constructed by the DM(s). In that way, different criteria are transformed into one common utility scale (with range 0–1). Summing the products of the sub-utilities multiplied with the corresponding weights of the criteria—which are defined by DMs—the final utility of each alternative is obtained (Schweier et al., 2019). The alternative with the highest utility value is the first-ranked alternative (the best one). The simplest form of MAUT method is SAW (previously described). MAUT demands more information from the DM than the SAW method, as the form of the utility function needs to be defined by the DM (Kangas et al., 2015). A detailed description of MAUT can be found in Keeney and Raiffa (1976).

9.2.2.5 SMART

In Simple Multi-Attribute Rating Technique (SMART) (Edwards, 1977), criteria and alternatives are both evaluated directly, using points between 0 and 100. The best alternative (for a given criterion) and the most important criterion will have 100 points. Then the weight of a criterion is defined to be the points assigned to it divided by the total points assigned to all criteria. Once all the partial scores and criteria weights are obtained, the overall score for each alternative is calculated using the weighted sum model (Ishizaka and Nemery, 2013):

$$S_i = \sum_k w_k p_{ik} \tag{9.9}$$

where p_{ik} is the partial score for alternative A_i with respect to criterion C_k and w_k is the weight of C_k.

9.2.2.6 SMARTER

Simple Multi-Attribute Rating Technique Exploiting Ranks (SMARTER) is a version of SMART that is primarily based on the ranking of elements. Within this method there are different techniques for calculating final weights, and here are presented rank order centroid (ROC) and rank reciprocal rule (RR). ROC (Edwards and Barron, 1994) provides the weights following the rule:

$$a_j = (1/m)\sum_{i=j}^{m}1/i \tag{9.10}$$

where a_j is the weight of the jth element, i is a rank of the jth element, and m is the number of elements. RR (Stillwell et al., 1981) computes final weights by dividing reciprocal of rank of the jth element with the sum of the reciprocals for all elements:

$$a_j = \frac{1/r_j}{\sum_i 1/r_i} \tag{9.11}$$

9.2.2.7 PROMETHEE

The PROMETHEE methods were developed in the 1980s (Brans et al., 1986). In PROMETHEE I and II, the outranking degree $\Pi\ (a_k, a_l)$ for each pair of alternatives (a_k, a_l) is calculated as

$$\Pi\left(a_k, a_l\right) = \sum w_j F_j\left(a_k, a_l\right) \tag{9.12}$$

where $F_j(a_k, a_l)$ is the preference function and w_j are the weights of the criteria, which cannot be interpreted as importance of the criteria (like in MAUT and SAW) but more like votes given to different criteria (Hokkanen and Salminen, 1997; Kangas et al., 2015). This method performs a pairwise comparison of alternatives—with respect to every criterion—using several generalized criteria functions (usual, U-shaped, V-shaped, level, linear, and Gaussian) (Brans et al., 1986). Values of $F_j(a_k, a_l)$ depend on the difference d_j between two alternatives for any criterion j and on the indifference and preference thresholds q and p. Two alternatives are indifferent for criterion j as long as d_j does not exceed the indifference threshold q. If d_j becomes greater than p, there is a strict preference (Pohekar and Ramachandran, 2004). Then, the outranking degrees Π are used to calculate the leaving flow for each alternative:

$$\phi^+\left(a_k\right) = \sum_{l \neq k} \frac{\Pi\left(a_k, a_l\right)}{n-1}, \tag{9.13}$$

and the entering flow:

$$\phi^-\left(a_k\right) = \sum_{l \neq k} \frac{\Pi\left(a_l, a_k\right)}{n-1}. \tag{9.14}$$

PROMETHEE I will produce full ranking of alternatives only in situations when one alternative is better than another with respect to both positive and negative flow; otherwise, they are incomparable. Contrary, PROMETHEE II will always produce complete ranking of alternatives because it uses the difference between positive and negative flow (net flow) (Hokkanen and Salminen, 1997; Kangas et al., 2015; Blagojevic et al., 2019b).

9.2.2.8 ELECTRE III

The ELECTRE methods were originally developed by Roy (1968). Similar to PROMETHEE, ELECTRE has an indifference threshold and a preference threshold, but it also has additional veto threshold—which is used to eliminate alternatives that perform excessively bad in any criteria. Therefore, the ELECTRE III method can be considered as a noncompensatory model (Rogers and Bruen, 1998), meaning that a bad score of any alternative with respect to one criterion cannot be compensated with good scores in other criteria.

9.2.2.9 Analytic Hierarchy Process (AHP)

The AHP (Saaty, 1980) method is very popular in decision practice and is often used for group decision-making. In a standard case, the AHP method enables decomposition of complex decision problem into a hierarchy, where the goal is at the top level, while criteria, subcriteria (if any), and alternatives are at the lower levels. However, there are many other cases when the purpose of AHP is only to define weights of criteria and/or subcriteria (i.e., Dragincic et al., 2015) without any alternatives (Blagojevic et al., 2019a). AHP is based on pairwise comparisons of the elements of the hierarchy at all levels. For comparisons, decision makers use Saaty's importance scale (Table 9.1). For instance, value 1 corresponds to the case in which two elements are equal, while value 9 corresponds to the case in which one of the two elements is absolutely dominant. The above process can be easily followed by a DM without prior mathematical knowledge or experience in decision-making, meaning that the goal of AHP is to solve complex problems through a simple procedure (Blagojevic et al., 2019a).

Numerical values that are equivalent to linguistic values are placed in an appropriate comparison matrix A:

$$
A = \begin{bmatrix}
a_{11} & a_{12} & . & . & a_{1n} \\
a_{21} & a_{22} & . & . & a_{2n} \\
. & & & & . \\
. & & & & . \\
a_{n1} & a_{n2} & . & . & a_{nn}
\end{bmatrix}
\tag{9.15}
$$

The matrix is symmetric and reciprocal; that is, $a_{ij} = 1/a_{ji}$ and all main diagonal elements are equal to 1. After all the judgments are made, the local priorities of the criteria and the alternatives are calculated. This procedure of computing a priority vector is commonly called prioritization. Relevant literature proposes a number of different prioritization methods (see Srdjevic, 2005). Also, the consistency of decision maker judgments is presented through the consistency ratio (CR), which is calculated for every comparison matrix.

Finally, the synthesis is performed by multiplying the criteria-specific priority vector of the alternatives with the corresponding criterion weight and then appraising the results to obtain the final composite alternatives' priorities with respect to the goal. The highest value of the priority vector indicates the best-ranked alternative. In a group decision-making context, aggregation of individual judgments (AIJ), aggregation of individual priorities (AIP) (Ramanathan and Ganesh, 1994; Forman and Peniwati, 1998), and "soft" consensus models (Dong et al., 2010; Blagojevic et al., 2016b) are the most used methods to obtain group priorities.

TABLE 9.1
Saaty's Importance Scale

Definition	Importance
Equal importance	1
Weak dominance	3
Strong dominance	5
Demonstrated dominance	7
Absolute dominance	9
Intermediate values	(2,4,6,8)

9.3   STUDY AREA

Vojvodina Province is located in the Republic of Serbia in the south-east part of the Carpathian (Pannonian) Basin in Central Europe and covers the area of about 21,506 km². The topographic, soil, and climate properties of the study area are described in detail in Bezdan et al. (2019a), Lalić et al. (2011), Hrnjak et al. (2014), Mihailović et al. (2015). The region extends between 44.6°N and 46.2°N, and 18.9°E and 21.5°E (Figure 9.1).

The relief of the area is mostly plain with an exception of two hilly regions (The Fruška Gora Mountain and the Vršac Mountains) with elevations mostly below 500–600 m.a.s.l. Vojvodina has a dense hydrographic network consisting of some of Europe's large rivers (the Danube, Tisa, and Sava) and more than 22,000 km of drainage canals (Savic et al., 2017). The region is predominantly an agricultural area with more than 17,500 km² of arable land, which is about 75% of the total area. Deep and fertile soil types (Chernozem and Phaeozem) with good water retention characteristics are prevalent. Vojvodina lies in the temperate continental climate zone (Lalić et al., 2011) characterized by cold winters and warm and humid summers, large temperature ranges during the year, and uneven distribution of precipitation in time (Hrnjak et al., 2014). The Köppen climate classification subtype is "Cfwbx," which corresponds to the moderate-warm rainy type, with warm summers, maximum precipitation in summer, and secondary maximum precipitation in late fall (Mihailović et al., 2015). Below is a summary of the climate elements in the Vojvodina region based on the data from nine meteorological stations of the Republic Hydrometeorological Institute of Serbia for the period from 1971 to 2016. The following meteorological stations were considered: Rimski Šančevi, Sremska Mitrovica, Bečej, Kikinda, Palić, Sombor, Vršac, Zrenjanin, and Beograd.

Air temperature is one of the most important climate elements. In Vojvodina, for the period from 1971 to 2016, the average annual temperature was 11.5°C. In this region, the average temperature in the nongrowing season is 4.7°C, and in the growing season it is 18.3°C. The large difference between the coldest and warmest days indicates the continental climate (Demirovic,

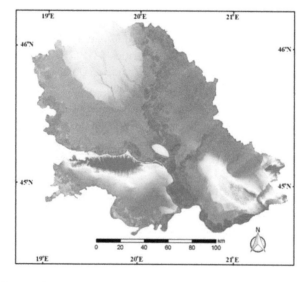

FIGURE 9.1   Vojvodina Province.

2016), which can be seen from the data that on average, the coldest month is January (0.4°C) and the warmest is July (21.9°C).

Precipitation is a very important climatic and bioclimatic element, crucial for the development of plants. Considering the atmospheric processes and relief features, precipitation in Vojvodina is spatially and temporally unevenly distributed (Josimov-Dundjerski, 2009). Vojvodina has the characteristics of the Danube regime of precipitation distribution with uneven distribution by months (Mihailović et al., 2004). Rainfall occurs mostly in June (82.4 mm) and is the lowest in February (33.6 mm). The highest monthly rainfall was recorded in July (326 mm), although months without precipitation were also recorded (April, October, November, and December) in the observed period. The average precipitation in the nongrowing season is 249 mm; in the growing season it is 366 mm, and the average annual precipitation in Vojvodina is 615 mm.

Relative humidity, as a significant climatic element, represents the amount of water vapor in the atmosphere and depends on the temperature of the air to which it is inversely proportional. In Vojvodina, the relative humidity in the spring and summer months (April–August) is the lowest and the average is around 66%, while the highest relative humidity occurs in December (85%) and January (84%). The average humidity in the nongrowing season is 79%; in the growing season it is 67%, and the annual average is 73%.

Insolation, or the duration of sunshine, represents the time interval of direct incoming sunlight (Josimov-Dundjerski, 2009). In the Vojvodina region, the longest average duration of sunshine occurs in the summer months (July, 295 h), and the shortest duration is in the winter months (December, 60 h). The annual average sunshine is 2091 h; in the nongrowing season it is 629 h, and in the growing season it is 1464 h.

Wind is a climate element that greatly affects evapotranspiration and evaporation. The average annual wind speed in Vojvodina is 2.5 m/s. The highest average wind speed occurs in March and April (about 3 m/s) and the lowest in August (1.9 m/s).

Evapotranspiration consists of complex, interdependent processes of movement of water from the soil and through plants into the atmosphere (Trajković, 2009). Potential evapotranspiration or plant water needs represents the amount of water that would be consumed under the conditions of optimal water supply to the plants, when the plants achieve maximum transpiration and maximum yields (Josimov-Dunđerski, 2009). The highest values of evapotranspiration occur in the summer months. The average daily ET_0 value in Vojvodina in July calculated by the FAO-56 Penman-Monteith method is 4.7 mm. The lowest values are in the winter months (December and January) averaging 0.5 mm per day. The average sum in the nongrowing season is 192 mm; in the growing season it is 694 mm, and the annual sum averages 886 mm.

9.4 MULTI-CRITERIA EVALUATION OF LAND SUITABILITY FOR IRRIGATION

To evaluate land suitability for irrigation and to define spatial priorities in installing new irrigation systems in Vojvodina Province, Blagojevic et al. (2016c) proposed a combination of AHP and the geographic information system (GIS) in group context. AHP was chosen because it has the capacity to integrate a large amount of heterogeneous data and because it can be used for group decision-making. Using a combination of AHP and GIS is very common in spatial decision-making related to land and water management (Huang et al., 2011). Here, GIS was used to present all the criteria spatially (as maps or raster layers) and to enable their combination (summarization) in the final suitability map for irrigation, while AHP was used for defining the corresponding criteria weights. The proposed methodology is described in more details in Blagojevic et al. (2016c).

9.4.1　Factors for Evaluating Land Suitability for Irrigation

The main focus in every land evaluation is to define factors for land evaluation, factor ratings, and factor significance (or weights) (Blagojevic et al., 2016c). According to Purnell (1979), an evaluation is of the land; not just the soil and all aspects of the environment need to be considered. The FAO (1985) proposed *Guidelines for Land evaluation for irrigated agriculture* in which 32 potential factors for land evaluation were presented (Table 9.2). "The reader should use these Guidelines selectively, as not all the factors listed will be relevant in a given evaluation, but at a level that is consistent with achieving practicable recommendations" (FAO, 1985). In the same document, factor ratings are defined as: S1, S2, S3, N1, and N2—indicating, in terms of a single factor or a single interaction of a group of factors—whether the land is highly suited, moderately suited, marginally suited, marginally not suited, or permanently not suited, respectively. For deciding about the land suitability class from factor ratings, the relative significance of each class-determining factor was defined with one of the following expressions: Very important, Moderately important, Less important, or Not important.

Sys et al. (1991) suggested two methods for land evaluation. The first one is the limitation method, where the evaluation is carried out by comparing the land characteristics with the limitation levels of the requirement tables. The second one is the parametric method for the evaluation of land characteristics, which requires a numerical scale (normally 100 for a maximum). If a land characteristic is optimal for the considered land utilization type, the maximum rating of 100 is attributed; if the same land characteristic is inadequate, a minimal rating is applied. An important characteristic is rated in a wide scale (100–25), a less important characteristic in a narrower scale (100–60).

TABLE 9.2
Factors for Land Evaluation for Irrigated Agriculture

Crop (agronomic)	1. Growing period; 2. Radiation; 3. Temperature; 4. Rooting; 5. Aeration; 6. Water requirements; 7. Nutritional requirements (NPK); 8. Water quality limitations; 9. Salinity limitations; 10. Sodicity limitations; 11. pH, micronutrients and toxicities; 12. Pest, disease, weed limitations; 13. Flood, storm, wind, frost, hail limitations
Management requirements and limitations	14. Location; 15. Water application management requirements; 16. Pre-harvest farm management; 17. Harvest and post-harvest requirements; 18. Requirements for mechanization
Land development or improvement	19. Land-clearing requirements; 20. Flood protection requirements; 21. Drainage requirements; 22. Land grading requirements; 23. Physical, chemical, organic aids, and amendments; 24. Leaching requirements; 25. Reclamation period; 26. Irrigation engineering needs
Conservation and **environmental** requirements	27. Long-term salinity, sodicity hazards; 28. Groundwater or surface water hazards; 29. Long-term erosion hazards; 30. Environmental hazards
Socioeconomic requirements	31. Farmers' attitudes to irrigation; 32. Others that are class-determining

Source: Food and Agricultural Organization of the United Nations (FAO), Guidelines: Land evaluation for irrigated agriculture, Rome, 1985.

This introduces the concept of weighting factors. The land suitability index can then be calculated from the individual ratings by the square root method, the results of which are then categorized through five suitability classes: S1-highly suitable (>80); S2-moderately suitable (60–80); S3-marginally suitable (45–59); N1-currently unsuitable (30–44); and N2-permanently unsuitable (<30). In this approach, the factors affecting the land suitability for irrigation can be subdivided into four groups (Albaji et al., 2008): physical soil properties, chemical soil properties, drainage properties, and environmental factors, such as slope. The main problem here is that Sys et al. (1991) haven't suggested using any other factor, e.g., socioeconomic ones.

Procedures (2004) proposes a land classification style for irrigation that involves physical and chemical considerations of soil (geological deposit, texture, salinity and sodicity, drainage, fertility) and topography (field size, shape and length of irrigation run; water distribution slope and erosion control; stoniness and brush/tree cover; surface drainage). These factors are divided based on permanent and changeable factors, where typical changeable factors include salinity, sodicity, stoniness, drainage, minor irregularities in relief, brush and tree cover, and flood hazards (Maletic and Hutchings, 1967).

Chen et al. (2010) used hydraulic conductivity of soil, percent slope, soil texture, depth to water tables, and the electrical conductivity of groundwater to identify the potential of expanding irrigated cropping land use in the Macintyre Brook catchment of Queensland in Australia. Criterion (factor) weights were derived by AHP within an individual context. The suitability classes (factor ratings), consisting of four levels used in this study, were adapted from the FAO system (FAO, 1976). They are stated as highly suitable (S1), moderately suitable (S2), marginally suitable (S3), and unsuitable (N). Tadic (2012) suggested using different data for evaluation of agricultural land suitability for irrigation such as physical planning data (structures, roads, land use), natural resource data (soil properties, water resources, hydrological data), meteorological data (precipitation, air temperature, humidity, wind isolation), environmental data (protected areas, biodiversity), and agricultural potential (cropping patterns, mechanization) but did not give any suggestions on how to define factor weights.

Albaji et al. (2008) used slope, drainage properties, electrical conductivity of soil solutions, $CaCO_3$ status, soil texture, and soil depth as factors, while Rabia et al. (2013) spatially evaluated the land suitability for surface and drip irrigation, taking into account seven parameters (factors): soil texture, soil depth, $CaCO_3$ status, salinity/alkalinity, drainage, slope, and surface stoniness. In both cases, the methodology used for the evaluation refers to the Sys et al. (1991) parametric method. According to Ali (2010), there are several important factors that are affecting irrigation planning and development: soil, climate, topography, water source, crop(s) to be cultivated, energy, labor, capital, commodity/product market, national policy and priority, institutional infrastructure, economic factor, environmental aspect, and sociocultural aspect.

Belic et al. (2011) suggested using climate, soil, hydrology, crop, ecology, and economic factors to evaluate land suitability for irrigation with corresponding weights of 0.1, 0.3, 0.2, 0.1, 0.1 and 0.2, but there is no explanation of how the authors got these particular factor weights. Anane et al. (2012) selected four criteria to rank suitable sites for irrigation, which were soil suitability criteria (texture, depth, salinity, and land slopes); resource conflict criteria (land use and freshwater availability); economic criteria (wastewater transfer and closeness to the roads); social criteria (farness from residential areas and closeness to the already irrigated districts); and environmental criteria (intrinsic vulnerability of the shallow aquifer and current quality of the shallow aquifer).

Based on previous information and experts' opinions, Blagojevic et al. (2016c) identified 16 factors that play an important role in land evaluation for irrigation in Vojvodina, and, according to their nature and role in the decision-making process, these factors were treated as criteria. Since Saaty (1980) suggested that the number of elements compared must be nine at maximum, all identified factors were clustered into five major factor groups: Soil (SO), Climate (CL), Economy (EC), Water infrastructure (IN), and Environmental (EN). They were sorted as follows:

- Land slope (SL), soil drainage (DR), soil suitability for irrigation (IR), geomorphology (GM), and the total available water in the root zone (AW) (soil factors group);
- Water deficit (WD) and drought vulnerability (DV) (climate factors group);
- Soil fertility and production potential (FP), land use (LU), proximity to markets (PM) and development of livestock (DL) (economy factors group);
- Distance from water bodies (DW), density of drainage network (DN), and land consolidation (LC) (water infrastructure factors group); and
- Surface water quality (SW) and subsurface water quality (UW) (environmental factors group).

9.4.2 Developing Land Suitability Map for Irrigation

Four irrigation and water management experts evaluated the selected factor groups and factors in AHP standards, and their group decision is presented in Table 9.3. The three most important factors were soil suitability for irrigation (0.162), surface water quality (0.141), and distance from water bodies (0.104), while the least important criterion was proximity to markets (0.010) (Blagojevic et al., 2016c).

Since factors for land suitability for irrigation are of heterogeneous types (qualitative and/or quantitative), different forms (continuous or discrete), and different domains of measurement, it is crucial to standardize all factors by bringing them into a common domain of measurement (Anane et al., 2012). Therefore, each of the 16 factors was assigned a different rating (value) on a scale from 1 (low priority for irrigation) to 5 (very high priority) according to experts' experiences and domestic and international references. This was done in the GIS environment (Figure 9.2). Finally, the cell values were multiplied by the corresponding group weights of the factors (Table 9.3) and summarized in the GIS software using a raster calculator. The final result is the map of the spatial priorities for irrigation development in Vojvodina Province (Figure 9.3).

TABLE 9.3
Group Weights of All Factors

Factor Groups	Factors	Weights (w)	Rank
Soil (SO)	Land slope (SL)	0.024	14
	Soil drainage (DR)	0.053	7
	Soil suitability for irrigation (IR)	0.162	**1**
	Geomorphology (GM)	0.081	5
	Total available water (AW)	0.097	4
Climate (CL)	Water deficit (WD)	0.045	9
	Drought vulnerability (DV)	0.060	6
Economy (EC)	Soil fertility and prod. pot. (FP)	0.043	11–12
	Land use (LU)	0.043	11–12
	Proximity to markets (PM)	0.010	16
	Development of livestock (DL)	0.015	15
Water infrastructure (IN)	Distance from water bodies (DW)	0.104	**3**
	Density of drainage network (DN)	0.044	10
	Land consolidation (LC)	0.029	13
Environmental (EN)	Surface water quality (SW)	0.141	**2**
	Subsurface water quality (UW)	0.047	8

Source: Blagojevic, B. et al., *J. Hydroinform.*, 18, 579–598, 2016c.

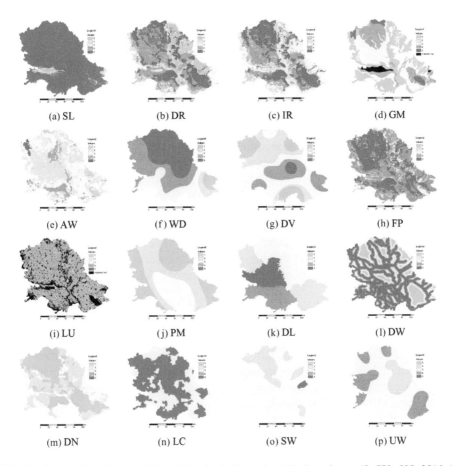

FIGURE 9.2 Standardized factors. (From Blagojevic, B. et al., *J. Hydroinform.*, 18, 579–598, 2016c.)

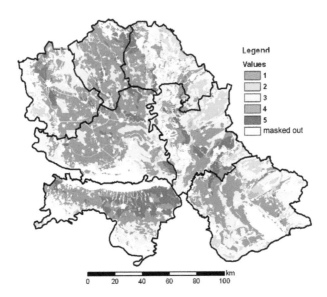

FIGURE 9.3 Spatial priorities for irrigation development in Vojvodina Province. (From Blagojevic, B. et al., *J. Hydroinform.*, 18, 579–598, 2016c.)

9.5 MULTI-CRITERIA SELECTION OF EVAPOTRANSPIRATION METHOD FOR DROUGHT MANAGEMENT

Bezdan et al. (2019a) present the standardized precipitation evapotranspiration index (SPEI)-based approach to agricultural drought monitoring (the ADM-SPEI approach). Although it was created for the Vojvodina region, the developed approach to agricultural drought monitoring based on the SPEI index (Vicente-Serrano et al., 2010) has been described and explained in detail, thereby allowing the ADM-SPEI approach to be modified and applied in any agroclimatic conditions (Bezdan, 2019). The proposed approach is based on the modified Standardized Precipitation and Evapotranspiration Index (SPEI). However, in this approach, the reference evapotranspiration (ET_0) is replaced by the potential crop evapotranspiration (ET_c) in order to enable relating the SPEI to a specific crop and monitor agricultural drought more precisely. The ADM-SPEI approach consists of three phases alongside the corresponding steps. The first phase of the presented approach is choosing the appropriate method for calculating the ET_0/ET_c. Here, AHP—as one of the most used MCDA methods in the field of agriculture, water management (e.g. Vojtek and Vojteková, 2019; Mir and Padma, 2016; Dragincic et al., 2015; Dragincic and Vranesevic, 2014; Bezdan et al., 2019a, 2019b etc.)—is used in order to select the most suitable method for the calculation of ET_0 (or ET_c) for using within the SPEI index in the Vojvodina region.

Phase two presents the calculation of the modified SPEI (AD-SPEI crop) adjusted to the area of interest and related to specific crops. In addition to the other steps, it is important to mention in this phase the possibility of using the AD-SPEI crop for the assessment of the capability of the irrigation systems to combat drought. In the last phase of the proposed approach, the validation was carried out from several different perspectives including examining the correlation of the index with the crop yields, comparing with the generally accepted drought indices (SPI, SPEI, and SC-PDSI), and taking into account the experts' feedback.

The second phase of the presented approach—related to multi-criteria selection of the most suitable method for the calculation of evapotranspiration for use within the SPEI index in the Vojvodina region—is explained in more detail in the following text, while results of other phases can be found in Bezdan et al. (2019a).

9.5.1 DEFINING ALTERNATIVES—SUITABLE METHODS FOR CALCULATING ET_0 (OR ET_c)

Reference evapotranspiration is a very important element of water balance. In the hydro-amelioration practices, the calculation of evapotranspiration is one of the basic steps in designing irrigation and drainage systems (Bezdan et al., 2017). When reliable lysimeter data are not available, the FAO-56 PM method is recommended as a standard procedure (Allen et al., 1998; Irmak et al., 2003; Utset et al., 2004; Trajković, 2005, 2007; Gavilán et al., 2006; Alexandris et al., 2006). This method is generally considered to be the most reliable in a wide range of different climatic conditions and locations because it is based on physical principles and takes into account the major climatic factors that influence evapotranspiration (Shahidian et al., 2012). The FAO-56 PM method is also recommended by the Food and Agriculture Organization (FAO) of the United Nations and the International Commission on Irrigation and Drainage (ICID) as the standard procedure. However, high data requirements characterizing this method complicate its application in the areas with unavailable, missing, or inaccurate meteorological data (Cruz-Blanco et al., 2014). In most regions, these data are limited, and experimental models have been developed for this purpose to estimate evapotranspiration (Valipour, 2015). Numerous equations for estimating reference evapotranspiration can be classified into temperature, radiation, evaporation, mass transfer, and combined (Gocic and Trajkovic, 2010; Tabari et al., 2013). Xystrakis and Matzarakis (2010) state that there is no rule with regard to choosing the optimal equation for estimating reference evapotranspiration in different climates, because even in the same climates, different studies give different results with respect to the performance of empirical models. Trajkovic (2009) states that the best results are obtained by regional formulas and that each method produces good results in the conditions for which it is developed.

Here, the list of the most suitable methods for calculating ET_0 or ET_c (representing alternatives) in the Vojvodina region is selected. The choice is based on consultations with local experts from the specific area of the Vojvodina region, scientific research, and relevant literature. As a result of this, the following methods have been selected as alternatives: FAO-56 PM (PM) (Allen et al., 1998), Turc (TU) (Turc, 1961), Thornthwaite (TH) (Thornthwaite, 1948), bioclimatic method (BM) based on hydrophytothermic indices (Vučić, 1971), and Hargreaves (HG) (Hargreaves and Samani, 1985).

9.5.1.1 FAO-56 Penman–Monteith

The FAO-56 Penman–Monteith (FAO-56 PM) equation was presented in Allen et al. (1998) and is expressed as:

$$ET_0 = \frac{0,408\Delta\left(R_n - G\right) + \gamma\dfrac{900}{T+273}u_2\left(e_s - e_a\right)}{\Delta + \gamma\left(1 + 0,34u_2\right)}$$

where ET_0 is the reference evapotranspiration [mm day^{-1}], R_n is the net radiation at the crop surface [MJ m^{-2} day^{-1}], G is the soil heat flux density [MJ m^{-2} day^{-1}], T is the air temperature at 2 m height [°C], u_2 is the wind speed at 2 m height [m s^{-1}], e_s is the saturation vapor pressure [kPa], e_a is the actual vapor pressure [kPa], e_s—e_a is the saturation vapor pressure deficit [kPa], D is the slope vapor pressure curve [kPa °C^{-1}], and γ is the psychrometric constant [kPa °C^{-1}].

9.5.1.2 Turc Equation

The Turc equation is one of the simplest empirical equations used for estimating reference evapotranspiration (ET_0) in humid conditions and is expressed as (Turc, 1961; Trajković, 2009):

$$ET_0 = 0.013 \cdot \left(23.88 \cdot R_s + 50\right) \cdot T \cdot \left(T + 15\right)^{-1} \text{ when } RH > 50\%$$

$$ET_0 = 0.013 \cdot \left(23.88 \cdot R_s + 50\right) \cdot T \cdot \left(T + 15\right)^{-1}\left(1 + \frac{50 - RH}{70}\right) \text{when } RH < 50\%$$

where ET_0 is the reference evapotranspiration [mm day^{-1}], T is the mean air temperature [°C], and R_s is the solar radiation [MJ m^{-2} day^{-1}].

9.5.1.3 Thornthwaite Method

The Thornthwaite method (Thornthwaite, 1948) is a temperature-based method for the estimation of ET_0 and is expressed as (Trajković, 2009):

$$ET_0 = 16 \times \left[\frac{10 \times T}{I}\right]^a \times k; \quad I = \sum_{i=1}^{12} i; \quad i = \left[\frac{T}{5}\right]^{1,514}; \quad a = 0.016\, I + 0.5$$

where ET_0 is the reference evapotranspiration [mm month^{-1}], T is the daily mean air temperature [°C], I is the heat index which is calculated as the sum of 12 monthly index values i, a is the exponent being a function of I, and k is the correction coefficient computed as a function of the latitude and month.

9.5.1.4 Hargreaves Method

The limitation of reliable data resulted in Hargreaves et al. (1985) developing a function to estimate reference evapotranspiration where only the data about mean maximum and mean minimum air temperature and extraterrestrial radiation are required (Droogers and Allen, 2002). The Hargreaves equation is expressed as:

$$ET_0 = 0.0023 \cdot 0.408 \cdot R_a \left(\frac{T_{max} + T_{min}}{2} + 17.8 \right) \sqrt{T_{max} - T_{min}}$$

where ET_0 is the reference evapotranspiration [mm day^{-1}], Tmax is the mean daily maximum air temperature [°C], T_{min} is the mean daily minimum air temperature [°C], and R_a is the extraterrestrial radiation [MJ m^{-2} day^{-1}].

Trajkovic (2007) modified the Hargreaves equation by adjusting the Hargreaves exponent and proposed a value of 0.424 instead of the original 0.5 to be used in the Western Balkans (Trajković, 2009; Trajkovic, 2007):

$$ET_0 = 0.0023 \cdot 0.408 \cdot R_a \cdot \left(T_{max} - T_{min} \right)^{0.424} \cdot \left(\frac{T_{max} + T_{min}}{2} + 17.8 \right)$$

9.5.1.5 The Bioclimatic Method

Since the bioclimatic method is not widely known, whereas in the Western Balkans it is frequently used, it is briefly described in the following paragraph. Vučić (1971) developed the bioclimatic method in the Vojvodina region, and it is based on the correlation between crop evapotranspiration and physical factors of evapotranspiration (temperature, humidity, solar radiation, vapor pressure, etc.). If mean daily air temperature is used for the calculation of the ET_C, the bioclimatic coefficient is called the hydrophytothermic index (K). The hydrophytothermic index represents the amount of water (mm) consumed through the evapotranspiration for every degree of mean daily air temperature. The method is described in literature (Pejić et al., 2008, 2009, 2011b, 2011c, 2011d; Pejić and Gajić, 2006; Marinković et al., 2009; Maksimović and Dragović, 2002; Aksić et al., 2008; Bezdan et al., 2019a). Based on experimental field research, numerous authors have determined hydrophytothermic indices for various fields, vegetables, and fruit crops. The values of the ET_C can be calculated on a monthly basis by multiplying the corresponding hydrophytothermic index by the sum of the mean daily air temperature in the month observed:

$$ET_C = K \cdot \sum_{i=1}^{n} T_i$$

where n is the number of days in a month, and T is the mean daily air temperature in the observed month. The values of the hydrophytothermic indices used for the calculation of the ET_c presented in the Table 9.4 have been collected from literature.

TABLE 9.4
The Hydrophytothermic Indices (K)

Month	Maize	Sugar b.	Alfalfa	Soybean	Potato	Beans	Cabbage	Hop	Onion	Tobacco	W. Wheat
Mar											0.22
Apr			0.17					0.12	0.19		0.24
May	0.11	0.15	0.25	0.11	0.19			0.15	0.19	0.13	0.26
Jun	0.18	0.20	0.25	0.17	0.19	0.22		0.18	0.19	0.20	0.21
Jul	0.18	0.21	0.25	0.18	0.19	0.24	0.20	0.21	0.19	0.21	
Aug	0.18	0.20	0.22	0.17	0.19		0.20	0.18	0.19	0.19	
Sep	0.11	0.12	0.15	0.11	0.19		0.20			0.14	

9.5.2 Defining the Criteria for the Selection of Method for the Calculation of ET_0, and ET_c

This choice is made by consulting local experts in this field and relevant literature. The following five criteria were selected.

Availability of reliable input data (AD). The methods for the calculation of evapotranspiration are based on various parameters and, therefore, require different input data. Many meteorological stations worldwide measure a limited number of meteorological data. Their public availability is often limited. Furthermore, the quality of the measured data is frequently disputable. Hence, it is important that the meteorological data be reliable. This criterion is an important prerequisite for the accuracy of the final results obtained within the application of this method.

Literature-based suggestions (LS). Literature offers numerous methods for the estimation of evapotranspiration based on different principles. Likewise, a lot of analysis and research conducted in diverse climatic conditions examine the possibilities of the application of these methods. Authors' approaches to the recommended methods for the particular regions vary. Being familiar with relevant literature and research about the particular region can have a huge impact on the appropriate selection of the most suitable method for the calculation of evapotranspiration.

Ease of calculation (EC)/the simplicity of method. A wide range of methods for the calculation of evapotranspiration vary from empirical to combined ones. In addition to the methods that are easy to use for the calculation, there are a large number of those that demand complex calculations. Oftentimes, the methods that enable simple and fast calculations are preferred. The complexity of the method does not guarantee its quality.

Common practice (CP). Many authors recommend comparing the results obtained using different methods with the results calculated by some of the sophisticated methods in the investigated area, such as lysimeters. Since the aforementioned research methods are demanding and expensive, this type of data is often not available in many locations. For that reason, it is presumed that the years long application of the method for the calculation of evapotranspiration and experts' practical experience have ensured the selection of the most suitable method for the agroclimatic conditions of the particular region. For the abovementioned reasons, the common practice or positive experience in local, specific conditions was taken into consideration as an appropriate criterion.

The Relevance of the input parameters for the local climatic conditions (RP). Evapotranspiration is conditioned by numerous climatic parameters. However, on the particular location, certain parameters can have a more dominant impact than the others. Therefore, it is important to know climatic conditions in the region of interest in order to select the adequate method for the calculation of evapotranspiration. The method that contains the most relevant climatic parameters of the particular region that affect the intensity of evapotranspiration probably produces the most reliable results.

9.5.3 Obtaining Individual Experts' and Group Decisions

Nine decision makers (DMs) from the Vojvodina region were interviewed and asked to express their opinion in decision matrices. In this process, experts participated from the University of Novi Sad-Faculty of Agriculture, the public water management company "Vode Vojvodine," the Institute of Field and Vegetable Crops from Novi Sad, and experts in the design and management of irrigation systems employed in the leading agricultural companies in this region were also involved. Each expert was interviewed separately by using Saaty's relative-importance scale (Saaty, 1980) shown in Table 9.1. They provided linguistic pairwise comparisons, and related numerical values were placed in AHP comparison matrices. Firstly, each decision maker (expert) compared criteria in relation to the goal—selecting the most appropriate method for calculating ET_0 or ET_c within the SPEI index in the Vojvodina region and then the alternatives were compared in relation to each criterion separately. For each decision maker, a total number of six matrices were obtained and then by using a standard AHP calculation procedure (Saaty, 1980), the final priority vector for each decision maker was separately obtained.

One final group decision was obtained from individual experts' decisions in order to get the rank of alternatives and select the most suitable one. The first-ranked alternative was considered to be the most suitable method for calculating evapotranspiration within the SPEI index. In this case, aggregation of individual priorities (AIP) (Ramanathan and Ganesh, 1994; Forman and Peniwati, 1998) and weighted arithmetic mean method (WAMM) (Ramanathan and Ganesh, 1994) were used to obtain the final group decision. There was also the possibility of using some other aggregation procedures or consensus models presented in numerous studies, but this method was selected because it is simple and one of the most commonly used mathematical procedures for the aggregation of individual decisions into a group one. Nine experts, who participated in the decision-making process, had equal input in the group decision. Literature offers a lot of methods, and different authors suggest various approaches to solving problems of the decision makers' input in a group decision. In this research, we decided to assign equal weight to all the experts in order to avoid further complicating the procedure. The results are shown in Table 9.5. The following rank of methods could be seen: the first ranked is bioclimatic method (BM); then Thornthwaite (TH), Hargreaves (HG), FAO-56 PM (PM); and the last ranked is Turc (TU). To summarize, the bioclimatic method based on hydrophytothermic indices is considered to be the most suitable method for calculating evapotranspiration within the SPEI index for the Vojvodina region.

9.5.4 Results—Calculation of the Modified SPEI (AD-SPEI Crop)

After the most suitable method for calculating evapotranspiration within the SPEI index for the Vojvodina region is selected, the calculation of the modified SPEI (AD-SPEI crop) adjusted to the area of interest and related to the crops is performed. More details about the steps in between can be found in Bezdan et al. (2019a).

TABLE 9.5
Individual and Group Decisions

Decision Makers (Experts)	Individual Decisions				
	PM	TU	TH	BM	HG
DM1	0.134	0.092	0.270	0.356	0.148
DM2	0.103	0.073	0.290	0.429	0.105
DM3	0.131	0.164	0.288	0.250	0.167
DM4	0.130	0.075	0.211	0.382	0.202
DM5	0.075	0.094	0.242	0.494	0.096
DM6	0.182	0.063	0.211	0.359	0.185
DM7	0.074	0.062	0.409	0.386	0.069
DM8	0.169	0.082	0.210	0.354	0.185
DM9	0.082	0.106	0.327	0.337	0.147
Group decision	**0.120**	**0.090**	**0.273**	**0.372**	**0.145**

The AD-SPEI crop index is based on the crop-specific climatic water balance. The crop-specific climatic water balance is a very important parameter because based on it, the crop irrigation requirements could be determined, which is a prerequisite for designing irrigation and drainage systems.

The significance of this modification of the SPEI index, differing from the indices such as the original SPEI or the SPI, is that the value of the drought index can be transformed backwards into the values of the crop-specific climatic water balance, as explained in Bezdan et al. (2019a).

Based on the value of crop-specific climatic water balance, it is possible to determine the irrigation hydromodule necessary for dimensioning the elements of the irrigation system. The irrigation hydromodule defines the required ability of irrigation equipment to compensate for water deficit. Based on it, the calculation of required capacities of all elements of the system can be performed. The hydromodule is expressed as the required amount of water per unit time per hectare (l/s/ha or mm/day):

$$q \text{ (mm/day)} = \frac{h \text{ (mm)}}{t \text{ (day)}} \text{ or } q \text{ (l/s/ha)} = \frac{h \text{ (mm)} \cdot 10000}{t \text{ (day)} \cdot n \cdot 3600}$$

where q is the irrigation hydromodule, h is the required amount of water, t is the number of days, and n is the effective working hours of the irrigation equipment during a day.

Irrigation systems are designed based on the crop irrigation requirements in critical months. In the Vojvodina region, those critical months are July and August. Therefore, the number of days equals 31. It is considered that with contemporary irrigation systems, the maximum effective working time of the machine is about 22 hours a day. In the Vojvodina region, the most common values of irrigation hydromodules are around 0.5–0.6 l/s/ha.

The following are examples of the application of the AD-SPEI crop index in analyzing the capabilities of irrigation systems to combat drought. If it is considered that the irrigation system is designed based on the value of the hydromodule of 0.6 l/s/ha, and if it is adopted that the efficiency of the system is 90% (contemporary automated machines—Center Pivot or Linear irrigation machinery), the monthly amount of water that the system can deliver is 133 mm. This value corresponds to the monthly water deficit in the water balance of the crops that can be recovered and for

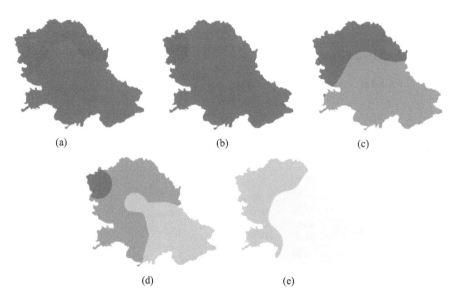

FIGURE 9.4   The drought categories in August representing the limits of the capability of irrigation systems designed using the hydromodule of 0.6 l/s/ha to successfully protect different crops: (a) Soybean, (b) maize, hop, (c) potato, onion and tobacco, (d) sugar beet and cabbage, and (e) alfalfa.

which the AD-SPEI crop values can be determined for each crop in a particular area. For example, the monthly value of the deficit in water balance of 133 mm in August corresponds to the value of the AD-SPEI1 crop for maize in Rimski Šančevi of −1.77. In July, for the same crop and the same location, the index value is −2.41. These results indicate that the irrigation system can effectively defend maize from the exceptional drought (AD-SPEI1crop = −2.41) in July, and in August from the extreme drought (AD-SPEI1crop = −1.77) at the observed site.

The maps (Figure 9.4) show the drought categories in August representing the limits of the capability of the irrigation systems designed using the hydromodule of 0.6 l/s/ha to successfully protect agricultural crops. The crops on the maps are grouped according to the values of hydrophytothermic coefficients in August. The lowest degree of protection is provided by irrigation systems to the crops with the highest water requirements expressed by hydrophytothermic coefficients, such as alfalfa. The irrigation systems designed based on the value of hydromodule of 0.6 l/s/ha can protect alfalfa from moderate drought. The irrigation systems can protect sugar beet and cabbage from moderate drought in southeastern Vojvodina to extreme drought in the northwest. The irrigation systems can protect potatoes, onions, and tobacco from severe droughts in central and southern regions to extreme drought in northern Vojvodina. The irrigation systems can protect maize and hops from extreme droughts in most of Vojvodina to exceptional drought in the northwest of Vojvodina. The irrigation systems provide the highest degree of protection to soybean, and they can protect soybean from extreme drought in the central and southern regions to exceptional drought in northern Vojvodina.

9.6   CONCLUSION

Decision-making related to irrigation management—as well as any other natural resources management—needs to be made in the context of multiple, usually conflicting criteria that are coming from different domains. For these problems, MCDA can be a suitable decision-making aid because it offers a process that leads to rational, justifiable, and explainable decisions; it can deal with mixed sets of data (quantitative and qualitative) including expert opinions; and last, but not least, MCDA allows participation of multiple experts and stakeholders (Mendoza and Prabhu 2003). We think that it is important for water and irrigation management professionals to become familiar with MCDA methods because all decisions in the future—related to water and land—will be based on multiple criteria (economic, environmental, and social). Therefore, basic ideas as well as the main steps of selected MCDA methods are presented in this chapter. Finally, the authors presented two real case studies from Serbia where MCDA was used to define spatial priorities for irrigation development and for drought monitoring and management.

REFERENCES

Aksić, M., N. Deletić, N., Gudžić, N., Gudžić S., & Stojković, S. (2008). Hidrofitometeorološki indeksi duvana tipa Virdžinija. *Journal of Agricultural Sciences, 53*(2), 91–97.

Albaji, M., Landi, A., Boroomand Nasab, S., & Moravej, K. (2008). Land suitability evaluation for surface and drip irrigation in Shavoor Plain Iran. *Journal of Applied Sciences, 8*(4), 654–659.

Alexandris, S., Kerkides, P., & Liakatas, A. (2006). Daily reference evapotranspiration estimates by the "Copais" approach. *Agricultural Water Management, 82*(3), 371–386.

Ali, H. (2010). Fundamentals of irrigation development and planning. In: *Fundamentals of Irrigation and On-farm Water Management* (Vol. 1). Springer Science & Business Media, New York.

Allen, R. G., Pereira, L. S., Raes, D., & Smith, M. (1998). Crop evapotranspiration: Guidelines for computing crop water requirements. FAO irrigation and drainage paper No. 56. *Rome: Food and Agriculture Organization of the United Nations, 56*(97), 156.

Anane, M., Bouziri, L., Limam, A., & Jellali, S. (2012). Ranking suitable sites for irrigation with reclaimed water in the Nabeul-Hammamet region (Tunisia) using GIS and AHP-multicriteria decision analysis. *Resources, Conservation and Recycling, 65*, 36–46.

Armenski, T., Stankov, U., Dolinaj, D., Mesaroš, M., Jovanović, M., Pantelić, M., Pavić D. et al. (2014). Social and economic impact of drought on stakeholders in agriculture. *Geographica Pannonica, 18*(2), 34–42.

Bachmair, S., Stahl, K., Collins, K., Hannaford, J., Acreman, M., Svoboda, M., Knutson C. et al. (2016). Drought indicators revisited: The need for a wider consideration of environment and society. *Wiley Interdisciplinary Reviews: Water, 3*(4), 516–536.

Belic, S., Belic, A., & Vranesevic, M. (2011). Suitability land for irrigation (Pogodnost lokaliteta za navodnjavanje). In: *Irrigation Water Quality* (Upotrebljivost voda za navodnjavanje), S. Belic, (Ed.). Faculty of Agriculture, Novi Sad, pp. 65–87 (in Serbian).

Belton, V., & Stewart, T. (2002). *Multiple Criteria Decision Analysis: An Integrated Approach*. Springer Science & Business Media, Boston, MA.

Bezdan, A. (2014). Water excess and water deficit risk assessment in landreclamation area. Doctoral dissertation, University of Novi Sad, Faculty of Agriculture, Novi Sad, Serbia.

Bezdan, A., Draginčić, J., Pejić, B., Blagojević, B., & Mesaroš, M. (2017). Poređenje metoda za izračunavanje referentne evapotranspiracije na području meteorološke stanice Rimski Šančevi. *Letopis naučnih radova Poljoprivrednog fakulteta, 41*(2), 61–67.

Bezdan, J. (2019). Standardized precipitation evapotranspiration based approach to agricultural drought monitoring in Vojvodina region. Doctoral dissertation, University of Novi Sad, Faculty of Agriculture, Novi Sad, Serbia.

Bezdan, J., Bezdan, A., Blagojević, B., Mesaroš, M., Pejić, B., Vranešević, M., Pavić D. & Nikolić-Đorić, E. (2019a). SPEI-based approach to agricultural drought monitoring in Vojvodina region. *Water, 11*(7), 1481.

Bezdan, A., Blagojevic, B., Vranesevic, M., Benka, P., Savic, R., & Bezdan, J. (2019b). Defining spatial priorities for irrigation development using the soil conservation and water use efficiency criteria. *Agronomy, 9*(6), 324.

Blagojevic, B., Athanassiadis, D., Spinelli, R., Raitila, J., & Vos, J. (2019a). Determining the relative importance of factors affecting the success of innovations in forest technology using AHP. *Journal of Multi-Criteria Decision Analysis*, 1–12.

Blagojević, B., Jonsson, R., Björheden, R., Nordström, E. M., & Lindroos, O. (2019b). Multi-Criteria Decision Analysis (MCDA) in Forest Operations–an Introductional Review. *Croatian Journal of Forest Engineering: Journal for Theory and Application of Forestry Engineering, 40*(1), 191–2015.

Blagojevic, B., Srdjevic, B., Srdjevic, Z., & Zoranovic, T. (2016a). Deriving weights of the decision makers using AHP group consistency measures. *Fundamenta Informaticae, 144*(3–4), 383–395.

Blagojevic, B., Srdjevic, B., Srdjevic, Z., & Zoranovic, T. (2016b). Heuristic aggregation of individual judgments in AHP group decision making using simulated annealing algorithm. *Information Sciences, 330,* 260–273.

Blagojevic, B., Srdjevic, Z., Bezdan, A., & Srdjevic, B. (2016c). Group decision-making in land evaluation for irrigation: A case study from Serbia. *Journal of Hydroinformatics, 18*(3), 579–598.

Bošnjak, Đ. (2001). *Problemi suše u Vojvodini i mere borbe protiv nje.* Zbornik radova Instituta za ratarstvo i povrtarstvo, 35, 391–401.

Brans, J. P., Vincke, P., & Mareschal, B. (1986). How to select and how to rank projects: The PROMETHEE method. *European Journal of Operational Research, 24*(2), 228–238.

Chen, Y., Yu, J., & Khan, S. (2010). Spatial sensitivity analysis of multi-criteria weights in GIS-based land suitability evaluation. *Environmental Modelling & Software, 25*(12), 1582–1591.

Cruz-Blanco, M., Lorite, I. J., & Santos, C. (2014). An innovative remote sensing based reference evapotranspiration method to support irrigation water management under semi-arid conditions. *Agricultural Water Management, 131,* 135–145.

Danielson, M., & Ekenberg, L. (2019). An improvement to swing techniques for elicitation in MCDM methods. *Knowledge-Based Systems, 168,* 70–79.

Demirović, D., 2016. Konkurentnost Vojvodine kao destinacije ruralnog turizma, Doctoral dissertation, University of Novi Sad, Faculty of Science, Novi Sad, Serbia, 2016.

Diakoulaki, D., Mavrotas, G., & Papayannakis, L. (1995). Determining objective weights in multiple criteria problems: The critic method. *Computers & Operations Research, 22*(7), 763–770.

Dong, Y., Zhang, G., Hong, W. C., & Xu, Y. (2010). Consensus models for AHP group decision making under row geometric mean prioritization method. *Decision Support Systems, 49*(3), 281–289.

Dragincic, J., & Vranešević, M. (2014). AHP-based group decision making approach to supplier selection of irrigation equipment. *Water Resources, 41*(6), 782–791.

Dragincic, J., Korac, N., & Blagojevic, B. (2015). Group multi-criteria decision making (GMCDM) approach for selecting the most suitable table grape variety intended for organic viticulture. *Computers and Electronics in Agriculture, 111,* 194–202.

Dragincic, J., Bezdan, A., Pejić, B., Mesaroš, M., & Blagojević, B. (2017). Analiza pojave suše na području Severnog Banata/Analysis of drought occurrence in North Banat. *Letopis naučnih radova Poljoprivrednog fakulteta, 41*(2), 77–84.

Dragović, S. (2001). *Potrebe i efekti navodnjavanja na povećanje i stabilizaciju prinosa u poljoprivrednim područjima Srbije.* Zbornik radova Instituta za ratarstvo i povrtarstvo, (35), 445–456.

Dragović, S. (2012). Effect of irrigation on field crops yield under the variable agroclimatic conditions of Serbia. *Agriculture & Forestry, 54*(8), 1–4.

Droogers, P., & Allen, R. G. (2002). Estimating reference evapotranspiration under inaccurate data conditions. *Irrigation and Drainage Systems, 16*(1), 33–45.

Edwards, W. (1977). How to use multiattribute utility measurement for social decisionmaking. *IEEE Transactions on Systems, Man, and Cybernetics, 7*(5), 326–340.

Edwards, W., & Barron, F. H. (1994). SMARTS and SMARTER: Improved simple methods for multiattribute utility measurement. *Organizational Behavior and Human Decision Processes, 60*(3), 306–325.

Fabian, J., & Zelenhasic, E. (2016). Modelling of meteo-droughts. *Water Resources Management, 30*(9), 3229–3246.

Food and Agricultural Organization of the United Nations (FAO) (1976). A framework for land evaluation. Rome.

Food and Agricultural Organization of the United Nations (FAO) (1985). Guidelines: Land evaluation for irrigated agriculture. Rome.

Forman, E., & Peniwati, K. (1998). Aggregating individual judgments and priorities with the analytic hierarchy process. *European Journal of Operational Research, 108*(1), 165–169.

Gavilán, P., Lorite, I. J., Tornero, S., & Berengena, J. (2006). Regional calibration of Hargreaves equation for estimating reference ET in a semiarid environment. *Agricultural Water Management, 81*(3), 257–281.

Gocic, M., & Trajkovic, S. (2010). Software for estimating reference evapotranspiration using limited weather data. *Computers and Electronics in Agriculture, 71*(2), 158–162.

Hajkowicz, S. A., McDonald, G. T., & Smith, P. N. (2000). An evaluation of multiple objective decision support weighting techniques in natural resource management. *Journal of Environmental Planning and Management, 43*(4), 505–518.

Hargreaves, G. H., & Samani, Z. A. (1985). Reference crop evapotranspiration from temperature. *Applied Engineering in Agriculture, 1*(2), 96–99.

Hargreaves, G. L., Hargreaves, G. H., & Riley, J. P. (1985). Irrigation water requirements for Senegal River basin. *Journal of Irrigation and Drainage Engineering, 111*(3), 265–275.

Heim Jr, R. R. (2002). A review of twentieth-century drought indices used in the United States. *Bulletin of the American Meteorological Society*, *83*(8), 1149–1166.

Hokkanen, J., & Salminen, P. (1997). Choosing a solid waste management system using multicriteria decision analysis. *European Journal of Operational Research*, *98*(1), 19–36.

Hrnjak, I., Lukić, T., Gavrilov, M. B., Marković, S. B., Unkašević, M., & Tošić, I. (2014). Aridity in Vojvodina, Serbia. *Theoretical and Applied Climatology*, *115*(1–2), 323–332.

Huang, I. B., Keisler, J., & Linkov, I. (2011). Multi-criteria decision analysis in environmental sciences: Ten years of applications and trends. *Science of the Total Environment*, *409*(19), 3578–3594.

Hwang, C. L., & Yoon, K. (1981). Methods for multiple attribute decision making. In *Multiple Attribute Decision Making*. Springer, Berlin, Germany, pp. 58–191.

Irmak, S., Allen, R. G., & Whitty, E. B. (2003). Daily grass and alfalfa-reference evapotranspiration estimates and alfalfa-to-grass evapotranspiration ratios in Florida. *Journal of Irrigation and Drainage Engineering*, *129*(5), 360–370.

Ishizaka, A., & Nemery, P. (2013). *Multi-criteria Decision Analysis: Methods and Software*. John Wiley & Sons, Chichester, UK, 310 p.

Janssen, R. (1991). *Multi-Objective Decision Support for Environmental Management*. Kluwer Academic Publishers, Boston, MA.

Josimov-Dunđerski, J. (2009). Eko-hidrološki uslovi za primenu mokrih polja u Vojvodini. Doctoral dissertation, University of Novi Sad, Faculty of Agriculture, Novi Sad, Serbia, 2009.

Kangas, A., Kurttila, M., Hujala, T., Eyvindsin, K., & Kangas, J. (2015). *Decision Support for Forest Management*, 2nd edition. Springer, Berlin, Germany, 307 p.

Keeney, R., & Raiffa, H. (1976). *Decision with Multiple Objectives: Preferences and Value Tradeoffs*. Wiley, New York, 569 p.

Keeney, R.L. (1982). Decision analysis: An overview. *Operations Research*, *30*(5), 803–838.

Keeney, R.L. (1992). *Value-Focused Thinking. A Path to Creative Decision Making*. Harvard University Press, Boston, MA, 416 p.

Lalić, B., Mihailović, D. T., & Podraščanin, Z. (2011). Future state of climate in Vojvodina and expected effects on crop production. *Ratarstvo i povrtarstvo*, *48*(2), 403–418.

Lessel, J., Sweeney, A., & Ceccato, P. (2016). An agricultural drought severity index using quasi-climatological anomalies of remotely sensed data. *International Journal of Remote Sensing*, *37*(4), 913–925.

Maksimović, L., & Dragović, S. (2002). Effect of sugar beet irrigation in different environmental growing conditions. *Zbornik radova Instituta za ratarstvo i povrtarstvo*, *36*, 43–56.

Maletic, J. T., & Hutchings, T. B. (1967). Selection and classification of irrigated land. In: *Irrigation of Agricultural Lands*, R. M. Hagan, H. R. Haise and T. W. Edminster (Eds.). American Society of Agronomists, Madison, WI.

Marinković, B., Crnobarac, J., Brdar, S., Antić, B., Jaćimović, G., & Crnojević, V., (2009). Data mining approach for predictive modeling of agricultural yield data. *Proc. First Int Workshop on Sensing Technologies in Agriculture, Forestry and Environment* (BioSense09), Novi Sad, Serbia, 1–5.

Mendoza, G. A., & Prabhu, R. (2003). Qualitative multi-criteria approaches to assessing indicators of sustainable forest resource management. *Forest Ecology and Management*, *174*(1–3), 329–343.

Mihailović, D. T., Lalić, B., Arsenić, I., & Malinović, S. (2004). Klimatski uslovi za proizvodnju semena. *U: M Milošević, M Malešević (ured.), Semenarstvo. Naučni institut za ratarstvo i povrtarstvo, Novi Sad, 1*, 243–266.

Mihailović, D. T., Lalić, B., Dršković, N., Mimić, G., Djurdjević, V., & Jančić, M. (2015). Climate change effects on crop yields in Serbia and related shifts of Köppen climate zones under the SRES-A1B and SRES-A2. *International Journal of Climatology*, *35*(11), 3320–3334.

Mir, S. A., & Padma, T. (2016). Evaluation and prioritization of rice production practices and constraints under temperate climatic conditions using Fuzzy Analytical Hierarchy Process (FAHP). *Spanish Journal of Agricultural Research*, *14*(4), 22.

Mishra, A. K., & Singh, V. P. (2010). A review of drought concepts. *Journal of Hydrology*, *391*(1–2), 202–216.

Nordström, E. M. Integrating multiple criteria decision analysis into participatory forest planning. Doctoral Dissertation, Swedish University of Agricultural Sciences, Umeå, Sweden, 2010.

Pejić, B., & Gajić, B. (2006). Water balancing-bioclimatic method as a functional approach to precise irrigation. *Journal of the Romanian National Society of Soil Science*, *40*(1), 45–50.

Pejić, B., Gvozdanović-Varga, J., Vasić, M., & Milić, S., (2008). Water balance, bioclimatic method as a base of rational irrigation regime of onion. *IV Balkan Symposium on Vegetables and Potatoes*, *830*, 355–362.

Pejić, B., Maksimović, L., Škorić, D., Milić, S., Stričević, R., & Ćupina, B. (2009). Effect of water stress on yield and evapotranspiration of sunwlower/Efecto del estrés hídrico sobre el rinde y apotranspiración de girasol/Effet du stress hydrique sur le rendement et l'évapotranspiration du tournesol. *Helia*, *32*(51), 19–32.

Pejić, B., Jaćimović, G., Latković, D., Bošnjak, Đ., Marinković, B., & Mačkić, K. (2011a). Indeks aridnosti kao osnova analize uticaja režima padavina i temperature vazduha na prinos kukuruza u Vojvodini. *Ratar. Povrt*, *48*(1), 195–202.

Pejić, B., Ćupina, B., Dimitrijević, M., Petrović, S., Milić, S., Krstić, Dj., & Jaćimović, G. (2011b). Response of sugar beet to water deficit. *Romanian Agricultural Research*, *28*, 151–155.

Pejic, B., Maksimovic, L., Cimpeanu, S., Bucur, D., Milic, S., & Cupina, B. (2011c). Response of soybean to water stress at specific growth stages. *Journal of Food Agriculture & Environment*, *9*(1), 280–284.

Pejic, B., Maheshwari, B. L., Šeremešić, S., Stričević, R., Pacureanu-Joita, M., Rajić, M., & Ćupina, B. (2011d). Water-yield relations of maize (Zea mays L.) in temperate climatic conditions. *Maydica*, *56*(4), 315–321.

Pohekar, S. D., & Ramachandran, M. (2004). Application of multi-criteria decision making to sustainable energy planning—a review. *Renewable and Sustainable Energy Reviews*, *8*(4), 365–381.

Procedures Manual for the Classification of Land for Irrigation in Alberta. (2004). Alberta Agriculture, Food and Rural Development, Resource Management and Irrigation Division, Irrigation Branch, Lethbridge, Alberta, Canada.

Purnell, M. F. (1979). The FAO approach to land evaluation and its application to land classification for irrigation. World Soil Resources Reports (FAO) 50: Land Evaluation Criteria for Irrigation. *Report of an Expert Consultation*, 27 February–2 March, Rome, Italy.

Quiring, S. M. (2009). Monitoring drought: An evaluation of meteorological drought indices. *Geography Compass*, *3*(1), 64–88.

Rabia, A. H., Figueredo, H., Huong, T. L., Lopez, B. A. A., Solomon, H. W., & Alessandro, V. (2013). Land suitability analysis for policy making assistance: A GIS based land suitability comparison between surface and drip irrigation systems. *International Journal of Environmental Science and Development*, *4*(1), 1.

Rajic, M., & Bezdan, A. (2012). Contribution to research of droughts in Vojvodina province. *Carpathian Journal of Earth and Environmental Sciences*, *7*(3), 101–107.

Ramanathan, R., & Ganesh, L. S. (1994). Group preference aggregation methods employed in AHP: An evaluation and an intrinsic process for deriving members' weightages. *European Journal of Operational Research*, *79*(2), 249–265.

Rogers, M., & Bruen, M. (1998). A new system for weighting environmental criteria for use within ELECTRE III. *European Journal of Operational Research*, *107*(3), 552–563.

Roy, B. (1968). Classement et choix en présence de points de vue multiples. *Revue française d'informatique et de recherche opérationnelle*, *2*(8), 57–75.

Ružica, S., & Nevenka, D. (2013). Determination of spatiotemporal distribution of Agricultural drought in Central Serbia (Šumadija). *Scientific Research and Essays*, *8*(11), 445–453.

Saaty, T. L. (1980). *The Analytic Hierarchy Process*. McGraw-Hill, New York, pp. 324.

Savić, R., Pejić, B., Ondrašek, G., Vranešević, M., & Bezdan, A. (2013). Iskorišćenost prirodnih resursa Vojvodine za navodnjavanje. *Агрознање*, *14*(1), 133–142.

Savic, R., Ondrasek, G., Letic, L., Nikolic, V., & Tanaskovik, V. (2017). Nutrients accumulation in drainage channel sediments. *International Journal of Sediment Research*, *32*(2), 180–185.

Schweier, J., Blagojević, B., Venanzi, R., Latterini, F., & Picchio, R. (2019). Sustainability assessment of alternative strip clear cutting operations for wood chip production in renaturalization management of pine stands. *Energies*, *12*(17), 3306.

Shahidian, S., Serralheiro, R., Serrano, J., Teixeira, J., Haie, N., & Santos, F. (2012). Hargreaves and other reduced-set methods for calculating evapotranspiration. *Evapotranspiration-Remote Sensing and Modeling*, A. Irmark (Ed.). In Tech, 4, pp. 59–80.

Srdjevic, B. (2005). Combining different prioritization methods in the analytic hierarchy process synthesis. *Computers & Operations Research*, *32*(7), 1897–1919.

Stillwell, W. G., Seaver, D. A., & Edwards, W. (1981). A comparison of weight approximation techniques in multiattribute utility decision making. *Organizational Behavior and Human Performance*, *28*(1), 62–77.

Stricevic, R., & Djurovic, N. (2013). Determination of spatiotemporal distribution of Agricultural drought in Central Serbia (Šumadija). *Scientific research and Essays*, *8*(11), 445–453.

Stricevic, R., Djurovic, N., & Djurovic, Z. (2011). Drought classification in Northern Serbia based on SPI and statistical pattern recognition. *Meteorological Applications*, *18*(1), 60–69.

Sys, C., Van Ranst, E., & Debaveye, J. (1991). Land Evaluation. Part I: Principles in land evaluation and crop production calculations. Agricultural Publications nr. 7, GADC, Brussels, Belgium.

Tabari, H., Grismer, M. E., & Trajkovic, S. (2013). Comparative analysis of 31 reference evapotranspiration methods under humid conditions. *Irrigation Science, 31*(2), 107–117.

Tadic, L. (2012). Criteria for evaluation of agricultural land suitability for irrigation in Osijek County Croatia. *Problems, Perspectives and Challenges of Agricultural Water Management*, 311–332.

Thornthwaite, C. W. (1948). An approach toward a rational classification of climate. *Geographical Review, 38*(1), 55–94.

Trajkovic, S. (2005). Temperature-based approaches for estimating reference evapotranspiration. *Journal of Irrigation and Drainage Engineering, 131*(4), 316–323.

Trajkovic, S. (2007). Hargreaves versus Penman-Monteith under humid conditions. *Journal of Irrigation and Drainage Engineering, 133*(1), 38–42.

Trajković, S. (2009). Metode proračuna potrebe za vodom u navodnjavanju. *Univerzitet u Nišu, Građevinsko-arhitektonski fakultet.*

Triantaphyllou, E., & Lin, C. T. (1996). Development and evaluation of five fuzzy multiattribute decision-making methods. *International Journal of Approximate Reasoning, 14*(4), 281–310.

Turc, L. (1961). Estimation of irrigation water requirements, potential evapotranspiration: A simple climatic formula evolved up to date. *Annales Agronomiques, 12*(1), 13–49.

Utset, A., Farre, I., Martínez-Cob, A., & Cavero, J. (2004). Comparing Penman–Monteith and Priestley–Taylor approaches as reference-evapotranspiration inputs for modeling maize water-use under Mediterranean conditions. *Agricultural Water Management, 66*(3), 205–219.

Valipour, M. (2015). Temperature analysis of reference evapotranspiration models. *Meteorological Applications, 22*(3), 385–394.

Vicente-Serrano, S. M., Beguería, S., & López-Moreno, J. I. (2010). A multiscalar drought index sensitive to global warming: The standardized precipitation evapotranspiration index. *Journal of Climate, 23*(7), 1696–1718.

Vojtek, M., & Vojteková, J. (2019). Flood susceptibility mapping on a national scale in Slovakia using the analytical hierarchy process. *Water, 11*(2), 364.

Vučić, N. (1971). Bioklimatski koeficijenti i zalivni režim biljaka-teorija i praktična primena. *Vodoprivreda*, 6–8.

von Winterfeldt, D., & Edwards, W. (1986). *Decision Analysis and Behavioral Research*. Cambridge University Press, Cambridge, 604, 6–8.

Xystrakis, F., & Matzarakis, A. (2010). Evaluation of 13 empirical reference potential evapotranspiration equations on the island of Crete in southern Greece. *Journal of Irrigation and Drainage Engineering, 137*(4), 211–222.

Zargar, A., Sadiq, R., Naser, B., & Khan, F. I. (2011). A review of drought indices. *Environmental Reviews, 19*(NA), 333–349.

Zeleny, M. (1982). *Multiple Criteria Decision Making*. McGraw-Hill, New York, pp. 563.

10 Appraisal of Agricultural Lands for Irrigation in Hot Subhumid Region of Jayakwadi Command Area, Parbhani District, Maharashtra, India

Bhaskara Phaneendra Bhaskar, Rajendra Hegde,
Sampura Chinnappa Ramesh Kumar, and
Venkataramappa Ramamurthy

CONTENTS

10.1 INTRODUCTION

The development of an irrigation network to feed the increasing human population is a global concern through choosing crops and cropping patterns based on land capabilities with adequate provision of conserving the natural resource (van Wambeke and Rossiter, 1987). Of late, researchers and planners have laid much emphasis on land irrigability classes to achieve sustainability in agricultural production (Alagh, 1990). It is emphasized that there is an urgent need for appropriate irrigated land-use policy for optimal water as well as land resources in fragile semiarid ecosystems of India (Khoshoo and Deekshatulu, 1992). The impact of land use and land cover on the hydrology, and thus irrigation, in tropical catchments has gained much interest since the last century. The advanced geospatial technologies are used in generation of land resource mapping using classified satellite images to understand the nexus between land use/cover change and their parameterization in the management of irrigation schemes, agriculture, and ecosystems (Akombo et al., 2014). The Food and Agriculture Organization (FAO) published *A Framework for Land Evaluation* with further modifications in later publications considering land qualities in each unit with the requirements of each land-use type (FAO, 1985, 1993). The earlier work done on land resource inventory and land evaluation for rural agricultural planning at a global level and an Indian scenario is discussed below.

10.1.1 LAND RESOURCE CHARACTERIZATION FOR IRRIGABILITY CLASSIFICATION: GLOBAL PERSPECTIVE

Site-specific soil mapping is often considered as Order 1 soil surveys. The best-known parametric systems for rating the quality of land are reported in the literature (Riquier et al., 1970; Sys and Vereheye, 1975; Sys et al., 1993). Productivity ratings are a reflection of real value for agriculture or forestry (Miller, 1984); but later recognition of the importance of management in obtaining economic yields prompted the use of crop equivalent rating (CER) values as an index for evaluating soils in Minnesota (Robert et al., 1995; Gajja et al., 2006). The soil salinity mapping was done using the electromagnetic induction (EMI) method in New South Wales, Australia (vander Lelij, 1983). Using soil–landscape model presentations of interpreted data (Drohan et al., 2010) with key soil indicators for crop production at a local level was proposed (Grealish et al., 2015). In land evaluation studies, the agricultural landscape photographs in conjunction with soil surveys were used to measure the perception of farmers about the changes in agri-landscapes (Eija et al., 2014; Tvet, 2009). The complexity in assessment of land suitability for agriculture needs multi-criteria to integrate land, topography, climate, water resources available for irrigation, soil capabilities, and current management practices including land use and land cover (Elsheikh et al., 2013). Multi-criteria decision-making (MCDM) is one of the most widely used methods in land suitability (Chen et al., 2010; Al-Mashreki et al., 2010). In Syria, Amel and Benni (2010) used Landsat images and worked out the salinity index to map salt-affected soils but later, Chandana et al. (2004) proposed a normalized differential salinity index and found it useful for mapping salt-affected soils in Hambantota District, Sri Lanka. The four soil-related criteria, namely salinity, depth, texture, and moisture followed by sequential assessments of the topography and water availability criteria were used in land suitability in the Emirate of Abu Dhabi (Abdelfattah, 2013), but the MCDM approach was used in land suitability assessment for date palm production (Abdelfattah et al., 2015). The study results show that 14.03% and 16.29% of total lands in the Emirate of Abu Dhabi are highly to moderately suitable for date palm plantation. The AHP-GIS integrated method is widely used as a powerful spatial decision support system in different fields, especially for land suitability assessment for irrigated agriculture (Feizizadeh and Blaschke, 2013). Six soil quality indicators such as pH, organic carbon, cation exchange capacity, texture, salinity, and slope were considered in the northeast area of Tadla Plain (Morocco) for generating suitable areas for agriculture using a geographic information system (GIS) and an analytical hierarchy process (AHP), which classified about 77% of cultivated soils suitable to agricultural production (Ennaji et al., 2018).

10.1.2 LAND RESOURCE CHARACTERIZATION FOR IRRIGATION: INDIAN PERSPECTIVE

The land resource surveys were made with the use of satellite images in combination with cadastral maps and evaluated *Vertisols* and *vertic* intergrades of basaltic terrain in the Jayakwadi command area. These soils have fair to poor productivity with severe limitations of sodicity, low permeability and effective rooting depth for irrigation (Bhaskar et al., 2014, 2017) and sodicity, erosion, drainage, organic carbon, and calcium carbonate in the Mula command area of Ahmednagar (Kharche and Pharande, 2010). In India, the salt-affected soils were mapped using Landsat imageries on a reconnaissance level (Sehgal et al.,1988), but Dwivedi (1992) mapped salt-affected soils in the Indo-Gangetic alluvial plains using Landsat and TM data. Later on, remote sensing (RS) data was mandatory and used widely for estimating biophysical parameters and indices (Rao et al., 1996; Panigrahy et al., 2006). With the availability of PAN data at 5.8 m spatial resolution from IRS-IC/ID, soil resource maps on 1:25,000 or a larger scale have been attempted using PAN merged LISS-III data. The high-resolution IKONOS data has the potential for farm-level soil mapping at the scale of 1:10,000 (Manchanda et al., 2002). The importance of systematic, large-scale land resource surveys is a prerequisite for evaluating their potential for a wide range of land-use options and in formulating land-use plans (Sathish and Niranjana, 2010). The large-scale mapping using IRS-1C PAN merged data of November and December was used to delineate a physiographic land-use map for basaltic terrain of Junewani village in Maharashtra (Srivastava and Saxena, 2004). Later on, similar approaches were used with IRS-P6-LISS-IV data in mini-watersheds of the Upper Subarnarekha subcatchment in Jharkhand, India (Londhe and Nathawat, 2010) and with IRS-Resourcesat-1 LISS-IV and Cartosat-1 fused data in the Mohammad village of the Nalgonda district, Andhra Pradesh (Wadodkar and Ravishankar, 2011). Boplab Mandal et al. (2017) used Cartosat DEM (30 m) and LISS-III data and reported that 31% of the area is estimated was moderately suitable (FAO approach) for surface irrigation in the Kansi watershed of the Purlia district of West Bengal. Several studies in India have applied the multi-criteria evaluation method for various applications including potential land suitability mapping for irrigation using groundwater (Adhikary et al., 2015; Jha et al., 2010). The profound implications of these developments for natural resources surveys and their applications are demonstrated in a case study taken from semiarid regions of Jayakwadi command of Parbhani district, Maharashtra.

10.1.3 IMPORTANCE OF STUDY

In India, agriculture and allied sectors accounted for 13.7% of the GDP (Gross Domestic Production, 2013) and stood second worldwide in farm output. Out of 160 million ha of cultivated land in India, 39 million ha are irrigated through underground water whereas 22 million ha are under canal irrigation, and the rest is under rainfed (Dhawan, 2017). Maharashtra has the irrigation potential of 13%, which is far below the national average of 33%. The state has the gross irrigated area of 1.93 m ha, out of which 21% is irrigated by canals, 14% by tanks, 58% by wells and 7% by other sources (Sawant et al., 1999). The Marathwada region in Maharashtra state has been more vulnerable and has often experienced severe drought over the last three decades. The distribution of soil subgroups of Vertisols and vertic intergrades in the Deccan Traps have high smectitic clay, low hydraulic conductivity, and high exchangeable magnesium (Mg) and sodium (Na) (Balpande et al., 1996). The formation of sodic soils in micro-highs along the side of non-sodic vertisols in micro-lows were reported in Purna Valley of Maharashtra (Vaidya and Pal, 2002). Further, the soil resource inventories in Maharashtra have shown that 1.5% of total geographical area (TGA) in command areas of the Godavari and Krishna river basins were moderately to strongly saline/sodic soils

(Challa et al., 1995). The drainage, salinity, and soil depth were the limitations in Vertisols of India for irrigation (Bhattacharjee, 1979). The Jayakwadi command area covers 0.276 m ha of cultivable land, having problems of shallow water table and salinization (Abhange et al., 1986). At present, the information on micro-topographical variations and their influence on irrigation in the command area is scanty. Therefore, the present study was carried out to characterize the soil-topography relationships and then to evaluate the soils for their suitability to surface irrigation for cotton- based cropping systems in the Jayakwadi command area.

10.2   MATERIALS AND METHODS

10.2.1   DESCRIPTION OF STUDY AREA

The study area is a part of Porward village (459 ha) of the Pabhani district covering under the Paithan left bank canal. This area lies between 18°50′N latitude and 76°50′E longitude, which is 24 km away from Parbhani (Figure 10.1). Deccan Traps are dominant and considered a result of a fissure type of lava eruption during the late Cretaceous to early Eocene period. The Deccan basalt has been assigned to Ajanta formations, having stratigraphic equivalents of Upper Ratangad formations of Western Maharashtra (Godbole et al., 1996). The red bole occurs as stringers or veins within basalt and has joints to serve as moderate to good aquifers (Singhal, 1997).

10.2.2   CLIMATE

The climate is hot, dry, and humid with a mean annual rainfall of 960 mm with 66 rainy days. Eighty percent of total rainfall is received during the rainy season (June to September), amounting to a total rainfall of 785 mm and water deficit of 124 mm (Ramakrishna Rao et al., 1986). The water balance (Thornthwaite and Mather, 1957) was calculated for 10 years (1984 to 1994, Table 10.1). The area receives 1,108 mm of rainfall with potential evapotranspiration (PE) of 1,759 mm. The rainy season

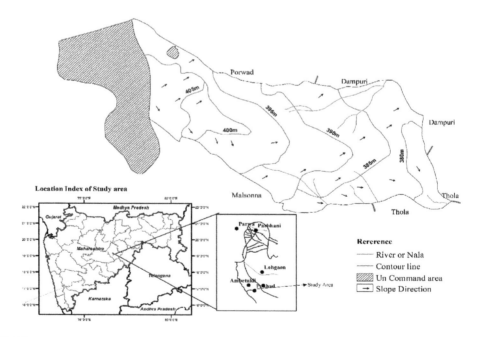

FIGURE 10.1   Location map of minor-4 in Jayakwadi command area, Parbhani district.

TABLE 10.1
Water Balance Data (1984 to 1994)

Weather Variables (mm)	January	February	March	April	May	June	July	August	September	October	November	December	Total
Rainfall	0.9	0.6	13.4	5.5	25.0	249.3	275.8	290.9	143.8	80.8	12.8	9.3	1108.0
PE	104.0	124.4	169.6	205.7	257.7	173.6	120.0	121.8	138.0	135.0	110.6	98.3	1758.8
P-PE	−103.2	−123.8	−156.2	−200.2	−232.7	75.7	155.0	169.1	5.8	−54.2	−97.8	−89.00	−650.6
AE	6.63	2.89	14.16	5.67	25.03	173.6	120.0	121.76	138.0	122.6	49.1	22.19	801.7
WD	97.4	121.5	155.4	200.0	232.67	0.0	0.0	0.0	0.0	12.4	61.5	76.1	957.0
WS	0.0	0.0	0.0	0.0	0.0	0.0	131.5	169.1	5.8	0.0	0.0	0.0	306.4

Note: PE = potential evaporation, AE = actual evapotranspiration, WD = water deficit, WS = water surplus, P = precipitation.

(June to September) showed that actual evapotranspiration (AE) is equal to PE, indicating optimal moisture supply to meet crop requirements during *kharif*. The area has *ustic* soil moisture regime and *isohyperthermic* soil temperature regime (Velayutham et al., 1999). It is further observed that the de Martonne aridity index (Id, 1926) values are 21 to 23 to define climate as Mediterranean with frequency of 20% to 70% of semidry conditions. The coefficient of variation is below 32% over a period of 1901 to 2002 (Table 10.2). Further, the BMDI values (Bhalme and Mooley, 1980) show that the region is subjected to incipient dry spells with values of −0.5 and −0.99 during crop season (Figure 10.2).

TABLE 10.2
de Martonne Aridity Index

| | I_d | | | % Frequency of |
| | | | Classification of | Semidry Conditions |
Year	Mean ± SD	CV (%)	Climate	Over a Decade
1901–1910	21.6 ± 5.1	23.7	Mediterranean climate	50
1911–1920	20.6 ± 6.8	32.9	with hot, dry	60
1921–1930	20.0 ± 6.4	31.9	summers and warm,	70
1931–1940	20.7 ± 6.6	31.9	wet winters	20
1941–1950	20.7 ± 6.5	31.4		40
1951–1960	21.2 ± 6.6	31.0		20
1961–1970	23.6 ± 5.3	22.5		30
1971–1980	20.6 ± 5.8	28.1		40
1981–1990	22.2 ± 5.4	24.5		50
1991–2002	21.0 ± 4.1	19.4		50

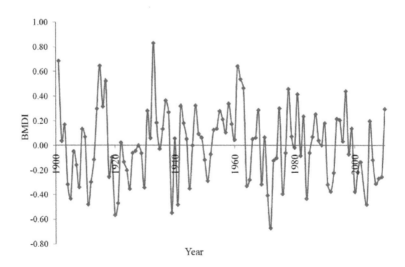

FIGURE 10.2   BMDI values computed from the hydrological years of 1901 to 2010 for Parbhani district.

10.2.3 Natural Vegetation and Cropping Pattern

The natural vegetation consists of trees such as *Azadirachta Indica* (neem), *Tamarindus Indica* (Tamarind), *Zizyphus jujube* (ber), *Acacia arabica* (babool), and *Ficus religiosa* (peepal); *Mangifera indica* (Mango), *Pyridium guava* (guava), *Punica granatum* (Pomegranate), and *Citrus reticulate* (citrus); and grasses such as *Cyprus rotendus* (doob), *Euphorbia dreaumculeidus* (dhudhi), *Cynodon dactylon* (doob), *Zizyphus numalaria* (Jherberi), and *Euphorbia hirta* (dhudhi). The major cropping pattern includes cotton (*Gossypium hirsutum* L; 200 ha) and groundnut (*Arachis hypogaea* L; 125 ha), wheat (*Triticum aestivum*; in 40 ha, sugar cane (*Saccharum officinarum* in 35 ha), and sorghum (*Sorghum bicolor*, in 25 ha).

10.2.4 Field Survey

The detailed soil survey was conducted using a *cadastral* map scaled at 1:8000 in minor–4 of Branch—67 of Paithan left bank canal near Porwad village. Intensive field traverse was made and field boundaries were checked with sufficient auger bores. Auger samplings were made at 1/4th to ½ m depending upon the variations in texture, erosion, and soil depth. Forty-five soil profiles were examined to a depth of 2 m or to the bedrock (if it is less than 2 m) with two to three master profiles for every 2 to 4 ha (All India Soil and Land Use Survey Organisation, 1970). The soil profiles were described as per the standard format (Schoeneberger et al., 2002) and classified in the subgroups of Vertisols, Inceptisols, and Entisols (Soil Survey Staff, 2014). Eight soil series were identified after soil correlation.

10.2.5 Laboratory Analysis

Soil samples from each horizon of eight soil series were collected and sieved to pass through 2 mm sieve for fine earth fraction for physical and chemical analysis. The particle-size analysis was done as per international pipette method with pretreatment of hydrogen peroxide to remove organic matter, dispersion, and shaking with sodium hexametaphosphate (Gee and Bauder, 1986). Soil reaction was determined in 1:2.5 soil/water suspension using a standard pH meter and electrical conductivity using a conductivity meter. Soil organic carbon (OC) was estimated using the wet digestion method and expressed in percentages. Exchangeable Ca^{2+} and Mg^{2+} in I1N NH_4OAc extracts were determined using atomic absorption spectrophotometer (AAS). Extractable potassium (K^+) and sodium (Na^+) were determined with 1N ammonium acetate extract (pH 7.0) using a flame photometer. The cation exchange capacity of soils was determined as per distillation method (Jackson, 1973). DTPA-extractable micronutrients viz. Fe, Mn, Zn, and Cu were extracted in 0.005 M DTPA at pH 7.3 (Lindsay and Norvell, 1978), and the concentration of the micronutrients was estimated using an atomic absorption spectrophotometer. The soil water content (in mass %) at 33 and 1,500 kPa was determined using a pressure membrane extractor (Richards, 1955). The hydraulic conductivity at saturation (Ks) of the soil sample was measured at the laboratory using the constant head method (Klute and Driksen, 1986).

10.2.6 Land Irrigability Classification

The methodology used in evaluating soil mapping units for gravity is given in Figure 10.3. The parametric approach from Sys et al. (1991, 1993) was applied (Table 10.3). The steps in irrigability analysis were given as under:

Step 1. Soil map with limiting symbol formula was used to define limitations of each series in the numerator and topography/drainage in the denominators.

Step 2. The capability index and soil units were rated by multiplying the proportion of each soil type by its respective soil rating. Based on the capability index, the soils' suitability was defined.

Step 3. Decision rules were proposed for irrigation and derived priority areas suitable for irrigated agriculture using ARC info version 13.

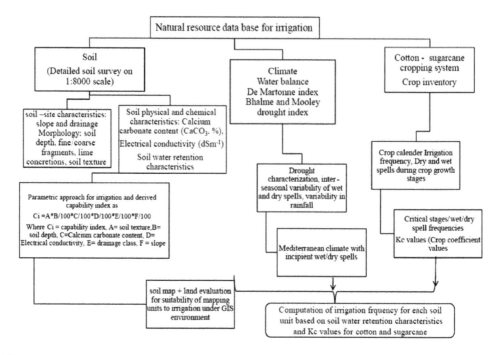

FIGURE 10.3   Methodology used in land evaluation for irrigation in Jayakwadi project.

TABLE 10.3
Rating of Textural Classes for Irrigation

Textural Class	Rating				
	<15%	15%–40%	40%–75%	15%–40%	40%–75%
Clay loam	100	90	80	80	50
Silty clay loam	100	90	80	80	50
Sandy clay loam	95	85	75	75	45
Loam	90	80	70	70	45
Silt loam	90	80	70	70	45
Silt	90	80	70	70	45
Silty clay	85	95	80	80	40
Clay	85	95	80	80	40
Sandy clay	80	90	75	75	35
Sandy loam	75	65	60	60	35
Loamy sand	55	50	45	45	25
Sand	30	25	25	25	25

| Soil Depth (cm) | Rating | Electrical Conductivity (dSm⁻¹) | Rating | | | CaCO₃ (%) | Rating | Drainage Classes | Rating | |
			Clay, Silty Clay, Sandy Clay	Other Textures					Clay, Silty Clay, Sandy Clay	Other Textures
<20	30	0–4	100	100		>50	80	Well-drained	100	100
20–50	60	4–8	90	95		20–50	90	Moderately well-drained	80	90
50–80	80	8–16	80	50		10–25	100	Imperfectly drained	70	80
80–100	90	16–30	70	35		0.3–10	95	Poorly drained	60	65
>100	100	>30	60	20		<0.3	90	Very poorly drained	40	65

The capability classes are defined according to the value of the capability (or suitability) index (Ci).

Capability Index	Class	Definition	Symbol
>80	I	Excellent	S1
60–80	II	Suitable	S2
45–60	III	Slightly suitable	S3
30–45	IV	Almost unsuitable	N1
<30	V	Unsuitable	N2

10.2.7 Irrigation Frequency

The irrigation interval of each soil mapping unit for cotton-based irrigation systems was calculated as per FAO (1979). The total available soil water (TAW) is the difference field capacity (FC) with permanent wilting point (PWC) and then multiplied with effective crop rooting depth (cm). After that, usable soil water (UW) was computed by multiplying maximum allowable depletion (MAD) with total available soil water (TAW). The irrigation frequency was calculated by dividing UW with crop water use rate.

The statistical analyses such as root mean sum square (RMSE), mean error (ME) for multiple regression analysis, correlations, and descriptive statistics were performed using Microsoft Excel and SPSS Windows version 10.

10.3 RESULTS AND DISCUSSION

10.3.1 Soil Morphology

Of the nine well-drained shrink–swell soil series on basalt capping over shale in minor–4 have a dark to very grey and brown matrix with uniformly clay texture throughout the profiles (Table 10.4). These soils have gradual wavy boundaries in subsoils. The nine soil series are grouped into three depth classes *viz.*, very deep (>90 cm): Parbhani series (P5), Brahmapuri (P6), Dhampuri (P7) and Thola (P9); deep (45–90 cm): Ambetak (P1), Pokhad (P3), Lohgaon (P4), and Malsonna (P8); and moderately deep (22.5 to 45 cm): Porwad (P2, All India soil and Land Use Survey Organisation, 1970). These soils have angular blocky structures in the B horizons and are very sticky and plastic when wet and very hard when dry. The prominent intersecting slickensides with wedge-shape aggregates—Lohgaon (P4), Parbhani (P5), Brahmapuri series (P6), and Thola series (P9)—were a result of shear displacement and vertical sliding of nonsheared units near the ground surface (Yaalon and Kalmar,

TABLE 10.4

Morphological Properties of the Cracking Clay Soils

Horizon	Depth (cm)	Matrix Color (moist)	Texture	Structure	Consistence	Effervescence	Other Features
P1. Ambatek Series							
Ap	0–16	10YR4/3	c	m2sbk	sh, fr, s,p	—	Surface cracks of 0.5 to 1.0 cm wide extended to 0.26 m.
AC	16–52	10YR4/2	c	m2sbk	sh, fr, s,p	—	Shiny pressure faces on ped surfaces
							Weathered basalt
P2. Porwad Series							
Ap	0–15	10YR3/2	c	c3abk	h, fi, s,p	—	
Bw1	15–34	10YR3/1	c	c3abk	h, fi, s,p	—	
Cr	34–73						
P3. Pokhad Series							
Ap	0–12	10YR3/2	c	m3sbk	vh, fi, s,p	es	Cracks of 0.5 to 1 cm wide extend to a depth of 0.5 m
Bw1	12–32	10YR3/1	c	m3abk	vh, fi, s,p	es	Shiny pressure faces on ped surfaces in 2nd and 3rd horizons
Bw2	32–56	10YR3/1	c	c3sbk	vh, fi, s,p		
Cr	56–125						Weathered basalt
P4. Lohgaon Series							
Ap	0–11	10YR4/2	c	m2sbk	vh, fi, s,p	ev	Cracks of 1.5 to 2 cm wide extend to a depth of 0.5 m
Bw1	11–40	10YR3/2	c	m2abk	vh, fi, s,p	ev	Shiny pressure faces on ped surfaces
Bss1	40–65	10YR3/1	c	c3sbk	vh, fi, s,p	ev	Intersecting slickensides forming parallelepipeds with wedge-shape aggregates in 3rd horizons
Cr	65–100					Weathered basalt	
P5. Parbhani Series							
Ap	0–10	2.5Y5/3	cl	m2sbk	h, fi, s,p	es	Cracks of 2.0 cm wide extend to a depth of 0.5 m
Bw1	10–44	10YR3/1	c	c2abk	h, fi, s,p	ev	Shiny pressure faces on ped surfaces
Bss1	44–72	10YR2/1	c	c2sbk	vh, fi, s,p	ev	Intersecting slickensides forming parallelepipeds with wedge-shape aggregates in 3rd, 4th, and 5th horizons
Bss2	72–102	10YR2/1	c	c2abk	vh, fi, s,p	ev	
Bss3	102–136	10YR3/1	c	c2sbk	vh, fi, s,p	ev	

(Continued)

TABLE 10.4 (Continued)
Morphological Properties of the Cracking Clay Soils

Horizon	Depth (cm)	Matrix Color (moist)	Texture	Structure	Consistence	Effervescence	Other Features
P6. Brahmapuri Series							
Ap	0–14	10YR3/2	c	c2sbk	h, fi, s,p	es	Cracks of 1.5 wide extend to a depth of 0.5 m
Bw1	14–43	10YR3/2	c	c3abk	vh, fi, s,p	es	Shiny pressure faces on ped surfaces
Bss1	43–81	10YR3/1	c	c3abk	vh, fi, s,p		Intersecting slickensides forming parallel epipeds with
Bss2	81–115	10YR3/1	c	c3abk	vh, fi, s,p		wedge-shape aggregates in 3rd and 4th horizons
P7. Dhampuri Series							
Ap	0–25	10YR4/3	c	m2sbk	h, fi, s,p	es	Surface cracks of 1.5 cm wide to a depth of 0.6 m
Bw1	25–52	10YR4/3	c	m2abk	h, fi, s,p	es	Presence of shiny pressure faces on ped surfaces
Bss1	52–73	10YR4/2	c	m2abk	h, fi, s,p		Intersecting slickensides and wedge-shape aggregates
Bss2	73–105	10YR5/3	c	m2abk	h, fi, s,p		forming parallel epipeds in 3rd and 4th horizons
P8. Malsonna Series							
Ap	0–16	10YR3/2	cl	m1sbk	h, fi, s,p	—	—
Bw1	16–53	10YR2/2	cl	m2sbk	sh, fi, s,p,	—	Presence of shiny pressure faces on ped surfaces
Cr	53–76	Weathered basalt					
P9. Thola Series							
Ap	0–9	10YR3/2	c	c3sbk	vh, vfi, vs, vp	es	Cracks of 0.5 to 1 cm wide extend to a depth of 0.5 m
Bw1	0–38	10YR3/1	c	c3sbk	vh, vfi, vs, vp	es	Shiny pressure faces on ped surfaces
Bss1	38–83	10YR3/2	c	c3abk	vh, vfi, vs, vp	es	Intersecting slickensides forming parallel epipeds with
Bss2	83–131	10YR3/2	c	c3abk	vh, vfi, vs, vp	es	wedge-shape aggregates in 3rd, 4th, and 5th horizons
Bss3	131–167	10YR4/3	c	c3abk	vh, vfi, vs, vp	es	

1978; Mermut and Acton, 1985). The structurally deformed B horizons have a direct relationship with linear frequencies of micro-highs and micro-lows of local relief (Bhaskar et al., 2001).

10.3.2 PARTICLE-SIZE DISTRIBUTION AND BULK DENSITY

These weighted means for sand are 10.3% (P9) to 21.7% (P7). The mean silt content is 21.4% with a coefficient of variation of 29% (P5&P6) and 26% silt for P9 with a variation of 7.9%. The soils over summit (P1 and P2) to middle slopes (from P3 to P7) have clay content of 51% to 58% to define particle-size class at the family level as *fine* and *very fine* in P1, P6, and P9 where clay content is more than 60% (Table 10.5).

TABLE 10.5
Particle-Size Distribution, Water Retention Characteristics, Bulk Density, Saturated Hydraulic Conductivity of Soils

Horizon	Particle-Size Distribution (<2 mm) Sand	Silt	Clay	Water Retention (%) −33 kPa	−1500 kPa	Bulk Density (Mg/m³)	Water Held at −1500 kPa/ clay (%)	Plant Available Water (%)	Ksat (cm/h)
	----------(%)----------								
P1. Ambatek Series									
Ap	15.2	25.8	58.9	34.7	25.9	1.62	0.44	8.8	0.23
AC	13.8	23.4	62.8	45.8	31.9	1.46	0.51	13.9	0.086
Weighted mean	14.14	24.04	61.76	42.84	30.3	1.50	0.49	12.54	0.126
P2. Porwad Series									
Ap	13.8	28.4	57.8	43.0	31.7	1.32	0.55	11.3	3.60
Bw1	12.8	26.4	60.8	44.1	31.8	1.52	0.52	12.3	2.01
Weighted mean	13.24	27.28	59.48	43.61	31.76	1.43	0.53	11.86	2.71
P3. Pokhad Series									
Ap	19.6	30.4	50.0	40.4	27.4	1.45	0.55	13.0	0.56
Bw1	16.2	29.4	54.4	44.8	30.0	1.62	0.55	14.8	4.19
Bw2	16.2	28.4	55.4	44.3	30.2	1.27	0.55	14.1	0.56
Weighted mean	16.82	29.22	53.96	43.81	29.60	1.46	0.55	14.21	2.21
P4. Lohgaon Series									
Ap	16.8	26.4	56.4	43.3	29.6	1.33	0.51	14.3	1.97
Bw1	16.8	23.4	59.8	44.0	30.7	1.45	0.51	13.3	0.57
Bss1	21.8	20.4	57.8	26.8	19.0	1.28	0.33	7.8	1.97
Weighted mean	18.72	22.75	58.45	37.26	26.01	1.36	0.44	11.35	1.35
P5. Parbhani Series									
Ap	20.6	30.0	49.4	31.2	20.6	1.26	0.42	10.6	4.35
Bw1	17.6	28.0	54.4	44.6	30.9	1.36	0.57	13.7	1.42
Bss1	16.6	27.0	56.4	45.4	31.2	1.47	0.55	14.2	0.43
Bss2	14.6	25.0	60.4	45.8	31.4	1.57	0.52	14.4	0.15
Bss3	12.6	25.0	62.4	46.4	32.6	1.27	0.53	13.6	3.49
Weighted mean	15.78	26.52	57.76	44.49	30.74	1.39	0.54	13.70	1.69
P6. Brahmapuri Series									
Ap	16.2	27.4	56.4	38.4	25.6	1.43	0.45	12.8	0.66
Bw1	12.2	23.4	60.4	40.7	28.4	1.62	0.47	12.3	0.04
Bss1	15.2	22.4	62.4	42.6	29.4	1.54	0.47	13.2	0.21
Bss2	13.8	22.0	63.8	43.7	30.1	1.26	0.47	13.6	4.21
Weighted mean	14.15	23.14	61.58	41.93	28.89	1.46	0.47	13.04	1.40

(Continued)

TABLE 10.5 (*Continued*)
Particle-Size Distribution, Water Retention Characteristics, Bulk Density, Saturated Hydraulic Conductivity of Soils

Horizon	Particle-Size Distribution (<2 mm) Sand	Silt	Clay	Water Retention (%) −33 kPa	−1500 kPa	Bulk Density (Mg/m³)	Water Held at −1500 kPa/ clay (%)	Plant Available Water (%)	Ksat (cm/h)
	----------(%)-----------								
P7. Dhampuri Series									
Ap	22.0	30.0	48.0	42.5	29.8	1.33	0.62	12.7	2.13
Bw1	21.5	28.0	50.5	43.6	31.1	1.56	0.62	12.5	0.18
Bss1	22.1	26.1	51.4	43.9	31.2	1.64	0.61	12.7	0.08
Bss2	21.4	24.0	54.8	44.6	32.4	1.71	0.59	12.2	0.04
Weighted mean	21.7	26.87	51.39	43.71	31.21	1.57	0.61	12.49	0.58
P8. Malsonna Series									
Ap	19.2	27.4	53.4	40.7	25.5	1.46	0.48	15.2	0.51
Bw1	18.2	26.4	55.4	37.4	24.9	1.52	0.45	15.2	0.26
Weighted mean	18.51	26.70	54.79	38.39	17.38	1.51	0.43	13.92	0.338
P9. Thola Series									
Ap	12.2	29.4	58.4	39.0	24.6	1.26	0.42	14.4	3.67
Bw1	11.2	28.4	60.4	41.7	26.8	1.37	0.44	14.9	1.12
Bss1	11.2	28.0	60.8	46.1	27.7	1.36	0.45	18.4	1.26
Bss2	10.2	28.4	61.4	47.1	33.3	1.39	0.54	13.8	0.89
Bss3	9.2	26.4	64.4	48.8	33.4	1.25	0.52	15.4	
Weighted mean	10.53	27.91	61.55	45.82	30.21	1.34	0.49	15.2	1.89

10.3.3 SOIL WATER CHARACTERISTICS AND SATURATED HYDRAULIC CONDUCTIVITY

These clay soils have field capacity (−33 kPa) of 37% to 45%. The field capacity shows increasing depth trends and is closely related with clay distribution. The clay has high specific area and, consequently, retains more water (Hillel, 1998). The permanent wilting point (−1500 kPa) is 25% to 30%. The plant available water holding capacity (AWC) is 11.4% (P4) to 15.7% (P9). The ratio between soil water held at −1500 kPa to total clay is 0.33 to 0.62 with a mean of 0.51 and percent of coefficient of variation of 12.98%. The available water holding capacity is 206 mm/m (P9) to 163 mm/m (P6), but other soils have less than 100 mm/m. The HC in the Parbhani series (P5) with an ESP of more than 15% was 4.35 cmh^{-1} in the Ap layer but decreased to 0.15 cmh^{-1} in the Bss horizon (Table 10.5). These soils have low to moderate hydraulic conductivity (Moore, 2001). The low hydraulic conductivity is due to dispersion of clay particles with increasing exchangeable sodium plus magnesium and influence on crop productivity (Yadav and Girdhar, 1981; Kadu et al., 2003).

The multiple regressions were worked out considering silt (%), clay (%), bulk density (Mgm^{-3}), and organic carbon (g/kg) with field capacity and permanent wilting point as given below:

- Field capacity (%, −33kPa) = −90.06 + 1.58X (silt, %) + 1.19X (clay, %) + 14.62X (bulk density, Mgm^{-3}) + 0.19 X (organic carbon, %) R^2 = 0.72 ** and F = 7.13
- Permanent wilting point (%, −1500 kPa) = −70.74 + 1.105X (silt, %) + 0.994 X (clay, %) + 8.25 X (bulk density, Mgm^{-3}) + 0.295 X (organic carbon, %) R^2 = 0.80** and F = 11.13

Both derived equations are significant at the 1% level. The Durbin Watson statistic is 2.105, indicating there is no autocorrelation detected in the sample. These equations are best fit for shrink-swell soils in the region.

10.3.4 CHEMICAL CHARACTERISTICS

These soils are slightly to strongly alkaline (pH of 8.1 to 9.2) with low electrical conductivity (0.08 to 0.87 dSm^{-1}). The organic carbon content, generally, decreases from summit (7.4 g kg^{-1}) to foot slopes (3.9 g kg^{-1}). Similar observations at lower landscape positions were reported under tropical environments (Papanicolaou et al., 2015). These soils are calcareous with calcium carbonate of 61.9 gkg^{-1} (P2) to 164.6 gkg^{-1} (P5). The content of CaCO$_3$ improves the water retention capacity and also promotes stable microaggregates (Shainberg and Letey, 1984). Calcium is the dominant cation on exchange complex with weighted mean of 11.51 cmol/kg in P5 to 50.33 cmol/kg in P8 (Table 10.6). The exchangeable magnesium shows increasing trends with depth and has a weighted

TABLE 10.6
Chemical Properties of Soils

Horizon	pH	EC (dSm^{-1})	Organic Carbon	CaCO$_3$	Ca	Mg	K	Na	Sum of Bases	ESP	ExCa/ ExMg
			g/kg		---------cmol(+)kg^{-1}--------						
P1. Ambatek Series											
Ap	8.5	0.20	7.0	100	38.0	18.0	0.70	0.65	52.5	0.9	2.1
AC	8.4	0.28	6.8	123	33.2	18.0	0.65	0.55	51.2	1.1	1.8
Weighted mean	8.43	0.26	6.9	116.9	34.5	18.0	0.66	0.58	51.5	1.0	1.8
P2. Porwad Series											
Ap	8.1	0.36	11.4	63	48.4	11.2	1.20	0.55	61.3	0.9	4.3
Bw1	8.2	0.25	8.7	61	48.4	10.0	0.70	0.65	59.7	1.1	4.8
Weighted mean	8.1	0.29	9.9	61.9	48.4	10.5	0.92	0.61	60.4	1.0	4.5
P3. Pokhad Series											
Ap	8.4	0.08	9.1	106	38.0	13.2	1.55	4.45	57.2	7.8	2.9
Bw1	8.8	0.65	7.2	102	33.2	16.8	0.70	6.20	56.9	10.9	2.0
Bw2	8.9	0.52	6.8	126	25.2	16.8	0.55	5.40	47.9	11.3	1.5
Weighted mean	8.7	0.49	7.4	111.5	31.2	16.1	0.80	5.59	53.7	10.5	1.9
P4. Lohangaon Series											
Ap	8.1	0.47	11.4	73	47.2	11.6	0.65	1.05	60.5	1.7	4.1
Bw1	8.2	0.54	8.7	91	44.0	14.0	0.50	1.25	59.7	2.1	3.1
Bss1	8.3	0.44	3.4	215	30.8	4.4	0.25	1.30	36.7	3.5	7.0
Weighted mean	8.2	0.49	7.1	135.6	39.5	9.9	0.43	1.24	50.9	2.5	4.7
P5. Parbhani Series											
Ap	8.2	0.56	6.4	191	13.6	7.6	2.6	8.8	32.6	26.9	1.8
Bw1	8.8	3.28	6.8	192	15.6	16.8	1.55	11.8	45.7	25.8	0.9
Bss1	9.1	—	5.3	141	10.4	21.6	1.70	12.2	45.8	26.6	0.5
Bss2	9.2	—	5.1	152	8.0	24.8	1.65	10.5	44.9	23.4	0.3
Bss3	9.2	—	4.8	160	10.8	24.8	1.55	8.9	46.1	19.3	0.4
Weighted mean	9.0	—	5.6	164.6	11.5	20.9	1.68	10.65	44.7	23.9	0.6
P6. Brahmapuri Series											
Ap	8.4	0.20	9.8	98	13.6	7.6	1.00	0.60	52.8	1.1	5.7
Bw1	8.4	0.20	7.0	100	38.0	15.8	0.65	0.75	55.0	1.4	2.4

(Continued)

TABLE 10.6 (*Continued*)
Chemical Properties of Soils

Horizon	pH	EC (dSm⁻¹)	Organic Carbon	CaCO₃	Ca	Mg	K	Na	Sum of Bases	ESP	ExCa/ ExMg
			g/kg		---------cmol(+)kg⁻¹---------						
Bss1	8.5	0.28	6.8	123	33.2	18.0	0.70	0.65	52.5	1.3	1.8
Bss2	8.4	0.28	6.0	137	32.0	18.0	0.65	0.55	51.2	1.1	1.8
Weighted mean	8.4	0.25	6.9	118.3	31.7	16.2	0.71	0.64	52.78	1.2	2.4
P7. Dhampuri series											
Ap	8.1	0.24	8.3	92	44.4	11.6	1.10	0.50	57.6	0.8	3.8
Bw1	8.2	0.18	8.0	72	49.2	11.2	0.65	0.55	61.6	0.9	4.4
Bss1	8.2	0.19	6.0	72	46.4	14.4	0.60	0.40	61.8	0.6	3.2
Bss2	8.2	0.27	5.2	57	46.0	15.6	0.75	0.35	62.7	0.6	2.9
Weighted mean	8.1	0.22	6.8	72.2	46.5	13.3	0.78	0.45	61.0	0.7	3.5
P8. Malsonna series											
Ap	7.9	1.38	9.8	54	54.8	8.4	1.10	0.65	64.9	1.0	6.5
Bw1	8.0	0.27	8.7	100	48.4	10.8	0.55	0.80	60.5	1.3	4.5
Weighted mean	7.9	0.61	9.0	86.1	50.3	10.0	0.71	0.75	61.82	1.21	5.1
P9. Thola series											
Ap	8.2	0.24	6.8	113	36.4	13.2	0.90	0.90	51.4	1.8	2.8
Bw1	8.7	0.25	5.3	124	34.8	12.4	0.65	1.85	49.7	3.7	2.8
Bss1	9.0	0.46	5.0	146	27.6	15.2	0.60	4.35	47.7	9.1	1.8
Bss2	9.1	0.55	1.5	139	24.4	16.4	0.55	5.35	46.7	11.5	1.5
Bss3	9.2	0.87	0.8	132	19.6	22.8	0.60	6.35	49.3	12.9	0.9
Weighted mean	8.9	0.53	3.24	135.37	26.7	16.6	0.61	4.5	48.3	9.3	1.8

mean of 9.9 cmol/kg ((P4) to 20.87 cmol/kg (P5). The Ca to Mg ratio is more than 3 P8, P7, P4, and P2. The exchangeable Ca and magnesium is high (Moore, 2001). The exchangeable sodium is high (23.9%) in P5 whereas potassium is 0.6 to 1.68 cmol/kg. Therefore, exchangeable Ca^{2+}/Mg^{2+} ratio > 2.5 (P2, P4, P7, and P8) was considered here as desirable, whereas 5 is considered as a critical limit for crop production (Kadu et al., 2003). Therefore, ESP > 5 as in P3 and P5 was considered as moderate limitation for all crops and >10 and >7.5 as severe limitation for cotton and pulses, respectively.

10.3.5 SOIL CLASSIFICATION

The eight shrink-swell soils in minor-4 of the Jayakwadi command area were classified in the orders of *Vertisols, Inceptisols, and Entisols* (Soil Survey Staff, 2014).

The Ambatek series (P1) on summits is classified as *clayey, mixed, isohyperthermic family of Typic Ustorthents* but the Porwad series (P2) and Malsonna series (P8) as *clayey, montmorillonitic, isohyperthermic family of Typic Haplustepts*. The Pokhad series (P3) with cambic horizons and wedge-shaped aggregates is classified as *fine, montmorillonitic, isohyperthermic family of Vertic Haplustepts*.

The Lohgaon series (P4) has cambic horizon and presence of slickensides within 100 cm and is classified as *fine, montmorillonitic, isohyperthermic family of Leptic Haplusterts*. The Parbhani series (P5) is classified as *very fine, montmorillonitic, isohyperthermic family of Sodic Haplusterts* with exchangeable sodium percentage more than 15% whereas the Brahmapuri (P6) and Thola series (P9) have clay more than 60% to place them as *very fine, montmorillonitic, isohyperthermic family of Typic Haplusterts*. Dhampuri (P7) is classified as *fine, montmorillonitic, isohyperthermic family of Chromic Haplusterts* as value moist of 4 and chroma of 3 within 30 cm.

10.3.6 Soil Map

In this section, the nine mapping units shown on the soil map (scale 1:10,000) and their hectarages are described in more detail (Figure 10.4). On upper slopes, three soil units—Ambetek, Porwad, and Pokhad—cover 71.6 ha (15.6% of minor–4). The Parbhani series (P5) have sodic phase (ESP more than 15%) and cause deflocculation and structural stability. The three units on middle slopes—Lohgaon, Parbhani, and Brahmapuri series—cover 139.55 ha (30.39% of total area). The soil units on lower slopes (Porwad, Malsonna, and Thola) cover 247.87 ha (53.97%).

10.3.7 The Land Suitability for Irrigation

Land suitability units were derived using a parametric system that defined these limitations in a qualitative way using limitation symbol formulae (Sys et al., 1991, Table 10.7). The Malasona (P2) and Pokhad (P7) units with heavy topsoil texture and moderate erosion on 3% to 5% slopes were evaluated as suitable. The Porwad (P8) and Ambetak (P9) were unsuitable with moderate subsoil permeability, heavy texture, and moderate salinity. The Thola, Dhampuri, Brahmapuri, and Lohgaon land units were slightly suitable because of strong subsoil alkalinity and strong microrelief. It was estimated that 233.4 ha land (50.81% TGA) on middle slopes (P4, P6, P7 and P8) required grading and leveling (t1) for good water management (w) whereas the soil units of the Parbhani, Porwad, and Ambetak series cover 10.3% of the area with sodicity (Y), long-term erosion hazard (e), and alkalinity (Z) limitations.

The suitability analyses were worked out to identify economically viable units for optimal water use in the command areas. The clay soils on middle slopes (122 ha, 26.7%) were suitable for irrigation, but clay soils on foot slopes (P9) were moderately suitable (24% of area, 110 ha, Figure 10.5).

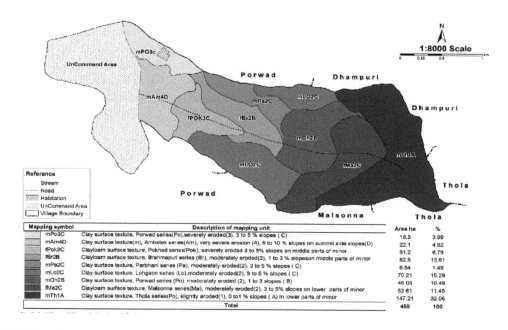

Mapping symbol	Description of mapping unit	Area ha	%
mPo3C	Clay surface texture, Porwad series(Po),severely eroded(3), 3 to 5 % slopes (C)	18.3	3.99
mAm4D	Clay surface texture(m), Ambetek series(Am), very severe erosion (4), 5 to 10 % slopes on summit side slopes(D)	22.1	4.82
fPok3C	Clayloam surface texture, Pokhad series(Pok), severely eroded 3 to 5% slopes on middle parts of minor	31.2	6.79
fBr2B	Clayloam surface texture, Brahmapuri series (Br), moderately eroded(2), 1 to 3 % slopeson middle parts of minor	62.5	13.61
mPa2C	Clay surface texture, Parbhani series (Pa), moderately eroded(2), 3 to 5 % slopes (C)	6.84	1.49
mLo2C	Clay surface texture, Lohgaon series (Lo),moderately eroded(2), 3 to 5% slopes (C)	70.21	15.29
mDh2B	Clay surface texture, Porwad series (Po), moderately eroded (2), 1 to 3 slopes (B)	48.05	10.46
fMe2C	Clayloam surface texture, Malsonna series(Me), moderately eroded(2), 3 to 5% slopes on lower parts of minor	52.61	11.45
mTh1A	Clay surface texture, Thola series(Po), slightly eroded(1), 0 to1 % slopes (A) in lower parts of minor	147.21	32.06
Total		**459**	**100**

FIGURE 10.4 Soil map of minor-4, Parbhani district.

TABLE 10.7

Suitability Subclass, Irrigation Frequency (Days) and Crop Yields

Soil series	Suitability Subclass/ Capability Index(Ci)	Symbol Formulae	Irrigation Frequency (days)			Yield (q ha⁻¹)			Soil Productivity Index
			Cotton	Jowar	Wheat	Cotton	Jowar	Wheat	
Ambetak	N2rt/29.5	**3H2S**/Cd3E3	9	11	11	11.5	12.9	15.4	107
Porwad	N1rt/34.5	**3H3S**/Ba3E3	7	9	9	10.9	13.5	15.4	104
Pokhad	S3r(e)/53.5	**3H2S-A2**/Ba2E3	11	14	14	11.7	10.0	14.3	102
Lohgaon	S2t''(e)/65.4	**3V2S**/Bc2E2	11	14	14	13.5	11.2	15.8	103
Parbhani	N1ZX'Y'a'/26.5	**4VSS2-A3**/Aa1E2	11	14	14	11.0	14.0	12.0	85
Brahmapuri	S2wt'/70.5	**4V1S**/Ac1E1	11	13	13	11.4	10.6	13.1	97
Dhampuri	S2wt'/73.5	**3H1S**/Bb2E2	11	14	14	10.0	10.6	13.1	98
Malsonna	S3r(t')/48.5	**3H2S**/Bb2E2-O1F1	12	15	15	10.9	13.5	15.4	104
Thola	S3d1(Y)/54.8	**5VSA3**/Ab1E1-W1P1OF1	14	17	17	11.4	10.6	13.1	97

Note: S2 = moderately suitable, S3 = marginally suitable, N1 = marginally not suitable, N2 = permanently not suitable.
Letter suffix: d' = Drainage, (Y) = Sodicity, r = rooting depth, t = land grading and leveling, w = water application management, Z = pH, micronutrient and toxicities, X' = leaching, a' = amendments, e = long-term erosion hazard.

a. Soil limitations (numinator):
  i. Subsoil permeability: 3 = moderate, 4 = slow, 5 = very slow
  ii. Top soil texture: V = very heavy(SC, SiC, C), H = heavy (CL, SiCL, SCL)
  iii. Soil salinity/alkalinity: A2 = moderate, A3 = severe, S2 = moderate
  iv. Depth limiting layers: S = soft weathered rock
  v. Soil depth classes: 1 = deep, 2 = moderately deep, 3 = shallow

b. Topographic, Erosion and Drainage limitations (Denominator)
  • Slope (%): A = 0, B = 2 – 5, C = 5 – 8
  • Transversal slope (%) = a = 1–2, b = 2–5, c = 5–8, d = 8–12
  • Microrelief: 1 = slight, 2 = moderate, 3 = strong
  • Erosion status: e1 = slight, e2 = moderate, e3 = severe
  • Flooding: f1 = slight
  • Groundwater depth: w1 = slight
  • Drainage: e1 = presence of mottling between 1.2 and 2 m
  • Ponding hazard: O1 = slight

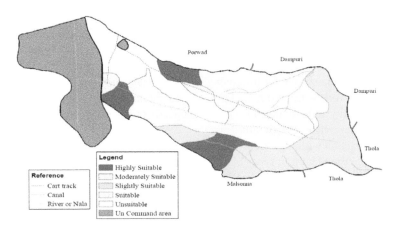

FIGURE 10.5 Land suitability map for surface irrigation in minor-4 of Jayakwadi project.

For each series, the irrigation frequencies were calculated considering an allowable soil water depletion of 50% for wheat and cotton, 65% for sugarcane, and daily water use was 0.6 mmday^{-1} for wheat, 0.75 mmday^{-1} for cotton, and 0.95 mmday^{-1} for sugarcane (Mohan and Arumugam, 1994). For these soil units, the irrigation intervals varied from 8.61 ± 1.35 days for cotton, to 8.9 ± 1.4 days for wheat, and 10 ± 1.64 days for sugarcane. It was reported in the literature that air-filled porosity <10% below 20 cm after five days of irrigation is considered a critical limit for plant root respiration, and it did not attain value less than 10% of air-filled porosity at a depth of 40 cm for 10 days (McGarry and Chan, 1984). This kind of irrigation trial in the command area of the Jayakwadi region is scanty and needed for identifying a threshold value of air-filled porosity for shrink-swell soils used for cotton and sugarcane. In the present study, the cotton yield varies from 1.0 t ha^{-1} in P7 to 1.3 t ha^{-1} in P4. The prolonged dry spells and excess wet periods in the Jayakwadi command area have resulted in the failure of *kharif* cotton (Ramakrishna Rao et al., 1986). The relative yield index showed that the *Leptic Haplusterts* and other vertic soil subgroups have above mean (99.3%), below the mean value in *Sodic Haplusterts*.

10.4   CONCLUSIONS

The land resource inventory (1:8000 scale) of minor–4 in the vicinity of the Jayakwadi irrigation project was made with an objective of evaluating the potentials of shrink-swell soils for irrigation. The clay soils on middle slopes were classified as suitable for irrigation and cover 122 ha (26.7% of area), but clay soils on foot slopes (P9) were moderately suitable, covering 24% of the area (110 ha). For each series, the irrigation intervals vary from 8.61 ± 1.35 days for cotton to 8.9 ± 1.4 days for wheat and 10 ± 1.64 days for sugarcane. It was estimated that 233.4 ha land (50.81% TGA) on middle slopes (P4, P6, P7 and P8) requires grading and leveling (t1) for efficient irrigation management (w) versus the soil units of the Parbhani, Porwad, and Ambetak series with sodicity (Y), long-term erosion hazard (e), and alkalinity (Z) limitations.

10.5   RECOMMENDATIONS

As the study is confined to minor-4 of the Jayakwadi command area (449 ha) with nine soil mapping units and evaluated for surface irrigation systems, the recommendations made from findings are indicative only.

- For the soil mapping unit mThiA (Thola series) and mPa2C (Parbhani series) with sodic phase that have exchangeable Ca^{2+}/Mg^{2+} ratio >2.5 and ESP >5 as in P3 and P5, this was considered as moderate limitation for all crops, and >10 and >7.5 was severe limitation for cotton and pulses, respectively. These land units are unsuitable for irrigation but upgraded and improved with continuous incorporation of organic matter and application of gypsum.
- The soil units such as mLo2C, fBr2B, and mDh2B are moderately suitable and cover 180.76 ha (39.36% of total area). These units need land leveling, removal of surface stones, and graded bunding system for improving water use efficiency.
- Future line of work is to be directed toward understanding the interactions between soil structural stability and also the indicators of reduced drainage. The other important aspects to explore under irrigated agriculture are:
  - The interaction with high bicarbonate groundwaters
  - Residence time factor of types of salts that remain in the soil longer
  - Implications for gypsum applications, and for interpreting SAR
  - The influence of high shrink/swell clays (smectites) on resisting dispersion of particles, bulk density, porosity, and internal drainage
  - Suggestion to have water use efficiency studies for cotton-based systems and make this study site as monitoring site for soil changes in the Marathwada region.

ACKNOWLEDGMENTS

The authors wish to express their sincere thanks to the S.G. Anantwar, chief technical officer, for helping in laboratory analysis and J. Zade, technical officer, for his support during field survey. The authors express their thanks to Deepak and Madiletti for GIS help.

REFERENCES

Abdelfattah, M.A. (2013). Integrated suitability assessment: A way forward for land use planning and sustainable development in Abu Dhabi, United Arab Emirates. *Arid Land Research and Management*, **27**: 41–64.

Abdelfattah, M.A. and Kumar, A.T. (2015). A web-based GIS enabled soil information system for the United Arab Emirates and its applicability in agricultural land use planning. *Arabian Journal of Geosciences*, **8**: 1813–1827.

Abhange, G.V., Pisolkar, S.S., More, S.D., Bharambhe, P.R. and Ghonsikar, C.P. (1986). Problems in Jayakwadi and Purna Commands and their management. Agricultural Development of Jayakwadi and Purna command areas. Marathwada Agricultural University Parbhani, pp. 155–165.

Adhikary, P.P., Chandrasekharan, H., Trivedi, S. and Dash, C.J. (2015). GIS applicability to assess spatio-temporal variation of groundwater quality and sustainable use for irrigation. *Arabian Journal of Geosciences*, **8(5)**: 2699–2711.

Akombo, R.A., Luwesi, C.N., Shisanya, C.A. and Obando, J.A. (2014). Green Water Credits for Sustainable Agriculture and Forestry in Arid and Semi-Arid Tropics of Kenya. *Journal of Agri-Food and Applied Sciences*, **2(4)**: 86–92.

Alagh, Y.K. (1990). Agro-climatic planning and regional development. *Indian Journal of Agricultural Economics*, **45(3)**: 244–268.

Al-Mashreki, M.H., Akhir, J.B.M., Rahim, S.A., Desa, K.M. and Rahman, Z.A. 2010. Remote sensing and GIS Application for assessment of land suitability potential for agriculture in the IBB Governorate, the Republic of Yemen. *Pakistan Journal of Biological Sciences*, **13**: 1116–1128.

All India Soil and Land Use Survey Organisation. (1970). *Soil Survey Manual*. IARI, New Delhi.

Amel, M.A. and Benni, J.T. (2010). Monitoring and evaluation of soil salinity in terms of spectral response using Land sat images and GIS in Mesopotamiam plain. *Journal of Iraqi Desert Studies: Special Issue of First Scientific Conference*, **2(2)**: 19–32.

Balpande, S.S., Deshpande, S.B. and Pal, D.K. (1996). Factors and processes of soil degradation in Vertisols of the Purna valley, Maharashtra, India. *Land Degradation & Development*, **7**: 313–324.

Bhalme, H.N. and Mooley, D.A. (1980). Large scale droughts/floods and monsoon circulation. *Monthly Weather Review*, **108**: 1197–1211.

Bhaskar, B.P., Anantwar, S.G., Challa, O., Bharambe, P.R. and Velayutham, M. (2001). Catenary variations in properties of shrink-swell soils in minor-4 of Jayakwadi irrigation project in Parbhani district, Maharashtra. *Journal of Geological Society of India*, **57**: 429–434.

Bhaskar, B.P., Prasad, J. and Tiwari, G. (2017). Evaluation of agricultural land resources for irrigation in the cotton growing Yavatmal district, Maharashtra, India. *Journal of Applied and Natural Science*, **9(1)**: 102–113.

Bhaskar, B. P., Sarkar, D., Bobade, S. V., Gaikwad, S. S. and Anantwar, S. G. (2014). Land evaluation for irrigation in cotton growing Yavatmal district, Maharashtra. *International Journal of Research of Agricultural Sciences*, **1(2)**: 128–136.

Bhattacharjee, J.C. (1979). Land evaluation criteria for irrigation in India (Paper 5.). Land evaluation criteria for irrigation. Report of expert. World Soil Resources Report. 50. Soil Resources, Management and Conservation Services, Land and Water Development Division, FAO, Rome, pp. 38–52.

Challa, O., Vadivelu, S. and Sehgal, J. (1995). *Soils of Maharashtra for Optimising Land Use*. NBSS Publications 54b, NBSS&LUP, Nagpur, India, 112p.

Chandana, P.G., Weerasinghe, K.D.N., Subsinghe, S. and Pathirana, S. (2004). Remote sensing approach to identify salt affected soils in Hambantota district. In *Proceedings of the Second Academic Sessions*, pp. 128–133.

Chen, Y., Yu, J. and Khan, S. (2010). Spatial sensitivity analysis of multi-criteria weights in GIS-based land suitability evaluation. *Environmental Modelling & Software*, 25, 1582–1591.

de Martonne, E. (1926). Une nouvelle fanction climatologique: l'indice d'aridité. *La Météorologie*, **2**: 449–458.

Dhawan, V. (2017). Water and Agriculture in India Background paper for the South Asia expert panel during the Global Forum for Food and Agriculture (GFFA) 2017. *Federal Ministry of food and Agriculture.* Teri. German Bundestag.

Dwivedi, R.S. (1992). Monitoring and the study of effects on image scale on delineation of salt affected soils in the Indo-Gangetic plains. *International Journal of Remote Sensing*, **13**: 1527–1536.

Drohan, P.J., Havlin, J.L., Megonigal, J.P. and Cheng, H.H. (2010). The "Dig It" Smithsonian soils exhibition: Lessons learned and goals for the future. *Soil Science Society of America Journal*, **74**: 697–705.

Eija, P., Grammatikopoulu, I., Timo Hurme, T., Soini, K. and Uusitalo, M. (2014). Assessing the quality of Agricultural landscape change with multi dimensions. *Land*, **3**: 598–616.

Elsheikh, R., Shariff, A.R.B.M., Amiri, F., Ahmad, N.B., Balasundram, S.K. and Soom, M.A.M. (2013). Agriculture land suitability evaluator (ALSE): A decision and planning support tool for tropical and subtropical crops. *Computers and Electronics in Agriculture*, **93**: 98–110.

Ennaji, W., Barakat, A., El Baghdadi, M., Oumenskou, H., Aadraoui, M., Karroum, L.A. and Hilali, A. (2018). GIS-based multi-criteria land suitability analysis for sustainable agriculture in the northeast area of Tadla plain (Morocco). *Journal of Earth System Science*, **127(79)**: 1–14.

FAO. (1979). Soil survey investigations for irrigation. FAO Soils Bulletin No. 42, FAO, Rome.

FAO. (1985). Guidelines: Land evaluation for irrigated agriculture. FAO Soils Bulletin No. 55. FAO, Rome.

FAO. (1993). The state of food and agriculture. Water policies and agriculture. Series 26. FAO, Rome.

Feizizadeh, B. and Blaschke, T. (2013). Land suitability analysis for Tabriz county, Iran: A multi-criteria evaluation approach using GIS. *Journal of Environmental Planning and Management*, **56**: 1–23.

Gajja, B.L., Khem Chand, K.and Singh, Y.V. (2006). Impact of land irrigability classes on crop productivity in canal command area of Gujarat: An economic analysis. *Agricultural Economics Research Review*, **19**: 83–94.

Gee, G.W. and Bauder, J.W. (1986). Particle size analysis. In: *Methods of Soil Analysis*, A. Klute (Ed.), 2nd ed., Vol. 9. American Society of Agronomy, Madison, WI, pp. 383–411.

Godbole, S.M., Rana, R.S. and Natu, S.R. (1996). Lava stratigraphy of Deccan basalts of Western Maharashtra. *Deccan flood basalts of India Gondwana Geological Society* 2: 125–134.

Grealish, G.J., Fitzpatrick, R.W. and Huston, J.L. (2015). Soil survey data rescued by means of user friendly soil identification keys and toposequence models to deliver soil information for improved soil management. *Geophysical Research Journal*, **6**: 81–91.

Hillel, D. (1998). *Environmental Physics: Fundamentals, Applications and Environmental Considerations.* Elsevier Science Publishing, Burlington, MA, 771p.

Jackson, M.L. (1973). *Soil Chemical Analysis.* Prentice Hall of India, New Delhi, India, pp. 38–56.

Jha, M. K., Chowdary, V. and Chowdhury, A. (2010). Groundwater assessment in Salboni Block, West Bengal (India) using remote sensing, geographical information system and multi-criteria decision analysis techniques. *Hydrogeology Journal*, **18(7)**: 1713–1728.

Kadu, P.R., Vaidya, P.H., Balpande, S.S., Satyavathi, P.L.A. and Pal, D.K. (2003). Use of hydraulic conductivity to evaluate the suitability of Vertisols for deep-rooted crops in semiarid parts of central India. *Soil Use and Management*, **19**: 208–216.

Kharche, V.K. and Pharande, A (2010). Land degradation assessment and land evaluation in Mula Command area of Irrigated agroecosystem of Maharashtra. *Journal of the Indian Society of Soil Science*, **58(2)**: 221–227.

Khoshoo, T.N. and Deekshatulu, B.L. (1992). *Land and Soils.* Indian Science Academy, Har-Anand Publications, New Delhi, India.

Klute, A. and Dirksen, C. (1986). Hydraulic conductivity and diffusivity: Laboratory methods. In: *Methods of Soil Analysis: Part 1, Physical and Mineralogical Methods*, A. Klute (Ed.). SSSA, ASA, Madison, WI, pp. 687–734.

Lindsay, W.L. and Norvell, W.A. (1978). Development of a DTPA soil test for zinc, iron, manganese, and copper. *Soil Science Society of America Journal*, **42**: 421–428.

Londhe, S. and Nathawat, M.S. (2010). Large scale mapping techniques for granitic terrain using high resolution satellite data. *Trends in Soil and Plant Sciences*, **1**: 19–31.

Manchanda, M.L., Kudrat, M. and Tiwari, A.K. (2002). Soil survey and mapping using remote sensing. *Tropical Ecology*, **43(1)**: 61–74.

Mandal, B., Dolui, G. and Satpathy, S. (2017). Land suitability assessment for potential surface irrigation of river catchment for irrigation development in Kansai watershed, Purulia, West Bengal, India. *Sustainable Water Resources Management.* doi:10.1007/s40899-017-0155-y.

McGarry, D. and Chan, K.Y. (1984). Preliminary investigation of clay soils' behaviour under furrow irrigated cotton. *Australian Journal of Soil Research*, **22**: 99–108.

Mohan, S. and Arumugam, N. (1994). Crop coefficient of major crops in South India. *Agricultural Water Management*, **26**: 67–80.

Mermut, A.R. and Acton, D.F. (1985). Surficial rearrangement and cracking in swelling clay soils of the Glacial Lake Regina Basin in Saskatchewan. *Canadian Journal of Soil Science*, **65**: 317–327.

Miller, G.A. (1984). Corn suitability rating: An index to soil productivity. Iowa State University. PM1168.

Moore, G. (2001). *Soil Guide: A Handbook for Understanding and Managing Agricultural Soils.* Bulletin 4343. A joint National Landcare and Agriculture Western Australia Project. Natural Resource Management Services Agriculture Western Australia, pp. 1–375.

Olsen, S.R. and Sommers, L.E. (1982). Phosphorus. In: *Methods of Soil Analysis*, A. L. Page et al., (Ed.), 2nd ed. American Society of Agronomy, Vol. 9, pp. 403–430.

Panigrahy, S., Manjunath, K.R. and Ray, S.S. (2006). Deriving cropping system performance indices using remote sensing data and GIS. *International Journal of Remote Sensing*, **26**: 2595–2606

Papanicolaou, A. N. T., Wacha, K. M. Abban, B. K., Wilson, C. G.,. Hatfield, J. L., Stanier, C. O. and Filley, T.R. (2015). From soilscapes to landscapes: A landscape-oriented approach to simulate soil organic carbon dynamics in intensively managed landscapes. *Journal of Geophysical Research: Biogeosciences*, **120**: 2375–2401.

Ramakrishna Rao, G., Sondge, V.D. and Bhonsle, S.S. (1986). Climatic conditions in Jayakwadi and Purnal commands with reference to water management. Agricultural Development in Jayakwadi and Purna Command Areas. Marathwada Agricultural University, Parbhani.

Rao, D.P., Gautam, N.C., Nagaraja, R. and Ram Mohan, P. (1996). IRSIC application in land use mapping and planning. *Current Science*, **70**: 575–578.

Richards, L.A. (1955). Retention and transmission of water in soil. *Yearbook of Agriculture (Water)*. United States Department of Agriculture, Washington, DC, pp. 144.

Riquier, J., Bramao, D.L. and Cornet, J.P. (1970). A New System of Soil Appraisal In Terms of Actual and Potential Productivity. AGL/TESR/70/6, FAO. Rome. 35p.

Robert, P.C., Rust, R.H. and Larson, W.E. (Editors). (1995). Site-specific Management for Agricultural Systems. American Society of Agronomy/Crop Science Society of America/Soil Science Society of America, Madison, WI.

Sathish, A. and Niranjana, K.V. (2010). Land suitability studies for major crops in Pavagadataluk, Karnataka using remote sensing and GIS techniques. *Journal of the Indian Society of Remote Sensing*, **38**(1): 143–151.

Sawant, S.D., Kulkarni, B.N., Achuthan, C.V. and Satya Sai, K.J.S. (1999). Agricultural development in Maharashtra-problems and prospects. National Bank for Agriculture and Rural development. Department of Economic analysis and Research. Karnataka Orion Press, Fort. Mumbai, pp. 1–173.

Schoeneberger, P.J., Wysocki, D.A., Benham, E.C. and Borderson, W.D. (2002). *Field Book–Describing and Sampling Soils.* version 2. NRSC, National Soil Survey Centre, Lincoln, NE.

Sehgal, J., Saxena, R.K. and Verma, K.S. (1988). Soil resources inventory of Indian using image interpretation techniques in remote sensing as a tool for soil scientist. In: *Proceedings of the Fifth Symposium of the Working Group on Remote Sensing* ISSS. Budapest, Hungary, 1988, pp. 17–31.

Shainberg, I. and Letey. J. (1984). Response of soils to sodic and saline conditions. *Hilgardia*, **52**(2): 1–57.

Singhal, B.B.S. (1997). Hydrogeological characteristics of Deccan trap formations of India. *Hard Rock Hydro Systems* (Proceedings of Rabat Symposium S2, May 1997. IAHS Publications no. 241.

Soil Survey Staff. (2014). *Keys to Soil Taxonomy.* 12th edition. USDA/NRSC, Washington, DC.

Srivastava, R. and Saxena, R.K. (2004). Technique of large scale soil mapping in basaltic terrain using satellite remote sensing data. *International Journal of Remote Sensing*, **25**(4): 679–688.

Sys, C. and Verheye, W. (1975). *Principles of Land Classification in Arid and Semi-Arid Areas* (Revised). Publ. Intern. Training Centre Post-Graduate Soil Scientists, Gent, Belgium, 42p.

Sys, C., Van Ranst, E. and Debaveye, J. (1991). *Land Evaluation, Part II.* Methods in land evaluation. General administration for development cooperation, Brussels, Belgium, pp. 24.

Sys, C., Van Ranst, E., Debaveye, J. and Beernaert, F. (1993). *Land Evaluation, Part 3: Crop Requirements.* Agricultural Publication 7. General Administration for Development Cooperation, Brussels, Belgium.

Thornthwaite, C.W. and Mather, J.R. (1957). Instructions and tables for computing potential évapotranspiration and the water balance. *Publications in climatology*, **10**: 3.

Tveit, M.S. (2009). Indicators of visual scale as predictors of landscape preference. A comparison between groups. *Journal of Environmental Management*, **90**: 2882–2888.

Vaidya, P.H. and Pal, D.K. (2002). Microtopography as a factor in the degradation of Vertisols in the Central India. *Land Degradation and Development*, **13**: 429–445.

Van der Lelij, A. (1983). Use of electromagnetic induction instrument (Type EM-38) for mapping soil salinity. Water Resources Commission, Murrumbidgee Division, New Southwales, Australia.

van Wambeke, A. and Rossiter, D. (1987). Automated land evaluation systems as a focus for soils research. IBSRAM Newsletter, 6, October 1987.

Velayutham, M., Mandal, D.K., Mandal, C. and Sehgal, J. (1999). Agroecological subregions of India for Planning and Development. NBSS&LUP, Pub. No. 35, 372p.

Wadodkar, M. and Ravi Shankar, T. (2011). Soil resource data base at village level for developmental planning. *Journal of the Indian Society of Remote Sensing*, **31(1)**: 43–57.

Yaalon, D.H. and Kalmar, D. (1978). Dynamics of cracking and swelling clay soils: Displacement of skeletal grains, optimum depth of slickensides, and rate of intra-pedonic turbation. *Earth Surface Processes*, **3**: 3142.

Yadav, J.S.P. and Girdhar, I.K. (1981). The effects of different magnesium: Calcium ratios and sodium absorption ratio values of leaching water on the properties of calcareous versus noncalcareous soils. *Soil Science*, **131**: 194–198.

Section IV

Case Studies
Irrigation System Selection in Western Asia

11 Appropriate Evaluation of Soils Delta of the Wadi Horan within the Province of Upper Euphrates, Iraq for Some Technologies of Irrigation Systems

AbdulKarem Ahmed Al-Alwany,
Farhan J. Mohamed Althyby, and Adel K. Salemn

CONTENTS

11.1 INTRODUCTION

The agricultural expansion in the desert lands is one of the most important objectives to be taken in the plans of agricultural expansion in Iraq, in order to bridge the gap between the population increase and the deterioration of land and food needs. The study area is one of the most important areas of future expansion in terms of agricultural potential (Investment map for promising areas – Anbar Agriculture Directorate, 2009).

Assessing land suitability for irrigation is an important tool not only in planning for agricultural development but also in overcoming the global problem of scarcity of water used for food production, especially because Iraq has for many reasons prevented it; this is due to Iraq not seriously planning to establish real agricultural development until it imports everything from crops. This has had a disastrous impact on the economy of the state and the deterioration of agricultural land due to the abandonment and lack of exploitation in an optimal manner, causing a state of deterioration, especially in the case of salinity; thus, there has been an increase in the proportion of land desertification.

Water is one of the important sources of growth and development for human societies. The water crisis is a real problem worldwide. Most of the world suffers from the scarcity of its water resources, whether it is for domestic use or for agriculture, industry, and development. Consequently, water management and investment are a challenge for workers in the field of agriculture and irrigation all in dry and semidry climatic conditions, and the optimal investment of this important resource for the development and transfer of modern technology and its dissemination in irrigation projects in conjunction with the study of soil characteristics and their suitability with irrigation methods, and the utilization of all available traditional and nontraditional water resources (Albaji et al., 2008).

FAO 1997 defines surface irrigation as holding water on the surface and allowing it to soak the soil to some depth. Drip irrigation is the flow of water droplets to a small part of the earth's surface and infiltration into the root region (Conway, 2003). The stability of global food security depends heavily on the management of natural resources. About 40% of all food production in the world is obtained from irrigated land. This result is not compatible with current rapid population growth. It has been identified that soil salinity and pollution as well as excessive urban development are also key factors affecting overall food production through irrigated agriculture. Good land resources management is one of the most important determinants of balanced food security in the world, and any imbalance in one of these resources leads to an imbalance in the ecosystem (Rabia, 2012a).

Rabia (2012b) also noted that studies of the appropriateness of the land for growing different agricultural crops are not effective if they are not taken into consideration. In order to determine the suitability of the land for different irrigation methods.

Yahia cultivar (1971), Anbar Governorate, according to its suitability for irrigated cultivation of crops, following a system (USBR, 1953). At the end of the seventies, the Soil Surveys in Iraq started applying the system (Sys and Verheye, 1972) for irrigation purposes in conjunction with the system of land suitability for irrigation proposed by Before (USBR, 1953).

Jassim (1981) assessed the land in Iraq by standard methods based on the characteristics of the land and the characteristics of the land proposed by Sys (1980; Khouri, 2000). The rationalization of water uses, the optimal investment of water resources, the introduction of modern irrigation technologies (Serhal, 1998). The Arab region is one of the most drought-prone areas in the world, with total water resources estimated at 330 km^3/year^{-1} (equivalent to 0.75% of global water resources).

(Al-Alwany, 2001) in his study on the assessment and classification of land suitability for the project of Saqlawiya in Anbar province to grow wheat and sorghum crops by adopting the standard method parametric approach for irrigated agriculture and preparing suitable land maps for these crops. The percentage of the project area is as follows: 28.76% of the study area is suitable for growing wheat, 28.86% for wheat, 12.48% for wheat and 20.35% for wheat, while maize for spring wheat was attributed to irrigated agriculture 54.55% of the total area. You are limited to maize cultivation and (35.9%) were unsuitable for maize cultivation.

Al-Alwany (2012) also noted in his geomorphic study of the Wadi Horan that the soil sediments in the bottom of the valley were coarse and moderately smooth, with low organic matter content with a clear rise in calcium carbonate content, ranging from 280 to 303 g·km^{-1} soil with homogeneity in its distribution in soils, and the rise of this latter component caused an increase in the values of the degree of soil interaction, which ranged between 7.6 and 8.1.

Akinbile et al. (2007) explained that the appropriateness of the land for irrigation requires a comprehensive assessment of soil characteristics, topography of the land within the field, and water quality for irrigation.

The main objective of this study is to evaluate the validity of the study area in Wadi Horan for the method of surface irrigation and drip irrigation in order to give recommendations to the policymakers so that the management of these lands is done in an optimal manner to sustain the resources of this region under current conditions and to rationalize water consumption.

11.2 MATERIALS AND METHODS OF WORK

11.2.1 DESCRIPTION OF THE STUDY AREA

The study area is located in the western part of the Iraqi desert, within the latitude between 33° 58′ 48.43″ and 33° 56′ 51.65″ north and longitude between 43° 31′ 48.24″ and 43° 33′ 37.71″. It is located 203 km from the city of Baghdad. It occupies an area of 10,480 dunums and is above sea level between 77 and 97 m. Figure 11.1 represents the location of the study area.

11.2.2 THE GEOLOGY OF THE STUDY AREA

The study area is located within the Stable Shelf, a stable structural unit that was not affected by the Alpine movement during the Mesozoic Era and the Tertiary Period. The characteristic features of these compositions are that the Alpine movement did not lead to the formation of folds but rather the formation of vertical displacements with some horizontal displacements in the base rocks, which in turn led to the formation of ripples for the layers of cover and formation of horst basins and elevations in different directions across the geological ages (Al-Dulaimi, 2008). It is clear from the geological map of the valley estuary (Figure 11.2) that the area was formed from the floodplain sediments, which descend on the northern side of the Wadi Horan valley. The deposits of the Euphrates formation are located on the southern side, and the deposits of the valley bottoms meet with the sediments of the Euphrates River downstream.

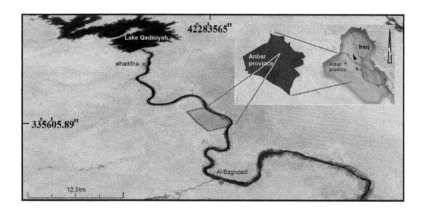

FIGURE 11.1 A satellite image representing the location of the study area.

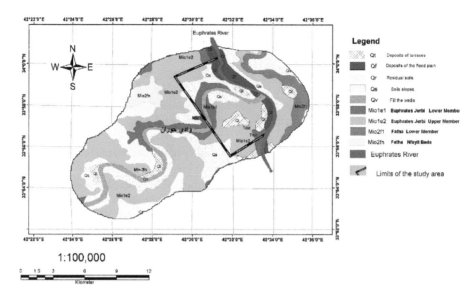

FIGURE 11.2 Geological map of Wadi Horan. (From General Company for Geological Survey and Mining, Plate Hit, 1997.)

11.2.3 TOPOGRAPHY AND PHYSIOGRAPHY STUDY AREA

In the region I have identified the following physical units:

11.2.4 SUMMIT

This is a physiographic unit that is very important in clarifying soil formation processes as well as reading the date of soil formation. In some sites belonging to this physiographic unit, the rocks of the original material appear close to the surface in shallow depths. The time is insufficient to form soil materials.

11.2.5 UPPER SLOPE

It is a physiographic unit with a moderate to very low slope. There are currently no manifestations of erosion, although it is slight on the soil surface and a depth of 30 cm. It is free of pebbles with a dark color, then a gravel layer appears at the depth of 60 cm with a very high content of gravel (60%) of the total volume. Soil and other physical characteristic of the soil of this physical unit is an increase in the clay content with depth, which reflects positively in the susceptibility of the soil to water retention.

11.2.6 LOWER SLOPE DOWN SLOPE

With very low to very few slopes, most of the soil of this physiographic unit is deep and free of gravel and deeper than the soil of the upper layer of the slope. The reason for the lack of gravel from the surface soil is the result of runoff of soil materials transferred from the soil of the upper slope unit. The base of the valley is usually one meter, and the dark color and high water retention can be identified in the soil of this physiographic unit.

11.2.7 TERRACES

They are almost flat parts from the perspective of the earth and are located along the course of the valley. The sediments in this unit have soft soil materials that are regularly present due to the sediments that are very slow and belong to the higher units transported by flood water and have a high content of the silt separation, where these joints are distinguished. With its ability to move by water over long distances, while sand separations are deposited at short distances, there is no gravel in the soil of this unit, while a decrease in clay content with depth and permeability characteristic is less than the soil of a unit (Lower slope).

11.2.8 FLOOD PLAIN

The soil of this unit is flat or near to flatness, as a ground perspective and without slope during periods of rain, this unit is a newly formed soil that contains sediments and weathered materials that move with the flood, and that most of its soil is sandy and has a low content of silt and mud, Figure 11.3. The cross section represents the beginning and end of the valley estuary with the Euphrates River.

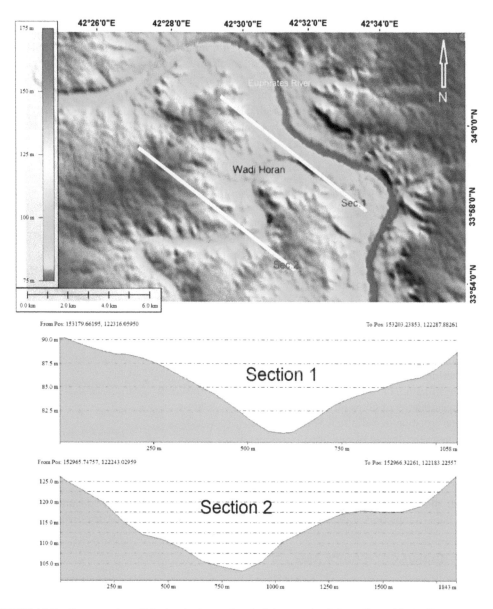

FIGURE 11.3 Cross section of the beginning and end of the mouth of the valley with the Euphrates River using the system (DEM-Mapper Gloppier).

11.2.9 Data Sources

- Topographic map 1: 50,000 from the US Defense Agency – 1990.
- Geological map 1: 250,000 from the General Authority for Geological Survey and Inquiry – Ministry of Industry and Minerals – Iraq 1995.
- Iraq's geological map: 1 million – General Establishment of Geological Survey and Mining – Ministry of Industry and Minerals – Iraq – Ministry of Water Resources – 1997.
- Exploration Soil Map (Buringh, 1960) Ministry of Agriculture – Iraq.

11.2.10 Field Work and the Collection and Analysis of Laboratory Samples

A semi-detailed soil survey was carried out in a free-lance, depending on the state of variance in the field, especially topography, regression, tissue, soil color, natural plant, land use, and other traits related to the irrigation process. The detection, dissection, and morphology of the pedons were described as morphology in SS Staf. (2003). Soil models were taken from each horizon for the purpose of laboratory analysis. The distribution of soil minutes volumes was estimated using the method described by Hesse (1976). Sedimentation method with acetone solution was done for gypsum, and then measuring was done of the electrical conductivity of the deposited sediment (Richards, 1954). Calcium carbonate was estimated by following the method described in Ryan et al. (2003); soils were classified into family level (Soil Survey Staff, 1993; USDA, 2004, 2006).

11.2.11 Analysis of Data for the Suitability of Irrigation Methods

The irrigation capacity index for both surface and drip irrigation was calculated according to the characteristics of the studied soils, to assess the suitability of these wetlands and to determine the appropriate ranges and determinants of the final grading. Based on the application of the equation (Sys et al., 1991) using soil characteristics and then classifying these properties and using them to calculate the irrigation potential index (Ci) according to the equation:

$$Ci = A*B/100* C/100 * G/100* D/100 * E/100 * F/100...n$$

where:

Ci: Capacity Capability Index
A: Determination of soil tissue
B: Determination of soil depth
C: Determination of Calcium Carbonate Condition
G: Determination of Gypsum Status
D: Determination of Salinity – Sodic
E: Estimate the degree of exchange
F: Estimation of the degree of regression
n: Of qualities

Table 11.1, shows the Final assessment of the suitability of irrigation of the soil units (Sys et al., 1991), and the Completion of the estimation process for surface irrigation or drip irrigation.

11.2.12 Climate of the Study Area

Monthly temperatures for July were 34.1°C, and winter temperatures dropped to the lowest monthly averages of 7.7°C in January. Rain is observed in winter and with a total lack of summer during the summer. To reach its peak in December, the annual rainfall recorded at a modern plant was

TABLE 11.1

Classes of the Suitability Index for Irrigation

Class Rank of Soil Suitability	The Description	Capability Index
S1	Highly suitable	>80
S2	Moderately suitable	60–80
S3	Marginally suitable	45–59
N1	Currently not suitable	30–44
N2	V Permanently not	<29

Source: Sys, C. et al., Land evaluation, Part l, principles in land evaluation and crop production calculations, International training centre for post-graduate Soil Scientists, University Ghent, 1991.

TABLE 11.2

Monthly Average of Some Elements of Climate for the Station of Modern Area and for the Period 1974–2010

The Month[a]	Monthly Rate of Temperature (Celsius)	Monthly Rainfall Rate (mm)	Monthly Rate of Evaporation (mm)
January	7.7	18.3	4.8
February	10.4	19.3	10.2
March	14.5	19.1	29.2
April	21.1	21.1	81.4
May	27.0	6.4	169.8
June	31.5	–	252.9
July	34.1	–	315.7
August	33.5	–	317.9
September	29.8	–	193.5
October	22.9	7.4	15.7
November	14.3	15.4	9.7
December	9.1	21.4	7.3
Annual rate	11.9	10.7	117.3

[a] General Organization for Meteorology and Seismic Monitoring, elevation of the station from the sea surface 146 m.

128.4 mm annually. Table 11.2 shows that the temperature has a significant impact on the loss of water by evaporation, so the highest loss of record during the summer months was from May to September. It reached 1249.9 mm, or 88.7% of annual evaporation of the plant, and Figure 11.4 shows the planned climate near the study area of the station.

11.2.13   USE OF THE LAND

Two types of land use are observed within the study area, the first for industrial use, as it includes limited parts of the delta as quarries for sand and gravel, while the other use is agricultural, whether irrigated or demi, as the land is used to wheat, corn, sesame and different vegetables for high soil fertility

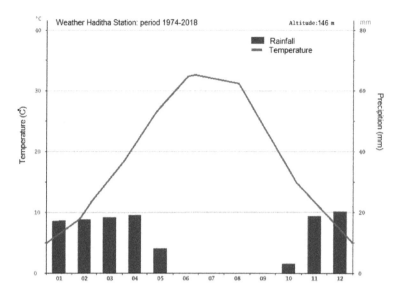

FIGURE 11.4   The climate plan of the air station in Haditha district.

FIGURE 11.5   The use of the land within the unit of the belly of the valley by planting wheat.

with a note of lack of interest And in using the methods of managing soils in the advanced state of lack of agricultural awareness on the one hand and the lack of financial support by the state (Figure 11.5).

11.2.14   NATURAL VEGETATION

The region is located within the subdesert region, according to the division of Guest (1966). The sources of natural vegetation in the mouth of the Horan Valley are the Western Desert, where the seeds are transmitted by floods during the winter season or by wind during the summer season. *Artemisia Herba-alba* and *Artemisia scoparia, Tamarix passorinoide, Zizyphus numalariae, haloxylon salicornicum, Cornulaca monacantha, Suaeda* spp., and *Peganum harmala*. The annual species recorded in the region are: *Malva parviflora*, wildebeest *lepornium Hordeum, Bromus sericeus, Bromus danthoniae, Centaurea sinaica, Numan decent, Anemone coronaria*, and *Brbssica arvensis* and *Anagallis arvensis* L. *Gypsophila capillaris.*

11.2.15 GEOMORPHOLOGY OF THE STUDY AREA

The area of the study is located in the lower valley unit according to the Parsons division (1955), which is characterized by various topographic features, where the flow of valleys that descend toward the Euphrates River or depressions to the west of it in the northern parts divide the surface of this unit and the formation of estuaries of these valleys at the confluence of the Euphrates. The unit of the Euphrates River is a contact area with the valley estuary. The expansion of this unit is influenced by the rock formations and the effect of the structures.

11.3 RESULTS AND DISCUSSION

11.3.1 SOIL CLASSIFICATION

The soil is defined as a dynamic natural body developed from different materials. It has chemical, physical, mineral, and biological properties and characteristics. It has a specific geographic distribution within the Earth's perspective. Its characteristics vary according to the heterogeneity of its composition factors. The study of the soil is of great importance in the subject of its suitability for irrigation, in terms of physical and chemical properties, because these qualities determine the adequacy. The most important soils prevailing in the study area, which was classified through the survey of semi-detailed soils, ranks the following:

11.3.2 DESERT SOILS (DRY) ARIDISOL

This is the predominant soil in the study area, covering large areas of its surface, and exploiting part of the desert soils in agriculture and grazing and exploiting in some parts the groundwater for irrigation. These soils are an important source of agricultural production and the best use of groundwater for the cultivation of wheat and barley in these soils, especially in the valleys in the rainy season, or watering from the Euphrates River through mechanical pumps established for this purpose. Gypsum soil prevails within these desert soils, consisting of the basic layers of gypsum with the presence of limestone and sandstone, a shallow soil. The soil is rich in calcite, and this type of soil receives low rainfall, which is reflected in the low production capacity. There are also gypsum soils, which are composed of gypsum, limestone, and sandstone, and their particles are disassembled with observation of water or wind erosion.

11.3.3 SOIL COMPOSITION (SEDIMENTARY SOILS) ENTISOLS

These include valleys, soils, alluvial fans and soils of the valleys, they are generally medium-to-soft soils formed due to the weak ability of waterways to transfer their loads, which led to their sedimentation, and depths range between (0.6 – 3 m), and new deposits are still added to them. In rainy seasons, these soils are tissues that vary in their tissues according to the sedimentation conditions within the valley of the valley, and the remaining soils are also observed in the region, which is one of the soils resulting from weathering operations, where the materials are sorted according to their size during the process of transporting them by wind or transporting water, as it transports the sand Coarse and Small sizes and reside coarse mixed with pieces of limestone, rocks and gypsum, and this soil is poor with organic materials, and it reflects the properties of the rocks of the parent materials that generate these soils, the study area covers an area of 10480 square dunums, and based on chemical, physical and morphological characteristics, Table 11.3, as shown on map Figure 11.6.

TABLE 11.3
Physical and Chemical Properties of the Soil of the Study Area

Physiographic Unit	No. Pedon	Classification of Soils	Slope %	Soil Depth (cm)	Texture	EC dS.m-1	pH	CaCO₃%	CaSO₄%	Drainage Class
(1) Plateau	1	Coarse loamy, mixed (gypsic), hyperthermic, Haplogypsids	0 to <3	50	SL	3.6	7.8	11.6	51.7	Well
(2) Floodplain	2	Fine, mixed (calcareous) hyperthermic, Typic Torrifluvents	0 to <2	120	SiL	2.3	7.7	22.3	1.3	Imperfect
(3) Wadi bottom	3	Coarse sandy, mixed, nonactive, hyperthermic, Torripsamments	0 to <2	>300	S+Greval +Boulder	2.8	8.2	23	1.8	Excessively
(4) Terraces	4	Coarse loamy, mixed (gypsic), hyperthermic, Haplogypsids	7 to <15	45	SL	3.9	7.6	12	18	Well

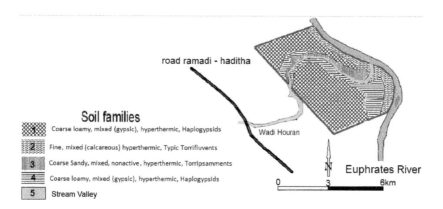

Soil families

1	Coarse loamy, mixed (gypsic), hyperthermic, Haplogypsids
2	Fine, mixed (calcareous) hyperthermic, Typic Torrifluvents
3	Coarse Sandy, mixed, nonactive, hyperthermic, Torripsamments
4	Coarse loamy, mixed (gypsic), hyperthermic, Haplogypsids
5	Stream Valley

FIGURE 11.6   Distributions of soil in the study area.

11.3.4 CLASSIFICATION OF SOIL UNITS AS SUITABILITY FOR SURFACE IRRIGATION AND DRIP

The results of Table 11.4 showed a difference in the values of the suitability of capability for surface irrigation. The soil units were classified as currently not suitable (N1) for family classification because of the severe determinants in the physical environment of these soils. This unit percentage (61.2%) of the study area and an area of 6,420 dunums². The unit of the floodplain unit and the unit of the sedimentation of the valley on the cultivar species is moderate (S2) because there are some determinants of irrigation that include the state. The physical composition of the verb In the case of Puncture and by 8.3, 4.2% and an area of 880.448 dunums² respectively, either unit terracing has obtained the appropriate degree of marginal (S3), to the presence of high determinants, especially the physical status of the soils and the situation topographic of degree gradient.

In the study area, the results showed that the unit of the floodplain was very suitable (S1) with an area of 8.3%, while the unit of the valley belly and river terraces showed a suitable marginal type. This is because of some determinants affecting drip irrigation, the physical environment of the root region, as well as the topography of the root region and the percentage of total area, 4.2 and 13.7%, respectively. The results in Table 11.4 showed that the plateau unit was unsuitable for drip irrigation at the present time. Figures 11.7 and 11.8 show suitable varieties for both methods.

TABLE 11.4

Soil Capability Index, Proportions and Areas Suitable for Surface Irrigation and Drip Irrigation

			Surface Irrigation		Drip Irrigation	
Map Unit	Area %	Areadunums²	Index Capability Ci	Suitability Classes	Index Capability Ci	Suitability Classes
1	61.2	6420	12.8	$N2_S$	14.9	$N2_S$
2	8.3	880	72.0	$S2_W$	76.9	$S1_W$
3	4.2	448	52.2	$S2_S$	52.2	$S3_S$
4	13.7	1444	36.0	$S3_{ts}$	42.3	$S3_{ts}$
5Stream valley	11.5	1208	—			

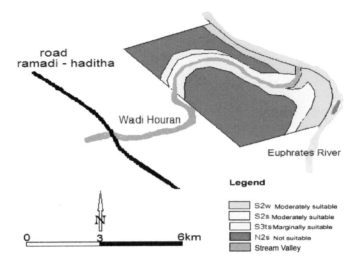

FIGURE 11.7 The study area according to its suitability for surface irrigation.

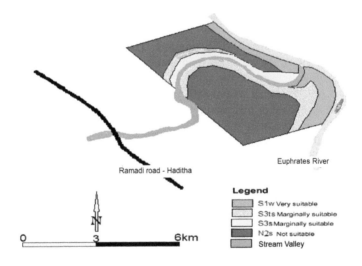

FIGURE 11.8 Area of study according to its validity for drip irrigation.

11.3.5 Sources and Quality of Irrigation Water

The main source of irrigation water in the study areas is the Euphrates River and the water wells located within the study area, whose water is characterized by a medium to high salt level and sometimes is not suitable for irrigation, with an electrical conductivity of 5.95 ds·m^{-1}.

11.4 CONCLUSIONS AND RECOMMENDATIONS

From field data, the suitability of irrigation systems (surface irrigation and drip irrigation) was analyzed and compared, and the study included analyzing the characteristics and characteristics of soil and land, and the results showed that the drip irrigation method is more appropriate than the surface irrigation system in the study area, however the main determinants were (limitation) for both methods due to the characteristics of the physical ocean and the topography of the study area, the drip irrigation method has proven studies that it keeps the soil within the field capacity water, and therefore, it will be more beneficial to use the irrigation method especially in arid climates and regions, and it is better to apply as the best method for your study area.

The effective management of river basins will greatly help in solving the problems of poor agricultural productivity in our country, Iraq, and many countries in the past decade have placed great importance on the development of river basins and drains of estuaries and dry valleys and decision-makers taking to achieve this fact and set up agricultural projects within this framework The scheme that involves the production of agricultural crops to secure food security instead of the immediate situation, which we see all agricultural products imported from outside the country despite the presence of two great rivers, the Tigris and Euphrates, as well as huge quantities in groundwater This was not achieved due to the difficult security conditions that rocked the country for decades and are still continuing.

REFERENCES

Akinbile, O.C., A.Y. Sangodoyin, I. Akintayo, F.E. Nwilene and Futakuchi. 2007. Growth and yield responses of upland Rice (NERICA 2) under different water Regimes in Ibadan, Nigeria. *Agricultural Journal* 1: 71–75.

Al-Alwany, AbdulKarem A. 2001. Evaluation of the land for the project of Saqlawiyah in Anbar province for the cultivation of wheat and yellow maize crops, Master thesis (unpublished), Faculty of Agriculture, Anbar University.

Al-Alwany, AbdulKarem A. 2012. Geomorphic study of the sediments of Wadi Horan in western Iraq, Ministry of Agriculture. *Journal of Iraqi Agriculture* 17(1): 35–46.

Albaji, M., A. Landi, N.S. Boroom and K. Moravej. 2008. Land suitability evaluation for surface and drip irrigation in Shavoor Plain Iran. *Journal of Applied Science* 8(4): 654–659.

Al-Dulaimi, A.S. 2008. A periodical bulletin issued by the Desert Studies Center. *Geology of Anbar Province.* University of Anbar, Center for Desert Studies.

Conway, K. 2003. Local solutions in the global water crisis. www.crdi.ca/en/ev-25649-201-1-DO_TOPIC. html.

Exploration Soil Map, Bayernek, 1960 Ministry of Agriculture, Iraq.

FAO. 2006. *Guidelines for Soil Profile Description.* Food and Agriculture Organization, Rome.

Geological Map. 1995. 1: 250000 from the General Authority for Geological Survey and Investigation of Metal, Ministry of Industry and Minerals, Iraq Year 1995.

Guest, E. (ed.). 1966. *Flora of Iraq,* Vol.1. Introduction Ministry of Agricultural and Agrarian Reforms, Baghdad. 213 pp.

Hesse, P.R. 1976. Particle size distribution in gypsic soils. *Plant and Soil* 44: 241–247.

Investment Map for Promising Areas. 2009. The Iraqi Ministry of Agriculture, Agriculture Directorate of Anbar province.

Iraq's Geomorphological Map. 1997. 1 million, General Establishment of Geological Survey and Mining, Ministry of Industry and Minerals, Iraq. Iraqi Ministry of Water Resources – Year 1997.

Jassim, H.F. 1981. Principles of regional soil survey land evaluation and land – use planning in Iraq, PhD Thesis. Ghent University Belgium. 547 pp.

Khoury, E.S. 2000. Modern irrigation methods and their role in reducing waste water and increasing production, Symposium on water resources in Syria and achievements in irrigation and land reclamation, Faculty of Civil Engineering, Tishreen University. Lathakia, May 3–4, 1998. Publications of the Supreme Council of Sciences.

Parsons, R.C. 1955. Ground water resources of Iraq. Vol. 8. Northern Desert, Report No. 415. Printing Office, Washington, DC.

Rabia, A.H. 2012a. Modeling of soil sealing by urban Sprawl in Wukro, Ethiopia Using Remote Sensing and GIS Techniques, Taza GIS-Days: the International conference of GIS Users, Fez, Morocco, SidiMohamed Ben Abdellah University, May 23–24, pp. 484–488.

Rabia, A.H. 2012b. A GIS based Land Suitability Assessment For Agricultural Planning In Kilte Awulaelo District, Ethiopia, *The 4th International Congress of ECSSS, EUROSOIL 2012 "soil science for the benefit of mankind and environment",* Bari, Italy. pp. 1257.

Richards, L.A. (ed.). 1954. Diagnosis and improvement of saline and alkaline soils. U.S.D.A. Agr. HB. No. 60.

Ryan, J., G. Estefan and A. Rashid. 2003. Soil and Plant Analysis Laboratory Manual International Center for Agricultural Research in the Dry Areas (ICARDA), Aleppo, Syria.

Serhal, Mi Damascene. 1998. Water in the Arab Countries between the Barriers to Scarcity and the Challenges of Agricultural Development, Scientific Symposium of the Federation of Syrian Chambers of Agriculture on Water in the Arab World, Damascus, May 11.

Soil Survey Staff. 1993. *Soil Survey Manual.* U.S.D.A. Hand book No. 18. U.S. Government.

Soil Survey Staff. 2003. *Key to Soil Taxonomy*, 9th edn. United State Department of Agriculture, Natural Resource Conservation Service, USA. pp. 332.

Sys, C. 1980. Land characteristics and qualities and methods of rating them. World Soil Resources Reports 52. FAO, Rome. pp. 23–49.

Sys, C. and W. Verheye. 1972. Principles of land evaluation in arid and semi-arid regions. ITC. Ghent. Belgium.

Sys, C., E. Van Ranst and J. Debaveye. 1991. Land evaluation, Part l, principles in land evaluation and crop production calculations. International Training Centre for Post-graduate Soil Scientists, University Ghent.

Topographic Map. 1990. 1: 50,000 from the US Defense Agency.

USBR. 1953. Land classification. Hand book. Vol. V. Part 2.

USDA. 2004. Soil Survey Laboratory Methods Manual, Soil Survey Investigation Report, No. 42, Version 4.0 November, Washington, DC.

USDA. 2006. *Key to Soil Taxonomy USDA*, United States Department of Agriculture Natural Resources Conservation Service, 10th edn. USA.

Yahia, H.M. 1971. Soil and soil condition in sediments of the Ramadi province, Iraq. PhD Thesis, University of Amsterdam, Holland.

12 The Best Irrigation System in Hamidih Plain, Iran

Mohammad Albaji, Abd Ali Naseri, and Saeid Eslamian

CONTENTS

12.1 INTRODUCTION

The scarcity of water in arid and semiarid regions such as Iran is a restrictive element for the agricultural sector; therefore, there is an urgent need to develop initiatives to save water in this particular sector (Abyaneh et al., 2017; Chegah et al., 2013). Among the proposed solutions is the use of pressurized irrigation systems. The present study was conducted in an area about 6,772 hectares in the Hamidih Plain, in the Khuzestan Province located in the southwest of Iran, during 2011–2012. The study area is located 20 km west of the city of Ahvaz, 31° 22′ to 31° 28′ N and 48° 21′ to 48° 28′ E. The average annual temperature and precipitation for the period of 1965–2011 were 25.4°C and 230 mm, respectively. Also, the annual evaporation of the area is 2035 mm. The Kharkhe River supplies the bulk of the water demands of the region. The application of irrigated agriculture has been common in the study area. Currently, the irrigation systems used by farmlands in the region are furrow irrigation, basin irrigation, and border irrigation schemes.

The area is composed of two distinct physiographic features, i.e., river terraces and river alluvial plains of which the river alluvial plains' physiographic unit is the dominant feature. Also, five different soil series were found in the area. The semi-detailed soil survey report of the Hamidih Plain (Khuzestan Water and Power Authority, 2011) was used in order to determine the soil characteristics.

Over much of the Hamidih Plain, the use of surface irrigation systems has been applied specifically for field crops to meet the water demand of both summer and winter crops. The major irrigated broad-acre crops grown in this area are wheat, barley, and maize, in addition to fruits, melons, watermelons, and vegetables such as tomatoes and cucumbers. But like other plains in Khuzstan Province, the important major crops are wheat and barley (Behzad et al., 2009; Naseri et al., 2009a; Albaji et al., 2012b; Albaji and Alboshokeh, 2017). There are very few instances of sprinkle and drip irrigation on large-area farms in the Hamidih Plain.

Five soil series and 26 series phases or land units were derived from the semi-detailed soil study of the area. The land units are shown in Figure 12.1 as the basis for further land evaluation practice. The soils of the area are of Aridisols and Entisols orders. Also, the soil moisture regime is aridic and torric, while the soil temperature regime is hyperthermic (Khuzestan Water and Power Authority, 2011). Therefore, the organic matter content is low due to high soil temperature and high decomposition ratio of this matter (Mahjoobi et al., 2010).

FIGURE 12.1   Soil map of the study area.

12.2   RESULTS AND DISCUSSIONS

As shown in Tables 12.1 and 12.2 for surface irrigation, only soil series coded 3 (1,094 ha – 16.15%) were classified as currently not suitable (N_1), and soil series coded 1, 2, 4, and 5 (5,578 ha – 82.37%) were classified as permanently not suitable (N_2) for any surface irrigation practices.

The analyses of the suitability irrigation map for surface irrigation (Figure 12.2) indicate that there was no highly suitable, moderately suitable, and marginally suitable land in this plain. The map also indicates that only some part of the cultivated area in this plain was evaluated as currently not suitable land because of severe limitations of drainage, salinity, and alkalinity. The major

TABLE 12.1

The Ci Values and Suitability Classes of Surface, Sprinkle, and Drip Irrigation for Each Land Unit

Codes of Land Units	Surface Irrigation		Sprinkle Irrigation		Drip Irrigation	
	Ci	Suitability Classes	Ci	Suitability Classes	Ci	Suitability Classes
1	17.55	N_2 n[a]	18	N_2 n[b]	16	N_2 sn[c]
2	15.79	N_2 nw	17.1	N_2 n	16	N_2 sn
3	42.99	N_1 nw	50.62	S_3 nw	48	S_3 snw
4	17.55	N_2 n	18	N_2 n	16	N_2 sn
5	23.03	N_2 ns	28.35	N_2 n	26.6	N_2 sn

[a,b] The limiting factors for surface and sprinkle irrigations: n: (Salinity & Alkalinity) and: w: (Drainage).
[c] The limiting factors for drip irrigation: n: (Salinity & Alkalinity) and s: (Calcium Carbonate).

TABLE 12.2

Distribution of Surface, Sprinkle, and Drip Irrigation Suitability

	Surface Irrigation			Sprinkle Irrigation			Drip Irrigation		
Suitability	Land Unit	Area (ha)	Ratio (%)	Land Unit	Area (ha)	Ratio (%)	Land Unit	Area (ha)	Ratio (%)
S_1	–	–	–	–	–	–	–	–	–
S_2	–	–	–	–	–	–	–	–	–
S_3	–	–	–	3	1,094	16.15	3	1,094	16.15
N_1	3	1,094	16.15	–	–	–	–	–	–
N_2	1, 2, 4, 5	5,578	82.37	1, 2, 4, 5	5,578	82.37	1, 2, 4, 5	5,578	82.37
[a]Mis Land		100	1.48		100	1.48		100	1.48
Total		6,772	100		6,772	100		6,772	100

[a] Miscellaneous Land: (River bed)

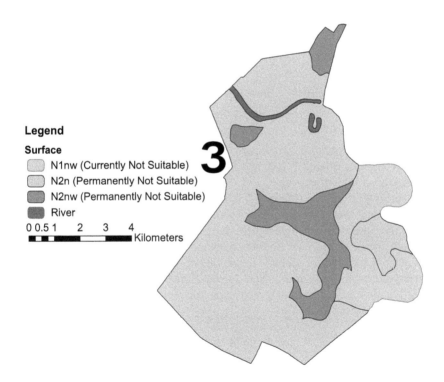

Legend

Surface
- N1nw (Currently Not Suitable)
- N2n (Permanently Not Suitable)
- N2nw (Permanently Not Suitable)
- River

0 0.5 1   2   3   4 Kilometers

FIGURE 12.2 Land suitability map for surface irrigation.

portion of this plain (located in the center, west, south, and north) is deemed as being permanently not suitable land due to very severe limitations of heavy soil texture, drainage, salinity, and alkalinity. For almost the total study area, elements such as soil depth, slope, and calcium carbonate were not considered as limiting factors.

In order to verify the possible effects of different management practices, the land suitability for sprinkle and drip irrigation was evaluated (Tables 12.1 and 12.2).

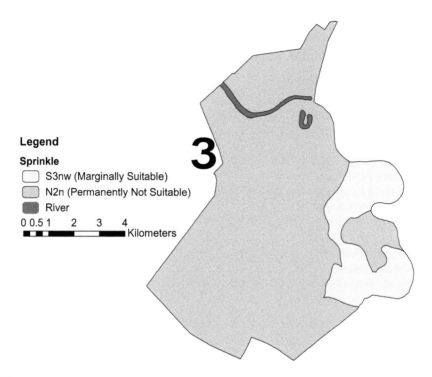

FIGURE 12.3   Land suitability map for sprinkle irrigation.

For sprinkle irrigation, only soil series coded 3 (1094 ha – 16.15%) were found to be marginally suitable (S_3), and soil series coded 1, 2, 4, and 5 (5578 ha – 82.37%) were classified as permanently not suitable (N_2) for sprinkle irrigation.

Regarding sprinkler irrigation (Figure 12.3), there was no highly suitable, moderately suitable, and currently not suitable land in this plain. The marginally suitable land can be observed in the smallest part of the cultivated zone in this plain (located in the east) due to high limitations of drainage, salinity, and alkalinity. As seen from the map, the largest part of this plain was evaluated as permanently not suitable land for sprinkle irrigation because of very severe limitations of salinity and alkalinity. Other factors such as drainage, depth, salinity, and slope never influence the suitability of the area. For almost the entire study area slope, soil depth, soil texture, and calcium carbonate were never taken as limiting factors.

For drip irrigation, only soil series coded 3 (1,094 ha, 16.15%) were found to be slightly suitable (S_3), and soil series coded 1, 2, 4, and 5 (5,578 ha – 82.37%) were classified as permanently not suitable (N_2) for drip irrigation.

Regarding drip irrigation (Figure 12.4), there was no highly suitable, moderately suitable, and currently not suitable land in this plain. The marginally suitable land can be observed in the smallest part of the cultivated zone in this plain (located in the east) due to high limitations of calcium carbonate, drainage, salinity, and alkalinity. As seen from the map, the largest part of this plain was evaluated as permanently not suitable land for drip irrigation because of very severe limitations of calcium carbonate, salinity, and alkalinity. Other factors such as drainage, depth, salinity, and slope never influence the suitability of the area. Slope, soil depth, soil texture, and drainage were never taken as limiting factors for almost the entire study.

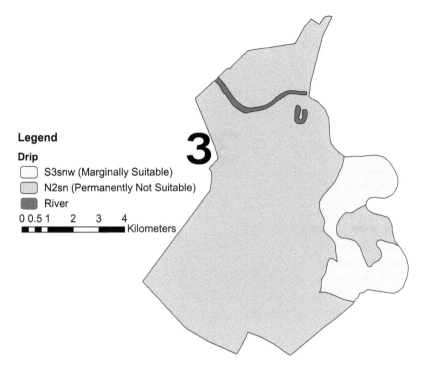

FIGURE 12.4 Land suitability map for drip irrigation.

TABLE 12.3

The Most Suitable Land Units for Surface, Sprinkle, and Drip Irrigation Systems by Notation to Capability Index (Ci) for Different Irrigation Systems

Codes of Land Units	The Maximum Capability Index for Irrigation(Ci)	Suitability Classes	The Most Suitable Irrigation Systems	Limiting Factors
1	18	$N_2 n$	Sprinkle	Salinity & Alkalinity
2	17.1	$N_2 n$	Sprinkle	Salinity & Alkalinity
3	50.62	$S_3 nw$	Sprinkle	Salinity & Alkalinity and Drainage
4	18	$N_2 n$	Sprinkle	Salinity & Alkalinity
5	28.35	$N_2 n$	Sprinkle	Salinity & Alkalinity

The comparison of the capability indices for surface, sprinkle, and drip irrigation (Tables 12.1 and 12.3) indicated that in soil series coded 1, 2, 3, 4, and 5, applying sprinkle irrigation systems was more suitable than surface and drip irrigation systems. Figure 12.5 shows the most suitable map for surface, sprinkle, and drip irrigation systems in the Hamidih Plain as per the capability index (Ci) for different irrigation systems. As seen from this map, all parts of this plain were suitable for sprinkle irrigation systems.

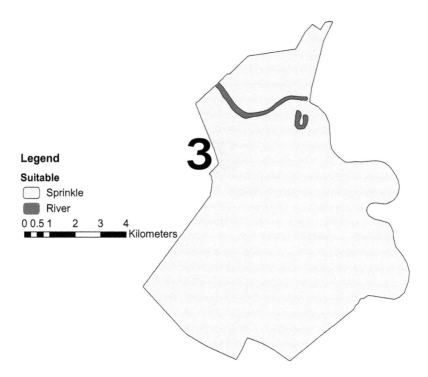

FIGURE 12.5   The most suitable map for different irrigation systems.

12.3   CONCLUSIONS

The results of Tables 12.1 and 12.3 indicated that by applying sprinkle irrigation instead of surface and drip irrigation methods, the land suitability of 6,672 ha (98.52%) of the Hamidih Plain's
land could be improved substantially. The comparison of the different types of irrigation revealed
that sprinkle irrigation was more effective and efficient than the drip and surface irrigation methods, and it improved the land suitability for irrigation purposes. Research carried out by Landi
et al. (2008); Rezania et al. (2009); Naseri et al. (2009b); Boroomand Nasab et al. (2010); Albaji
et al. (2012a); Jovzi et al. (2012); Albaji et al. (2008, 2014a, 2014b, 2014c, 2015, 2016) confirm
these results.

Moreover, the main limiting factors in using surface and sprinkle irrigation methods in this area
were drainage, salinity, and alkalinity, and the main limiting factors in using drip irrigation methods were the calcium carbonate, salinity, and alkalinity.

ACKNOWLEDGMENTS

We are grateful to the Research Council of Shahid Chamran University of Ahvaz for financial support (GN: SCU.WI98.280).

REFERENCES

Abyaneh, H.Z., Jovzi, M., Albaji, M. 2017. Effect of regulated deficit irrigation, partial root drying and N-fertilizer levels on sugar beet crop (*Beta vulgaris* L.). *Agricultural Water Management* 194, 13–23.

Albaji, M., Boroomand Nasab, S., Kashkuli, H.A., Naseri, A.A., Sayyad, G., Jafari, S. 2008. Comparison of different irrigation methods based on the parametric evaluation approach in North Molasani Plain, Iran. *Journal of Agronomy* 7 (2), 187–191.

Albaji, M., Boroomand Nasab, S., Hemadi, J. 2012a. Comparison of different irrigation methods based on the parametric evaluation approach in West North Ahvaz Plain. In: *Problems, Perspectives and Challenges of Agricultural Water Management*. M. Kumar, (ed.), InTech, Croatia, pp. 259–274.

Albaji, M., Papan, P., Hosseinzadeh, M., Barani, S. 2012b. Evaluation of land suitability for principal crops in the Hendijan region. *International Journal of Modern Agriculture* 1 (1), 24–32.

Albaji, M., Golabi, M., Boroomand Nasab, S., Jahanshahi. M. 2014a. Land suitability evaluation for surface, sprinkler and drip irrigation systems. *Transactions of the Royal Society of South Africa* 69 (2), 63–73.

Albaji, M., Golabi, M., Piroozfar, V.R., Egdernejad, A., Nazari Zadeh, F. 2014b. Evaluation of agricultural land resources for irrigation in the Ramhormoz Plain by using GIS. *Agriculturae Conspectus Scientificus* 79 (2), 93–102.

Albaji, M., Golabi, M., Egdernejad, A., Nazarizadeh, F. 2014c. Assessment of different irrigation systems in Albaji Plain. *Water Science and Technology: Water Supply* 14 (5), 778–786.

Albaji, M., Boroomand Nasab, S., Golabi, M., Sorkheh Nezhad, M., Ahmadee, M. 2015. Application possibilities of different irrigation methods in Hofel Plain. *YYÜ TAR BİL DERG (YYU J AGR SCI)* 25 (1), 13–23.

Albaji, M., Golabi, M., Hooshmand, A.R., Ahmadee, M. 2016. Investigation of surface, sprinkler and drip irrigation methods using GIS. *Jordan Journal of Agricultural Sciences* 12 (1), 211–222.

Albaji, M., Alboshokeh, A. 2017. Assessing agricultural land suitability in the Fakkeh region, Iran. *Outlook on Agriculture* 46 (1), 57–65.

Behzad, M., Albaji, M., Papan, P., Boroomand Nasab, S. 2009. Evan region qualitative soil evaluation for wheat, barley, alfalfa and maize. *Journal of Food, Agriculture & Environment* 7 (2), 843–851.

Boroomand Nasab, S., Albaji, M., Naseri, A.A. 2010. Investigation of different irrigation systems based on the parametric evaluation approach in Boneh Basht plain, Iran. *African Journal of Agricultural Research* 5 (5), 372–379.

Chegah, S., Chehrazi, M., Albaji, M. 2013. Effects of drought stress on growth and development frankenia plant (*Frankenia leavis*). *Bulgarian Journal of Agricultural Science* 19 (4), 659–666.

Jovzi, M., Albaji, M., Gharibzadeh, A. 2012. Investigating the suitability of lands for surface and under-pressure (drip and sprinkler) irrigation in Miheh Plain. *Research Journal of Environmental Sciences* 6 (2), 51–61.

Khuzestan Water and Power Authority (KWPA). 2011. Semi-detailed soil study report of Hamidih Plain, Iran (in Persian).

Landi, A., Boroomand-Nasab, S., Behzad, M., Tondrow, M.R., Albaji, M., Jazaieri, A. 2008. Land suitability evaluation for surface, sprinkle and drip irrigation methods in Fakkeh Plain, Iran. *Journal of Applied Sciences* 8 (20), 3646–3653.

Mahjoobi, A., Albaji, M., Torfi, K. 2010. Determination of heavy metal levels of Kondok soills-haftgel. *Research Journal of Environmental Sciences* 4 (3), 294–299.

Naseri, A.A., Albaji, M., Boroomand Nasab, S., Landi, A., Papan, P., Bavi, A. 2009a. Land suitability evaluation for principal crops in the Abbas Plain, Southwest Iran. *Journal of Food, Agriculture & Environment* 7 (1), 208–213.

Naseri, A.A., Rezania, A.R., Albaji, M. 2009b. Investigation of soil quality for different irrigation systems in Lali Plain, Iran. *Journal of Food, Agriculture & Environment* 7 (3&4), 955–960.

Rezania, A.R., Naseri, A.A., Albaji, M. 2009. Assessment of soil properties for irrigation methods in North Andimeshk Plain, Iran. *Journal of Food, Agriculture & Environment* 7 (3&4), 728–733.

13 The Best Irrigation System in Shahid Chamran Plain, Iran

Mohammad Albaji, Saeed Boroomand Nasab,
Saeid Eslamian, and Abd Ali Naseri

CONTENTS

13.1 INTRODUCTION

The scarcity of water in arid and semiarid regions such as Iran is a restrictive element for the agricultural sector; therefore there is an urgent need to develop initiatives to save water in this particular sector (Abyaneh et al., 2017; Chegah et al., 2013). Among the proposed solutions is the use of pressurized irrigation systems. The present study was conducted in an area about 35,500 hectares in the Shahid Chamran Plain, in Khuzestan Province located in the southwest of Iran, during 2011–2012. The study area is located 20 km southwest of the city of Ahvaz, 31° 5′ to 31° 21′ N and 47° 56′ to 48° 23′ E. The average annual temperature and precipitation for the period of 1966–2011 were 25.4°C and 230 mm, respectively. Also, the annual evaporation of the area is 2,350 mm (Khuzestan Water and Power Authority, 2011). The Kharkhe River supplies the bulk of the water demands of the region. The application of irrigated agriculture has been common in the study area. Currently, the irrigation systems used by farmlands in the region are furrow irrigation, basin irrigation, and border irrigation schemes.

The area is composed of two distinct physiographic features, i.e., river alluvial plains, and plateaus, of which the river alluvial plains' physiographic unit is the dominant feature. Also, nine different soil series were found in the area. The semi-detailed soil survey report of the Shahid Chamranplain (Khuzestan Water and Power Authority, 2011) was used in order to determine the soil characteristics.

Over much of the Shahid Chamran Plain, the use of surface irrigation systems has been applied specifically for field crops to meet the water demand of both summer and winter crops.

The major irrigated broad-acre crops grown in this area are wheat, barley, and maize, in addition to fruits, melons, watermelons, and vegetables such as tomatoes and cucumbers. But like other plains in Khuzstan Province, the important major crops are wheat and barley (Behzad et al., 2009; Naseri et al., 2009a; Albaji et al., 2012b; Albaji and Alboshokeh, 2017). There are very few instances of sprinkle and drip irrigation on large-area farms in the Shahid Chamran Plain.

Nine soil series and 41 series phases or land units were derived from the semi-detailed soil study of the area. The land units are shown in Figure 13.1 as the basis for further land evaluation practice. The soils of the area are of Aridisols orders. Also, the soil moisture regime is aridic, while the soil temperature regime is hyperthermic (Khuzestan Water and Power Authority, 2011). Therefore, the organic matter content is low due to soil high temperature and decomposition high ratio of this matter (Mahjoobi et al., 2010).

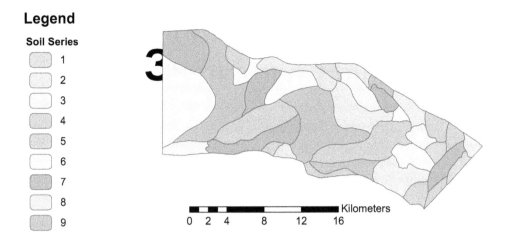

FIGURE 13.1   Soil map of the study area.

13.2   RESULTS AND DISCUSSIONS

As shown in Tables 13.1 and 13.2 for surface irrigation, only soil series coded 6 (6,320 ha – 17.80%) were found to be marginally suitable (S_3). Soil series coded 4, 5, 7, and 8 (16,800 ha – 47.32%) were classified as currently not suitable (N_1), and soil series coded 1, 2, 3, and 9 (12,380 ha – 34.88%) were classified as permanently not suitable (N_2) for any surface irrigation practices.

The analyses of the suitability irrigation map for surface irrigation (Figure 13.2) indicate that there was not highly suitable land and moderately suitable land in this plain. Some portion of the cultivated area in this plain (located in the west) is deemed as being marginally suitable land due to high limitations of calcium carbonate, salinity, and alkalinity. The currently not suitable area is located to the center and east of this area due to severe limitations of calcium carbonate, drainage, salinity, and alkalinity. Other factors such as slope, depth, and soil texture have no influence on the

TABLE 13.1
The Ci Values and Suitability Classes of Surface, Sprinkle, and Drip Irrigation for Each Land Unit

	Surface Irrigation		Sprinkle Irrigation		Drip Irrigation	
Codes of Land Units	Ci	Suitability Classes	Ci	Suitability Classes	Ci	Suitability Classes
1	26.32	N_2 snw[a]	34.2	N_1 sn[b]	33.25	N_1 sn[c]
2	26.32	N_2 snw	34.2	N_1 sn	33.25	N_1 sn
3	14.04	N_2 snw	15.2	N_2 sn	14	N_2 sn
4	35.1	N_1 sn	36	N_1 sn	31.5	N_1 sn
5	37.12	N_1 snw	45.9	S_3 snw	44.62	N_1 sn
6	46.41	S_3 sn	51	S_3 sn	44.62	N_1 sn
7	37.44	N_1 snw	46.8	S_3 snw	45.5	S_3 sn
8	37.12	N_1 snw	45.9	S_3 snw	44.62	N_1 sn
9	27.3	N_2 sn	28	N_2 sn	24.5	N_2 sn

[a,b] The Limiting Factors for Surface and Sprinkle Irrigation: s: (Calcium Carbonate), n: (Salinity & Alkalinity), and
     w: (Drainage).
[c]   The Limiting Factors for Drip Irrigations: s: (Calcium Carbonate) and n: (Salinity & Alkalinity).

TABLE 13.2

Distribution of Surface, Sprinkle, and Drip Irrigation Suitability

Suitability	Surface Irrigation			Sprinkle Irrigation			Drip Irrigation		
	Land Unit	Area (ha)	Ratio (%)	Land Unit	Area (ha)	Ratio (%)	Land Unit	Area (ha)	Ratio (%)
S_1	–	–	–	–	–	–	–	–	–
S_2	–	–	–	–	–	–	–	–	–
S_3	6	6,320	17.80	5, 6, 7, 8	22,345	62.94	7	5,500	15.49
N_1	4, 5, 7, 8	16,800	47.32	1, 2, 4	3,220	9.07	1, 2, 4, 5, 6, 8	20,065	56.52
N_2	1, 2, 3, 9	12,380	34.88	3, 9	9,935	27.99	3, 9	9,935	27.99
Total		35,500	100		35,500	100		35,500	100

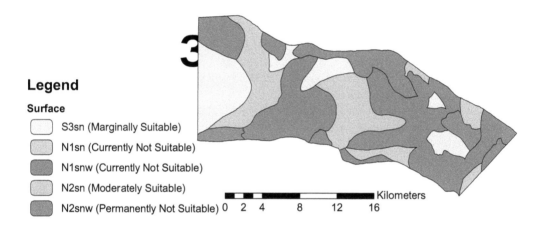

Legend

Surface

- S3sn (Marginally Suitable)
- N1sn (Currently Not Suitable)
- N1snw (Currently Not Suitable)
- N2sn (Moderately Suitable)
- N2snw (Permanently Not Suitable)

Kilometers
0 2 4   8   12   16

FIGURE 13.2 Land suitability map for surface irrigation.

suitability of the area whatsoever. The permanently not suitable land can be observed in the north and south of the plain because of very severe limitations of calcium carbonate, drainage, salinity, and alkalinity. For almost the total study area, elements such as soil depth, slope, and soil texture were not considered as limiting factors.

In order to verify the possible effects of different management practices, the land suitability for sprinkle and drip irrigation was evaluated (Tables 13.1 and 13.2).

For sprinkle irrigation, soil series coded 5, 6, 7, and 8 (22,345 ha – 62.94%) were found to be marginally suitable (S_3). Soil series coded 1, 2, and 4 (3,220 ha – 9.07%) was classified as currently not suitable (N_1), and soil series coded 3 and 9 (9,935 ha – 27.99%) were classified as permanently not suitable (N_2) for sprinkle irrigation.

Regarding sprinkler irrigation (Figure 13.3), there was not highly suitable land and moderately suitable land in this plain. The marginally suitable area can be observed in the largest part of the cultivated zone in this plain (located in the west, center, and the east) due to high limitations of calcium carbonate, drainage, salinity, and alkalinity. As seen from the map, some part of this plain was evaluated as currently not suitable land for sprinkle irrigation because of severe limitations of calcium carbonate, salinity, and alkalinity. Other factors such as drainage, depth, and slope never influence the suitability of the area. The permanently not suitable lands are located in the center of the plain, and their non-suitability of the land is due to very severe limitations of calcium carbonate, salinity, and alkalinity. Slope, soil depth, and soil texture were never taken as limiting factors for almost the entire study area.

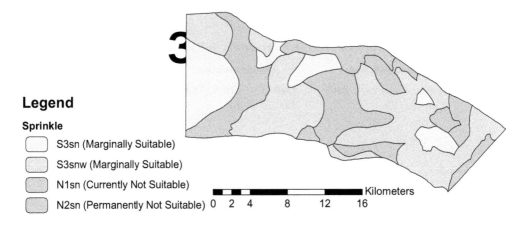

Legend

Sprinkle

☐ S3sn (Marginally Suitable)

☐ S3snw (Marginally Suitable)

☐ N1sn (Currently Not Suitable)

☐ N2sn (Permanently Not Suitable)

Kilometers
0   2   4   8   12   16

FIGURE 13.3   Land suitability map for sprinkle irrigation.

For drip irrigation, only soil series coded 7 (5,500 ha – 15.49%) were found to be slightly suitable (S_3). Soil series coded 1, 2, 4, 5, 6, and 8 (20,065 ha – 56.52%) were classified as currently not suitable (N_1), and soil series coded 3 and 9 (9,935 ha – 27.99%) were classified as permanently not suitable (N_2) for drip irrigation.

Regarding drip irrigation (Figure 13.4), there was not highly suitable land and moderately suitable land in this plain. The marginally suitable area can be observed in the some part of the cultivated zone in this plain (located in the west, center, and the east) due to high limitations of calcium carbonate, salinity, and alkalinity. As seen from the map, the largest part of this plain was evaluated as currently not suitable land for drip irrigation because of severe limitations of calcium carbonate, salinity, and alkalinity. Other factors such as drainage, depth, and slope never influence the suitability of the area. The permanently not suitable lands are located in the center and west of the plain, and their non-suitability of the land is due to very severe limitations of calcium carbonate, salinity, and alkalinity. Slope, soil depth, and soil texture were never taken as limiting factors for almost the entire study area.

The mean capability index (Ci) for surface irrigation was 34.54 (currently not suitable), while for sprinkle irrigation it was 40.20 (currently not suitable). Moreover, for drip irrigation it was 37.47 (currently not suitable). The comparison of the capability indices for surface, sprinkle, and

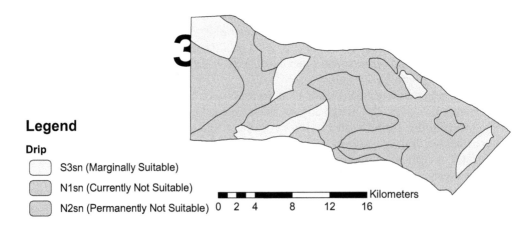

Legend

Drip

☐ S3sn (Marginally Suitable)

☐ N1sn (Currently Not Suitable)

☐ N2sn (Permanently Not Suitable)

Kilometers
0   2   4   8   12   16

FIGURE 13.4   Land suitability map for drip irrigation.

TABLE 13.3

The Most Suitable Land Units for Surface, Sprinkle, and Drip Irrigation Systems by Notation to Capability Index (Ci) for Different Irrigation Systems

Codes of Land Units	The Maximum Capability Index for Irrigation (Ci)	Suitability Classes	The Most Suitable Irrigation Systems	Limiting Factors
1	34.2	N_1 sn	Sprinkle	$CaCO_3$ and Salinity & Alkalinity
2	34.2	N_1 sn	Sprinkle	$CaCO_3$ and Salinity & Alkalinity
3	15.2	N_2 sn	Sprinkle	$CaCO_3$ and Salinity & Alkalinity
4	36	N_1 sn	Sprinkle	$CaCO_3$ and Salinity & Alkalinity
5	45.9	S_3 snw	Sprinkle	$CaCO_3$, Salinity & Alkalinity and Drainage
6	51	S_3 sn	Sprinkle	$CaCO_3$ and Salinity & Alkalinity
7	46.8	S_3 snw	Sprinkle	$CaCO_3$, Salinity & Alkalinity and Drainage
8	45.9	S_3 snw	Sprinkle	$CaCO_3$, Salinity & Alkalinity and Drainage
9	28	N_2 sn	Sprinkle	$CaCO_3$ and Salinity & Alkalinity

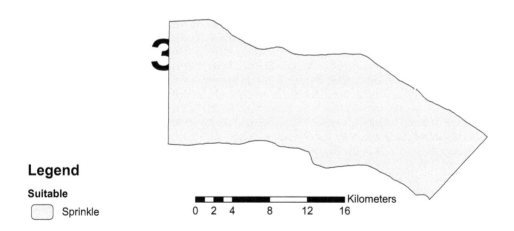

Legend

Suitable

Sprinkle

0 2 4   8   12   16 Kilometers

FIGURE 13.5   The most suitable map for different irrigation systems.

drip irrigation (Tables 13.1 and 13.3) indicated that in soil series coded 1, 2, 3, 4, 5, 6, 7, 8, and 9, applying sprinkle irrigation systems was more suitable than surface and drip irrigation systems. Figure 13.5 shows the most suitable map for surface, sprinkle, and drip irrigation systems in the Shahid Chamran Plain as per the Ci for different irrigation systems. As seen from this map, all of this plain was suitable for sprinkle irrigation systems.

13.3 CONCLUSIONS

The results of Tables 13.1 and 13.3 indicated that by applying sprinkle irrigation instead of surface and drip irrigation methods, the land suitability of 35,500 ha (100%) of the Shahid Chamran Plain's land could be improved substantially. The comparison of the different types of irrigation revealed that sprinkle irrigation was more effective and efficient than the drip and surface irrigation methods, and it improved the land suitability for irrigation purposes. Landi et al. (2008); Rezania et al. (2009); Naseri et al. (2009b); Albaji (2010); Boroomand Nasab et al. (2010); Albaji and Hemadi (2011); Albaji et al. (2008, 2009, 2012a, 2014a, 2014b, 2014c, 2015 and 2016) found similar results.

Moreover, the main limiting factor in using surface and sprinkle irrigation methods in this area were calcium carbonate, drainage, salinity, and alkalinity, and the main limiting factors in using drip irrigation methods were calcium carbonate, salinity, and alkalinity.

ACKNOWLEDGMENTS

We are grateful to the Research Council of Shahid Chamran University of Ahvaz for financial support (GN: SCU.WI98.280).

REFERENCES

Abyaneh, H.Z., Jovzi, M., Albaji, M. 2017. Effect of regulated deficit irrigation, partial root drying and N-fertilizer levels on sugar beet crop (*Beta vulgaris* L.). *Agricultural Water Management* 194, 13–23.

Albaji, M., Boroomand Nasab, S., Kashkuli, H.A., Naseri, A.A., Sayyad, G., Jafari, S. 2008. Comparison of different irrigation methods based on the parametric evaluation approach in North Molasani Plain, Iran. *Journal of Agronomy* 7 (2), 187–191.

Albaji, M,. Boroomand-Nasab, S., Naseri, A.A., Jafari, S. 2009. Comparison of different irrigation methods based on the parametric evaluation approach in Abbas Plain: Iran. *Journal of Irrigation and Drainage Engineering* 136 (2), 131–136.

Albaji, M., 2010. *Land Suitability Evaluation for Sprinkler Irrigation Systems.* Khuzestan Water and Power Authority (KWPA), Ahvaz, Iran (in Persian).

Albaji, M., Hemadi, J. 2011. Investigation of different irrigation systems based on the parametric evaluation approach on the Dasht Bozorg Plain. *Transactions of the Royal Society of South Africa* 66 (3), 163–169.

Albaji, M., Boroomand Nasab, S., Hemadi, J. 2012a. Comparison of different irrigation methods based on the parametric evaluation approach in West North Ahvaz Plain. In: *Problems, Perspectives and Challenges of Agricultural Water Management.* M. Kumar, (ed.), InTech, Croatia, pp. 259–274.

Albaji, M., Papan, P., Hosseinzadeh, M., Barani, S. 2012b. Evaluation of land suitability for principal crops in the Hendijan region. *International Journal of Modern Agriculture* 1 (1), 24–32.

Albaji, M., Golabi, M., Boroomand Nasab, S., Jahanshahi. M. 2014a. Land suitability evaluation for surface, sprinkler and drip irrigation systems. *Transactions of the Royal Society of South Africa* 69 (2), 63–73.

Albaji, M., Golabi, M., Piroozfar, V.R., Egdernejad, A., Nazari Zadeh, F. 2014b. Evaluation of agricultural land resources for irrigation in the Ramhormoz Plain by using GIS. *Agriculturae Conspectus Scientificus* 79 (2), 93–102.

Albaji, M., Golabi, M., Egdernejad, A., Nazarizadeh, F. 2014c. Assessment of different irrigation systems in Albaji Plain. *Water Science and Technology: Water Supply* 14 (5), 778–786.

Albaji, M., Boroomand Nasab, S., Golabi, M., Sorkheh Nezhad, M., Ahmadee, M. 2015. Application possibilities of different irrigation methods in Hofel Plain. *YYÜ TAR BİL DERG (YYU J AGR SCI)* 25 (1), 13–23.

Albaji, M., Golabi, M., Hooshmand, A.R., Ahmadee, M. 2016. Investigation of surface, sprinkler and drip irrigation methods using GIS. *Jordan Journal of Agricultural Sciences* 12 (1), 211–222.

Albaji, M., Alboshokeh, A. 2017. Assessing agricultural land suitability in the Fakkeh region, Iran. *Outlook on Agriculture* 46 (1), 57–65.

Behzad, M., Albaji, M., Papan, P., Boroomand Nasab, S. 2009. Evan region qualitative soil evaluation for wheat, barley, alfalfa and maize. *Journal of Food, Agriculture & Environment* 7 (2), 843–851.

Boroomand Nasab, S., Albaji, M., Naseri, A.A. 2010. Investigation of different irrigation systems based on the parametric evaluation approach in Boneh Basht plain, Iran. *African Journal of Agricultural Research* 5 (5), 372–379.

Chegah, S., Chehrazi, M., Albaji, M. 2013. Effects of drought stress on growth and development frankenia plant (*Frankenia Leavis*). *Bulgarian Journal of Agricultural Science* 19 (4), 659–666.

Khuzestan Water and Power Authority (KWPA). 2011. Semi-detailed soil study report of Shahid chamran Plain, Iran (in Persian).

Landi, A., Boroomand-Nasab, S., Behzad, M., Tondrow, M.R., Albaji, M., Jazaieri, A. 2008. Land suitability evaluation for surface, sprinkle and drip irrigation methods in Fakkeh Plain, Iran. *Journal of Applied Sciences* 8 (20), 3646–3653.

Mahjoobi, A., Albaji, M., Torfi, K. 2010. Determination of heavy metal levels of Kondok soills-haftgel. *Research Journal of Environmental Sciences* 4 (3), 294–299.

Naseri, A.A., Albaji, M., Boroomand Nasab, S., Landi, A., Papan, P., Bavi, A. 2009a. Land suitability evaluation for principal crops in the Abbas Plain, Southwest Iran. *Journal of Food, Agriculture & Environment* 7 (1), 208–213.

Naseri, A.A., Rezania, A.R., Albaji, M. 2009b. Investigation of soil quality for different irrigation systems in Lali Plain, Iran. *Journal of Food, Agriculture & Environment* 7 (3&4), 955–960.

Rezania, A.R., Naseri, A.A., Albaji, M. 2009. Assessment of soil properties for irrigation methods in North Andimeshk Plain, Iran. *Journal of Food, Agriculture & Environment* 7 (3&4), 728–733.

14 The Best Irrigation System in Ezeh and Baghmalek Plain, Iran

Mohammad Albaji, Mona Golabi, Abd Ali Naseri, and Saeid Eslamian

CONTENTS

14.1 INTRODUCTION

The scarcity of water in arid and semiarid regions such as Iran is a restrictive element for the agricultural sector; therefore, there is an urgent need to develop initiatives to save water in this particular sector (Abyaneh et al., 2017; Chegah et al., 2013). Among the proposed solutions is the use of pressurized irrigation systems. The present study was conducted in an area about 25,000 hectares in the Ezeh and Baghmalek Plain in Khuzestan Province, southwest of Iran (Figure 14.1), during 2010–2012. The study area is located 5 km southwest of the city of Zeh, 31° 10′ to 32° 21′ N and 49° 31′ to 50° 34′ E. The average annual temperature and precipitation for the period of 1959–2011 were 24°C and 655 mm, respectively. Also, the annual pan evaporation of the area is 1,685 mm (Khuzestan Water and Power Authority, 2012). The Karun River supplies the bulk of the water demands of the region. The application of irrigated agriculture has been common in the study area. Currently, the irrigation systems used by farmlands in the region are furrow irrigation, basin irrigation, and border irrigation schemes.

The area is composed of two distinct physiographic features, i.e., river alluvial plains and terraces, of which the river alluvial plains' physiographic unit is the dominant feature. Also, five different soil series were found in the area. The semi-detailed soil survey report of the Ezeh and Baghmalek Plain (Khuzestan Water and Power Authority, 2012) was used in order to determine the soil characteristics.

Over much of the Ezeh and Baghmalek Plain, the use of surface irrigation systems has been applied specifically for field crops to meet the water demand of both summer and winter crops. The major irrigated broad-acre crops grown in this area are wheat, barley, and maize, in addition to fruits, melons, watermelons, and vegetables such as tomatoes and cucumbers. But like other plains in Khuzstan Province, the important major crops are wheat, barley, and maize (Behzad et al., 2009; Naseri et al., 2009; Albaji et al., 2012a; Albaji and Alboshokeh, 2017). There are very few instances of sprinkler and drip irrigation on large-area farms in the Ezeh and Baghmalek Plain.

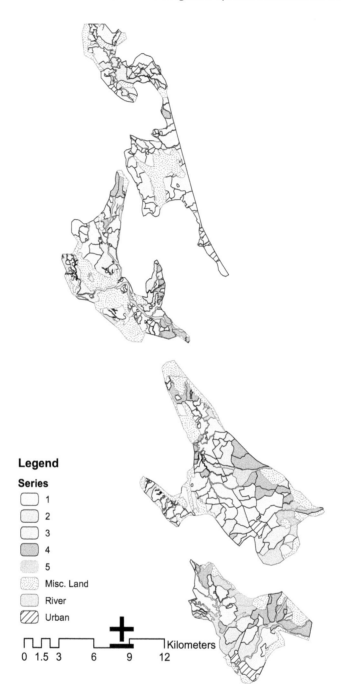

FIGURE 14.1 Soil map of the study area.

Five soil series and 16 series phases were derived from the semi-detailed soil study of the area. The soil series are shown in Figure 14.1 as the basis for further land evaluation practice. The soils of the area are of Mollisols, Inceptisols, and Entisols orders. Also, the soil moisture regime is ustic, while the soil temperature regime is hyperthermic (Khuzestan Water and Power Authority, 2012). Therefore, the organic matter content is low due to high soil temperature and high decomposition ratio of this matter (Mahjoobi et al., 2010).

14.2   RESULTS AND DISCUSSIONS

As shown in Tables 14.1 and 14.2 and Figure 14.2, for surface irrigation, the soil series coded 1, 2, and 5 (5952.4 ha – 23.8%) were moderately suitable (S_2); soil series coded 3 (4388.6 ha – 17.6%) were classified as marginally suitable (S_3); and soil series coded 4 (6116.1 ha – 24.5%) were found to be currently not suitable (N_1) for any surface irrigation practices. The highly suitable lands and permanently not suitable lands for surface irrigation did not exist in this plain. The main limiting factors in using surface irrigation methods in this area were soil depth, stone, gravel, and slope. For almost the total study area, elements such as salinity, alkalinity, drainage, and calcium carbonates content were not considered as limiting factors.

In order to verify the possible effects of different management practices, the land suitability for sprinkler and drip irrigation was evaluated (Tables 14.1 and 14.2).

For sprinkler irrigation (Figure 14.3), soil series coded 1, 2, and 5 (5952.4 ha – 23.8%) were moderately suitable (S_2); soil series coded 3 (4388.6 ha – 17.6%) were classified as marginally suitable (S_3); and soil series coded 4 (6116.1 ha – 24.5%) were found to be currently not suitable (N_1) for sprinkler irrigation. The highly suitable lands and permanently not suitable lands for sprinkler irrigation did not exist in this plain. The main limiting factors in using sprinkler irrigation methods

TABLE 14.1
Ci Values and Suitability Classes of Surface, Sprinkler, and Drip Irrigation for Each Soil Series

Codes of Soil Series	Surface Irrigation		Sprinkler Irrigation		Drip Irrigation	
	Ci	Suitability Classes	Ci	Suitability Classes	Ci	Suitability Classes
1	63	S_2 st[a]	64	S_2 st[b]	56	S_3 st[c]
2	65	S_2 st	65	S_2 st	68	S_2 s
3	58	S_3 st	57	S_3 st	63	S_2 s
4	32	N_1 st	33	N_1 st	37	N_1 s
5	65	S_2 st	65	S_2 st	72	S_2 s

[a] Limiting Factors for Surface Irrigation: t (Slope), s: (Soil depth, Stony Soil Texture, and Gravel Soil Texture).
[b] Limiting Factors for Sprinkler Irrigation: t (Slope), s: (Calcium Carbonate, Stony Soil Texture, and Gravel Soil Texture).
[c] Limiting Factors for Drip Irrigation: s (Calcium Carbonate, Stony Soil Texture, and Gravel Soil Texture).

TABLE 14.2
Distribution of Surface, Sprinkler, and Drip Irrigation Suitability

Suitability	Surface Irrigation			Sprinkler Irrigation			Drip Irrigation		
	Soil Series	Area (ha)	Ratio (%)	Soil Series	Area (ha)	Ratio (%)	Soil Series	Area (ha)	Ratio (%)
S_2	1,2,5	5952.4	23.8	1,2,5	5952.4	23.8	2,3,5	8372.3	33.5
S_3	3	4388.6	17.6	3	4388.6	17.6	1	1968.7	7.9
N_1	4	6116.1	24.5	4	6116.1	24.5	4	6116.1	24.5
[a]Mis Land		8542.9	34.2		8542.9	34.2		8542.9	34.2
Total		25000	100		25000	100		25000	100

[a] Miscellaneous Land: (Hill, Riverbed, Rock, Bad Lands, and Urban).

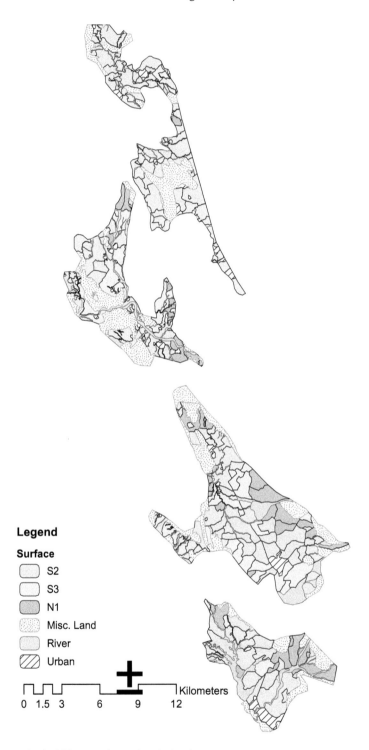

FIGURE 14.2   Land suitability map for surface irrigation.

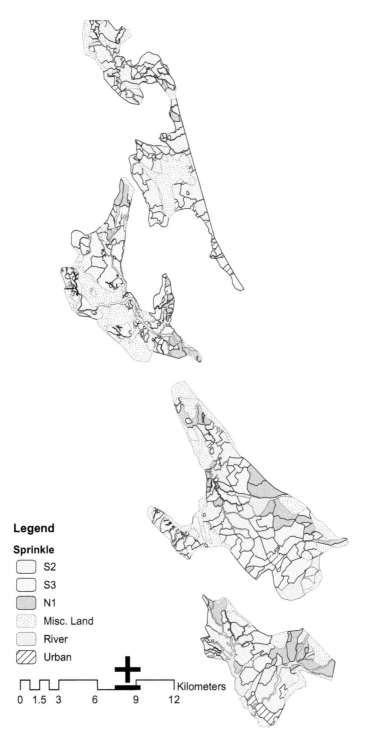

FIGURE 14.3 Land suitability map for sprinkle irrigation.

in this area were calcium carbonates, stone, gravel, and slope. For almost the total study area, elements such as salinity, alkalinity, drainage, and soil depth were not considered as limiting factors.

For drip irrigation (Figure 14.4), soil series coded 2, 3, and 5 (8372.3 ha – 33.5%) were moderately suitable (S_2); soil series coded 1 ((Figure 14.5) ha – 7.9%) were classified as marginally suitable (S_3); and soil series coded 4 (6116.1 ha – 24.5%) were found to be currently not suitable (N_1) for drip irrigation. The highly suitable lands and permanently not suitable lands for drip irrigation did not exist in this plain. Research carried out by Landi et al. (2008); Rezania et al. (2009); Boroomand Nasab et al. (2010); Albaji et al. (2008, 2012a); and Jovzi et al. (2012) confirm these results.

The main limiting factors in using drip irrigation methods in this area were calcium carbonates, stone, and gravel. For almost the total study area, elements such as salinity, alkalinity, drainage, soil depth, and slope were not considered as limiting factors.

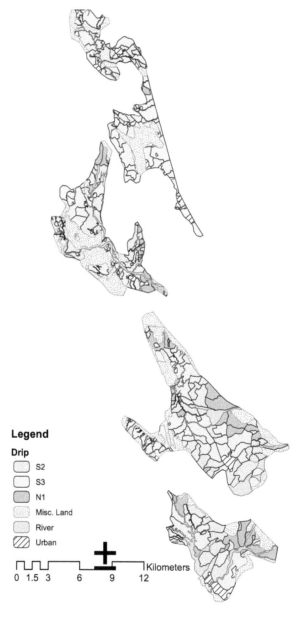

Legend

Drip
- S2
- S3
- N1
- Misc. Land
- River
- Urban

0 1.5 3   6   9   12 Kilometers

FIGURE 14.4   Land suitability map for drip irrigation.

14.3 CONCLUSIONS

Figure 14.5 shows the most suitable map for surface, sprinkler, and drip irrigation systems in the Ezeh and Baghmalek Plain as per the capability index (Ci) for different irrigation systems (Table 14.3). The comparison of the different types of irrigation revealed that drip irrigation was more effective and efficient than the sprinkler and surface irrigation methods, and it improved land suitability for

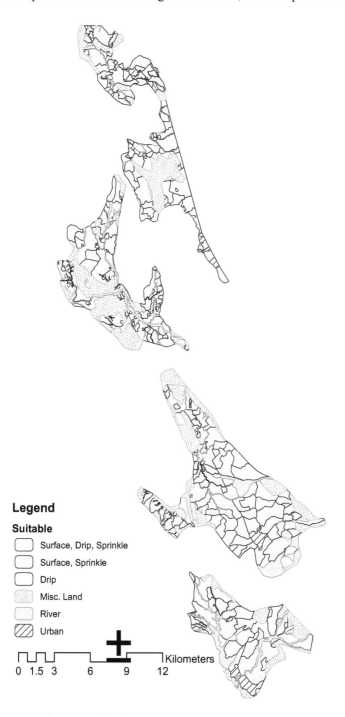

Legend

Suitable

☐ Surface, Drip, Sprinkle
☐ Surface, Sprinkle
☐ Drip
▨ Misc. Land
☐ River
▨ Urban

0 1.5 3   6   9   12   Kilometers

FIGURE 14.5   The most suitable map for different irrigation systems.

TABLE 14.3

The Most Suitable Soil Series for Surface, Sprinkler, and Drip Irrigation Systems by Notation to Capability Index (Ci) for Different Irrigation Systems

Codes of Soil Series	The Maximum Capability Indexfor Irrigation (Ci)	Suitability Classes	The Most Suitable Irrigation Systems	Limiting Factors
1	64	S_2 st	Sprinkler and surface	Slope and $CaCo_3$
2	68	S_2 st	drip, sprinkler and surface	Slope, calcium carbonate, stony soil texture, and gravel soil texture
3	63	S_2 s	Drip	Calcium carbonate, stony soil texture, and gravel soil texture
4	37	N_1 st	Drip, sprinkler and surface	Soil depth, slope, calcium carbonate, stony soil texture, and gravel soil texture
5	72	S_2 s	Drip	Calcium carbonate, stony soil texture, and gravel soil texture

irrigation purposes. The main limiting factors in using sprinkler and surface irrigation methods in this area were stone, gravel, and slope, and the main limiting factors in using drip irrigation methods were calcium carbonates, stone, and gravel.

ACKNOWLEDGMENTS

We are grateful to the Research Council of Shahid Chamran University of Ahvaz for financial support (GN: SCU.WI98.280).

REFERENCES

Abyaneh, H.Z., Jovzi, M., Albaji, M. 2017. Effect of regulated deficit irrigation, partial root drying and N-fertilizer levels on sugar beet crop (*Beta vulgaris* L.). *Agricultural Water Management* 194, 13–23.

Albaji, M., Boroomand Nasab, S., Kashkuli, H.A., Naseri, A.A., Sayyad, G., Jafari, S. 2008. Comparison of different irrigation methods based on the parametric evaluation approach in North Molasani Plain, Iran. *Journal of Agronomy* 7 (2), 187–191.

Albaji, M., Boroomand Nasab, S., Hemadi, J. 2012a. Comparison of different irrigation methods based on the parametric evaluation approach in West North Ahvaz Plain. In: *Problems, Perspectives and Challenges of Agricultural Water Management*. M. Kumar, (ed.), InTech, Croatia, pp. 259–274.

Albaji, M., Papan, P., Hosseinzadeh, M., Barani, S. 2012b. Evaluation of land suitability for principal crops in the Hendijan region. *International Journal of Modern Agriculture* 1 (1), 24–32.

Albaji, M., Alboshokeh, A. 2017. Assessing agricultural land suitability in the Fakkeh region, Iran. *Outlook on Agriculture* 46 (1), 57–65.

Behzad, M., Albaji, M., Papan, P., Boroomand Nasab, S. 2009. Evan region qualitative soil evaluation for wheat, barley, alfalfa and maize. *Journal of Food, Agriculture & Environment* 7 (2), 843–851.

Boroomand Nasab, S., Albaji, M., Naseri, A.A. 2010. Investigation of different irrigation systems based on the parametric evaluation approach in Boneh Basht plain, Iran. *African Journal of Agricultural Research* 5 (5), 372–379.

Chegah, S., Chehrazi, M., Albaji, M. 2013. Effects of drought stress on growth and development frankenia plant (*Frankenia Leavis*). *Bulgarian Journal of Agricultural Science* 19 (4), 659–666.

Jovzi, M., Albaji, M., Gharibzadeh, A. 2012. Investigating the suitability of lands for surface and under-pressure (drip and sprinkler) irrigation in Miheh Plain. *Research Journal of Environmental Sciences* 6 (2), 51–61.

Khuzestan Water and Power Authority (KWPA). 2012. Semi-detailed soil study report of Ezeh & Baghmalek Plains, Iran (in Persian).

Landi, A., Boroomand-Nasab, S., Behzad, M., Tondrow, M.R., Albaji, M., Jazaieri, A. 2008. Land suitability evaluation for surface, sprinkle and drip irrigation methods in Fakkeh Plain, Iran. *Journal of Applied Sciences* 8 (20), 3646–3653.

Mahjoobi, A., Albaji, M., Torfi, K. 2010. Determination of heavy metal levels of Kondok soills-haftgel. *Research Journal of Environmental Sciences* 4 (3), 294–299.

Naseri, A.A., Albaji, M., Boroomand Nasab, S., Landi, A., Papan, P., Bavi, A. 2009. Land suitability evaluation for principal crops in the Abbas Plain, Southwest Iran. *Journal of Food, Agriculture & Environment* 7 (1), 208–213.

Rezania, A.R., Naseri, A.A., Albaji, M. 2009. Assessment of soil properties for irrigation methods in North Andimeshk Plain, Iran. *Journal of Food, Agriculture & Environment* 7 (3&4), 728–733.

15 The Best Irrigation System in Dezful Plain, Iran

Mohammad Albaji, Abd Ali Naseri, and Saeid Eslamian

CONTENTS

15.1 INTRODUCTION

The scarcity of water in arid and semiarid regions such as Iran is a restrictive element for the agricultural sector; therefore, there is an urgent need to develop initiatives to save water in this particular sector (Abyaneh et al., 2017; Chegah et al., 2013). Among the proposed solutions is the use of pressurized irrigation systems. The present study was conducted in an area about 260,510 hectares in the Dezful Plain in Khuzestan Province, located in the southwest of Iran, during 2010–2011. The study area is located 90 km north of the city of Ahvaz, 31° 50′ to 32° 40′ N and 48° 10′ to 48° 35′ E. The average annual temperature and precipitation for the period of 1961–2011 were 24.6°C and 404 mm, respectively. Also, the annual evaporation of the area is 2,895 mm (Table 15.1) (Khuzestan Water and Power Authority, 2010). The Dez River supplies the bulk of the water demands of the region. The application of irrigated agriculture has been common in the study area. Currently, the irrigation systems used by farmlands in the region are furrow irrigation, basin irrigation, and border irrigation schemes.

The area is composed of three distinct physiographic features, i.e., river alluvial plains, piedmont alluvial plains, and colluvial fans, of which the piedmont alluvial plains' physiographic unit is the dominant feature. Also, 18 different soil series were found in the area (Table 15.2). The semidetailed soil survey report of the Dezful Plain (Khuzestan Water and Power Authority, 2010) was used in order to determine the soil characteristics. Table 15.3 shows some of the physicochemical characteristics for reference profiles of different soil series in the plain.

In the Dezful Plain, farmers are becoming increasingly aware of irrigation as a tool for optimizing production. When all other management practices are carried out efficiently, irrigation can help the farmers achieve the top yields and quality demanded for food self-sufficiency and even to the market. In the study area, irrigation is practiced from many water sources: surface water such as the Dez River, water harvesting, and digging wells from the groundwater. During the field work, a good observation in the Dezful Plain is that there is soil and water conservation practiced on the hillsides that enhance the increment of the water table level at the foot slope field, encouraging farmers to dig a well for irrigation practice. In the Dezful Plain, farmers are becoming increasingly aware of irrigation as a tool for optimizing production. When all other management practices are carried out efficiently, irrigation can help the farmers achieve the top yields and quality demanded for self food security and even to the market. Over much of the Dezful Plain, the use of surface irrigation systems has been applied specifically for field crops to meet the water demand of both summer and

TABLE 15.1
Mean Air Temperature, Relative Humidity, and Total Monthly Rainfall and Evaporation (1961–2011) at Dezful

Parameter	Jan	Feb	Mar	Apr	May	Jun	Jul	Aug	Sep	Oct	Nov	Dec	Average
Temperature (°C)	12.1	13.5	18.2	23.1	29.6	34.6	36.6	36.2	31.8	27.3	18.7	13.7	24.6
Relative humidity (%)	72	66.2	53.4	42.1	28.3	22.2	23.3	25.1	25.9	34.9	52.7	68.7	42.9
													Total
Rainfall (mm)	101.8	58.8	61.4	34.4	7.7	0.2	0.1	0	0	9.4	44.8	86.1	404.6
Evaporation (mm)	49.6	65.4	96	179.8	301.3	432.8	453	449.2	374.4	272.4	145.6	75.6	2895.1

TABLE 15.2
Soil Series of the Study Area

Characteristics Description	Series No.
Soil texture "Heavy: [a]SICL" without salinity and alkalinity limitation, depth 150 cm, level to very gently sloping: 0% to 2%, well drained.	1
Soil texture "Heavy: CL," slight salinity and alkalinity limitation, depth 150 cm, level to very gently sloping: 0% to 2%, moderately drained	2
Soil texture "Medium: SIL" without salinity and alkalinity limitation, depth 150 cm, level to very gently sloping: 0% to 2%, moderately drained.	3
Soil texture "Medium: L" without salinity and alkalinity limitation, depth 100 cm, level to very gently sloping: 0% to 2%, well drained.	4
Soil texture "Medium: SL" without salinity and alkalinity limitation, depth 150 cm, gently sloping: 2% to 5%, well drained.	5
Soil texture "Very Heavy: C," moderate salinity and alkalinity limitation, depth 150 cm, level to very gently sloping: 0% to 2%, poorly drained	6
Soil texture "Heavy: CL" without salinity and alkalinity limitation, depth 100 cm, level to very gently sloping: 0% to 2%, imperfectly drained.	7
Soil texture "Heavy: SCL" without salinity and alkalinity limitation, depth 150 cm, level to very gently sloping: 0% to 2%, well drained.	8
Soil texture "Medium: SL" without salinity and alkalinity limitation, depth 150 cm, level to very gently sloping: 0% to 2%, well drained.	9
Soil texture "Heavy: SICL," slight salinity and alkalinity limitation, depth 150 cm, level to very gently sloping: 0% to 2%, poorly drained.	10
Soil texture "Medium: L" without salinity and alkalinity limitation, depth 70 cm, gently sloping: 2% to 5%, well drained.	11
Soil texture "Medium: L," slight salinity and alkalinity limitation, depth 150 cm, level to very gently sloping: 0% to 2%, poorly drained.	12
Soil texture "Heavy: SCL" without salinity and alkalinity limitation, depth 150 cm, level to very gently sloping: 0% to 2%, moderately drained.	13
Soil texture "Medium: SIL," slight salinity and alkalinity limitation, depth 150 cm, gently sloping: 2% to 5%, moderately drained.	14
Soil texture "Heavy: CL" without salinity and alkalinity limitation, depth 150 cm, level to very gently sloping: 0% to 2%, moderately drained.	15
Soil texture "Medium: SL" without salinity and alkalinity limitation, depth 75 cm, level to very gently sloping: 0% to 2%, well drained.	16
Soil texture "Medium: SL" without salinity and alkalinity limitation, depth 150 cm, level to very gently sloping: 0% to 2%, well drained.	17
Soil texture "Heavy: CL," slight salinity and alkalinity limitation, depth 150 cm, level to very gently sloping: 0% to 2%, imperfectly drained.	18
Bad Lands, River Wash, and Urban	[a]Mis Land

[a] Texture symbols: C: Clay, CL: Clay Loam, L: Loam, SL: Sandy Loam, SIL: Silty Loam, SCL: Sandy Clay Loam, SICL: Silty Clay Loam, SIC: Silty Clay.

winter crops. The major irrigated broad-acre crops grown in this area are wheat, barley, alfalfa, and maize, in addition to fruits, melons, watermelons, and vegetables such as tomatoes and cucumbers. But like other plains in Khuzstan Province, the important major crops are wheat, barley, alfalfa, and maize (Behzad et al., 2009; Naseri et al., 2009; Albaji et al., 2012b; Albaji and Alboshokeh, 2017). There are very few instances of sprinkler and drip irrigation on large-area farms in the Dezful Plain.

Eighteen soil series and 98 series phases were derived from the semi-detailed soil study of the area. The soil series are shown in Figure 15.1 as the basis for further land evaluation practice.

TABLE 15.3

Some of Physicochemical Characteristics for Reference Profiles of Different Soil Series

CaCO$_3$ (%)	CEC (meq/100 g)	OM (%)	pH	ECe (ds·m^{-1})	Soil Texture	Depth (cm)	Soil Series Name	Soil Series No.
49	9.21	0.46	8.60	1.45	SICL	150	Andimeshk	1
45	11.24	0.32	8.20	6.23	CL	150	Bonvar	2
40	10.13	0.78	8.80	1.67	SIL	150	Chichali	3
45	12.29	0.41	8.90	1.14	L	100	Damaneh	4
40	11.98	0.39	8.50	0.95	SL	150	Davood	5
48	12.64	0.67	8.70	14.39	C	150	Dehinow	6
42	12.27	0.35	8.50	1.98	CL	100	Dezful	7
41	11.96	0.46	8.10	2.11	SCL	150	Haft Tapeh	8
38	15.87	0.81	8.80	2.37	SL	150	Kaab	9
44	10.26	0.31	8.50	6.91	SICL	150	Mianab	10
42	13.43	0.39	8.20	1.54	L	70	Mukhtar	11
45	12.83	0.47	8.40	7.42	L	150	Rokni	12
54	11.56	0.53	8.50	0.97	SCL	150	Salahabad	13
47	10.27	0.48	8.80	5.87	SIL	150	Samandi	14
46	14.51	0.85	8.50	1.37	CL	150	Shahabad	15
44	11.63	0.52	8.40	2.34	SL	75	Shahoon	16
45	10.74	0.69	8.20	1.05	SL	150	Shalgai	17
35	14.85	0.36	8.60	5.84	CL	150	Shush	18

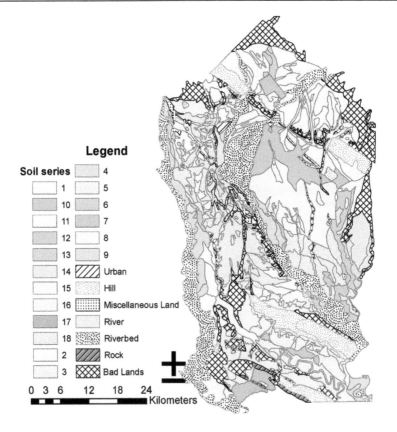

Legend

Soil series 4
1 5
10 6
11 7
12 8
13 9
14 Urban
15 Hill
16 Miscellaneous Land
17 River
18 Riverbed
2 Rock
3 Bad Lands

0 3 6 12 18 24 Kilometers

FIGURE 15.1 Soil map of the study area.

The soils of the area are of Entisols and Inceptisols orders. Also, the soil moisture regime is ustic and aquic, while the soil temperature regime is hyperthermic (Khuzestan Water and Power Authority, 2010b).

15.2 RESULTS AND DISCUSSIONS

As shown in Tables 15.4 and 15.5 for surface irrigation, the soil series coded 1 and 8 (41,531 ha – 15.95%) were highly suitable (S_1); soil series coded 2, 3, 4, 5, 7, 9, 13, 14, 15, 17, and 18 (96,780 ha – 37.14%) were classified as moderately suitable (S_2); and soil series coded 10, 11, and 12 (28,669 ha – 11%) were found to be marginally suitable (S_3). Only soil series coded 6 (2,084 ha – 0.8%) were classified as currently not suitable (N_1), and only soil series coded 16 (440 ha – 0.17%) were classified as permanently not suitable (N_2) for any surface irrigation practices.

The analysis of the suitability irrigation maps for surface irrigation (Figure 15.2) indicates that some portion of the cultivated area in this plain (located to the south and north) is deemed as being highly suitable land due to deep soil, good drainage, texture, salinity, and proper slope of the area. The moderately suitable area is located in the center of this area due to light limitations of slope and gravel soil texture. Other factors such as soil depth, salinity, and alkalinity have no influence on the suitability of the area whatsoever. The map also indicates that some part of the cultivated area in this plain was evaluated as marginally suitable because of the high limitations of drainage, salinity,

TABLE 15.4
Ci Values and Suitability Classes of Surface, Sprinkler, and Drip Irrigation for Each Soil Series

Codes of Soil Series	Surface Irrigation		Sprinkler Irrigation		Drip Irrigation	
	Ci	Suitability Classes	Ci	Suitability Classes	Ci	Suitability Classes
1	87.75	S_1[a]	90	S_1[b]	80	S_1[c]
2	75.02	S_2 sw	81.22	S_1	76	S_{2S}
3	78.97	S_2 sw	85.5	S_1	80	S_1
4	63.18	S_{2S}	68.4	S_{2S}	64	S_{2S}
5	60.75	S_2 ts	76.95	S_{2S}	76	S_{2S}
6	35.80	N_1 snw	42.26	N_1 snw	40.46	N_1 snw
7	69.25	S_2 sw	81.22	S_1	80	S_1
8	83.36	S_1	85.5	S_1	76	S_{2S}
9	65.81	S_{2S}	81	S_1	76	S_{2S}
10	47.38	S_3 snw	55.57	S_3 sw	53.2	S_3 sw
11	58.32	S_3 ts	65.40	S_2 s	64.8	S_2 s
12	48.76	S_3 sw	53.86	S_3 sw	54.72	S_3 sw
13	66.69	S_2 sw	72.2	S_{2S}	66.5	S_{2S}
14	69.25	S_2 tsw	77.16	S_{2S}	76	S_{2S}
15	78.97	S_2 sw	85.5	S_1	80	S_1
16	24.57	N_2 s	26.77	N_2 s	25.2	N_{2S}
17	65.81	S_{2S}	81	S_1	76	S_{2S}
18	75.02	S_2 sw	81.22	S_1	76	S_{2S}

[a] Limiting Factors for Surface Irrigation: t (Slope), s: (Calcium Carbonate, Gravel Soil Texture), n (Salinity and Alkalinity), and w: (Drainage).

[b] Limiting Factors for Sprinkler Irrigation: s: (Calcium Carbonate, Gravel Soil Texture) and w: (Drainage).

[c] Limiting Factors for Drip Irrigation: s: (Calcium Carbonate), n (Salinity and Alkalinity), and w: (Drainage).

TABLE 15.5

Distribution of Surface, Sprinkler, and Drip Irrigation Suitability

	Surface Irrigation			Sprinkler Irrigation			Drip Irrigation		
Suitability	Soil Series	Area (ha)	Ratio (%)	Soil Series	Area (ha)	Ratio (%)	Soil Series	Area (ha)	Ratio (%)
S_1	1, 8	41,531	15.95	1, 2, 3, 7, 8, 9, 15, 17, 18	131,253	50.39	1, 3, 7, 15	44,027	16.9
S_2	2, 3, 4, 5, 7, 9, 13, 14, 15, 17, 18	96,780	37.14	4, 5, 11, 13, 14	26,343	10.1	2, 4, 5, 8, 9, 11, 13, 14, 17, 18	113,569	43.59
S_3	10, 11, 12	28,669	11	10, 12	9,384	3.6	10, 12	9,384	3.6
N_1	6	2,084	0.8	6	2,084	0.8	6	2,084	0.8
N_2	16	440	0.17	16	440	0.17	16	440	0.17
[a]Mis Land		91,006	34.94		91,006	34.94		91,006	34.94
Total		260,510	100		260,510	100		260,510	100

[a] Miscellaneous Land: (Hill, Riverbed, Rock, Bad Lands, and Urban).

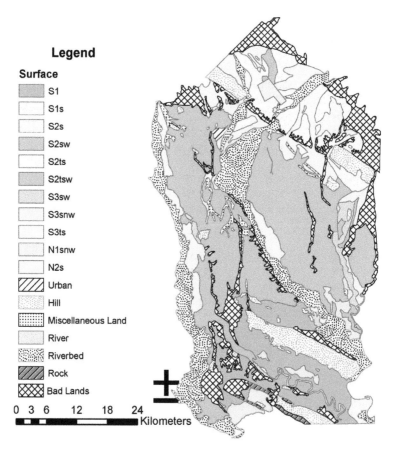

FIGURE 15.2　Land suitability map for surface irrigation.

and alkalinity. The currently not suitable land and permanently not suitable land can be observed only in the southeast and center of the plain because of severe limitations of calcium carbonate, drainage, salinity, and alkalinity. For almost the total study area, elements such as soil depth and calcium carbonate were not considered as limiting factors.

In order to verify the possible effects of different management practices, the land suitability for sprinkler and drip irrigation was evaluated (Tables 15.4 and 15.5).

For sprinkler irrigation, soil series coded 1, 2, 3, 7, 8, 9, 15, 17, and 18 (131,253 ha – 50.39%) were highly suitable (S_1), while soil series coded 4, 5, 11, 13, and 14 (26,343 ha – 10.1%) were classified as moderately suitable (S_2). Further, soil series coded 10 and 12 (9,384 ha – 3.6%) were found to be marginally suitable (S_3). Only soil series coded 6 (2,084 ha – 0.8%) were classified as currently not suitable (N_1), and only soil series coded 16 (440 ha – 0.17%) were classified as permanently not suitable (N_2) for sprinkler irrigation.

Regarding sprinkler irrigation (Figure 15.3), the highly suitable area can be observed in the largest part of the cultivated zone in this plain (located in the center, south, and the north) due to deep soil, good drainage, texture, salinity, and proper slope of the area. As seen from the map, some part of the cultivated area in this plain was evaluated as moderately suitable for sprinkler irrigation because of the light limitation of gravel soil texture. Other factors such as drainage, depth, salinity, and slope never influence the suitability of the area. The marginally suitable lands are located in the south and east of the plain, and their non-suitability of the land is due to the high limitations

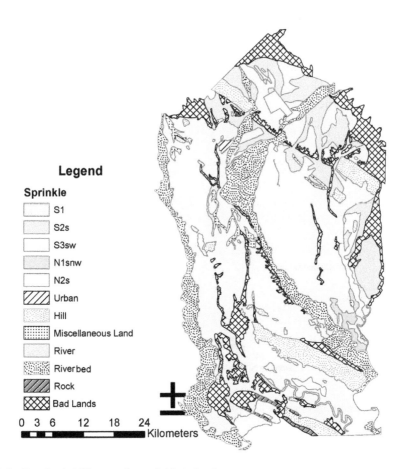

FIGURE 15.3 Land suitability map for sprinkle irrigation.

of drainage and calcium carbonate. The currently not suitable land and permanently not suitable land can be observed only in the southeast and center of the plain because of severe limitations of calcium carbonate, drainage, salinity, and alkalinity. For almost the total study area, elements such as soil depth and calcium carbonate were not considered as limiting factors.

For drip irrigation, soil series coded 1, 3, 7, and 15 (44,027 ha –16.9%) were highly suitable (S_1), while soil series coded 2, 4, 5, 8, 9, 11, 13, 14, 17, and 18 (113,569 ha – 43.59%) were classified as moderately suitable (S_2). Further, soil series coded 10 and 12 (9,384 ha, 3.6%) were found to be slightly suitable (S_3). Only soil series coded 6 (2,084 ha – 0.8%) were classified as currently not suitable (N_1), and only soil series coded 16 (440 ha – 0.17%) were classified as permanently not suitable (N_2) for drip irrigation.

Regarding drip irrigation (Figure 15.4), the highly suitable lands covered some part of the plain (located in the northeast). The slope, soil texture, soil depth, calcium carbonate, salinity, and drainage were in good conditions. The moderately suitable lands could be observed over a large portion of the plain (west, east, center, and south parts) due to the medium content of calcium carbonate. The marginally suitable lands were found in the south and the east of the area. The limiting factors for this soil series were high limitations of drainage and the medium content of calcium

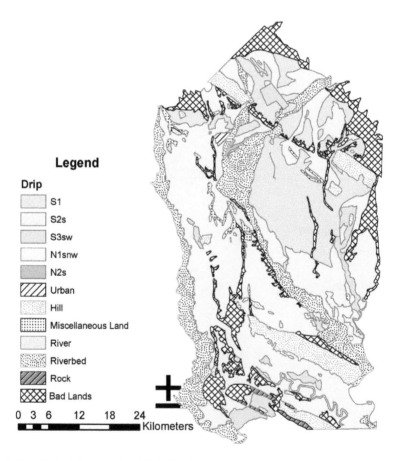

FIGURE 15.4   Land suitability map for drip irrigation.

TABLE 15.6

The Most Suitable Soil Series for Surface, Sprinkler, and Drip Irrigation Systems by Notation to Capability Index (Ci) for Different Irrigation Systems

Codes of Soil Series	The Maximum Capability Index for Irrigation (Ci)	Suitability Classes	The Most Suitable Irrigation Systems	Limiting Factors
1	90	S_1	Sprinkler	None exist
2	81.22	S_1	Sprinkler	None exist
3	85.5	S_1	Sprinkler	None exist
4	68.4	S_{2s}	Sprinkler	Gravel soil texture
5	76.95	S_{2s}	Sprinkler	Gravel soil texture
6	42.26	N_1 snw	Sprinkler	Gravel soil texture, salinity, & alkalinity and drainage
7	81.22	S_1	Sprinkler	None exist
8	85.5	S_1	Sprinkler	None exist
9	81	S_1	Sprinkler	None exist
10	55.57	S_3 sw	Sprinkler	Gravel soil texture, & drainage
11	65.40	S_{2s}	Sprinkler	Gravel soil texture
12	54.72	S_3 sw	Drip	CaCo₃ and drainage
13	72.2	S_{2s}	Sprinkler	Gravel soil texture
14	77.16	S_{2s}	Sprinkler	Gravel soil texture
15	85.5	S_1	Sprinkler	None exist
16	26.77	N_{2s}	Sprinkler	Gravel soil texture
17	81	S_1	Sprinkler	None exist
18	81.22	S_1	Sprinkler	None exist

carbonate. The currently not suitable land and permanently not suitable land can be observed only in the southeast and center of the plain because of severe limitations of calcium carbonate, drainage, salinity, and alkalinity. For the entire study area, slope, soil depth, and salinity were never considered as limiting factors.

The comparison of the capability indexes for surface, sprinkler, and drip irrigation (Tables 15.4 and 15.6) indicated that only in soil series coded 12, applying drip irrigation systems was the most suitable option as compared to surface and sprinkler irrigation systems. In soil series coded 1, 2, 3, 4, 5, 6, 7, 8, 9, 10, 11, 13, 14, 15, 16, 17, and 18, applying sprinkler irrigation systems was more suitable than surface and drip irrigation systems. Figure 15.5 shows the most suitable map for surface, sprinkler, and drip irrigation systems in the Dezful Plain as per the capability index (Ci) for different irrigation systems. As seen from this map, the largest part of this plain was suitable for sprinkler irrigation systems, and some parts of this area were suitable for drip irrigation systems.

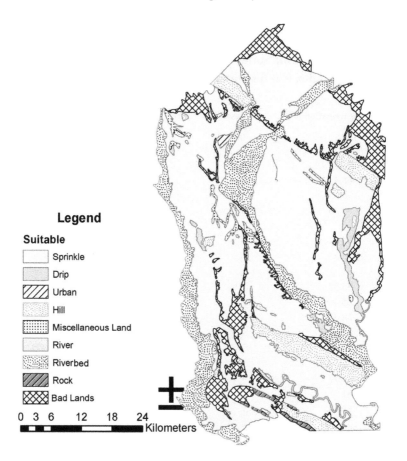

FIGURE 15.5   The most suitable map for different irrigation systems.

15.3   CONCLUSIONS

The results shown in Tables 15.4 and 15.6 indicate that by applying sprinkler irrigation instead of surface and drip irrigation methods, the land suitability of 165,498 ha (63.52%) of the Dezful Plain's land could be improved substantially. However, by applying drip irrigation instead of surface and sprinkler irrigation methods, the suitability of 4,006 ha (1.54%) of this plain's land could be improved. The comparison of the different types of irrigation revealed that sprinkler irrigation was more effective and efficient than the drip and surface irrigation methods, and it improved land suitability for irrigation purposes. The second best option was the application of drip irrigation, which was considered as being more practical than the surface irrigation method. To sum up, the most suitable irrigation systems for the Dezful Plain were sprinkler irrigation, drip irrigation, and surface irrigation, respectively. Landi et al. (2008); Naseri et al. (2009b); Rezania et al. (2009); Boroomand Nasab et al. (2010); Albaji (2010); Albaji and Hemadi (2011); Albaji et al. 2008, 2012a, 2014a, 2014b, 2014c, 2015 and 2016) found similar results.

Moreover, the main limiting factors in using surface irrigation method in this area were slope, calcium carbonate, gravel soil texture, salinity, alkalinity, and drainage. The major limiting factors in using the sprinkler irrigation method in this plain were calcium carbonate, gravel soil texture, and drainage. Also, the main limiting factors in using drip irrigation methods were the calcium carbonate salinity, alkalinity, and drainage.

ACKNOWLEDGMENTS

We are grateful to the Research Council of Shahid Chamran University of Ahvaz for financial support (GN: SCU.WI98.280).

REFERENCES

Abyaneh, H.Z., Jovzi, M., Albaji, M. 2017. Effect of regulated deficit irrigation, partial root drying and N-fertilizer levels on sugar beet crop (*Beta vulgaris* L.). *Agricultural Water Management* 194, 13–23.

Albaji, M., Boroomand Nasab, S., Kashkuli, H.A., Naseri, A.A., Sayyad, G., Jafari, S. 2008. Comparison of different irrigation methods based on the parametric evaluation approach in North Molasani Plain, Iran. *Journal of Agronomy* 7 (2), 187–191.

Albaji, M. 2010. *Land Suitability Evaluation for Sprinkler Irrigation Systems*. Khuzestan Water and Power Authority (KWPA), Ahvaz, Iran (in Persian).

Albaji, M., Hemadi, J. 2011. Investigation of different irrigation systems based on the parametric evaluation approach on the Dasht Bozorg Plain. *Transactions of the Royal Society of South Africa* 66 (3), 163–169.

Albaji, M., Boroomand Nasab, S., Hemadi, J. 2012a. Comparison of different irrigation methods based on the parametric evaluation approach in West North Ahvaz Plain. In: *Problems, Perspectives and Challenges of Agricultural Water Management*. M. Kumar, (ed.), InTech, Croatia, pp. 259–274.

Albaji, M., Papan, P., Hosseinzadeh, M., Barani, S. 2012b. Evaluation of land suitability for principal crops in the Hendijan region. *International Journal of Modern Agriculture* 1 (1), 24–32.

Albaji, M., Golabi, M., Boroomand Nasab, S., Jahanshahi. M. 2014a. Land suitability evaluation for surface, sprinkler and drip irrigation systems. *Transactions of the Royal Society of South Africa* 69 (2), 63–73.

Albaji, M., Golabi, M., Piroozfar, V.R., Egdernejad, A., Nazari Zadeh, F. 2014b. Evaluation of agricultural land resources for irrigation in the Ramhormoz Plain by using GIS. *Agriculturae Conspectus Scientificus* 79 (2), 93–102.

Albaji, M., Golabi, M., Egdernejad, A., Nazarizadeh, F. 2014c. Assessment of different irrigation systems in Albaji Plain. *Water Science and Technology: Water Supply* 14 (5), 778–786.

Albaji, M., Boroomand Nasab, S., Golabi, M., Sorkheh Nezhad, M., Ahmadee, M. 2015. Application possibilities of different irrigation methods in Hofel Plain. *YYÜ TAR BİL DERG (YYU J AGR SCI)* 25 (1), 13–23.

Albaji, M., Golabi, M., Hooshmand, A.R., Ahmadee, M. 2016. Investigation of surface, sprinkler and drip irrigation methods using GIS. *Jordan Journal of Agricultural Sciences* 12 (1), 211–222.

Albaji, M., Alboshokeh, A. 2017. Assessing agricultural land suitability in the Fakkeh region, Iran. *Outlook on Agriculture* 46 (1), 57–65.

Behzad, M., Albaji, M., Papan, P., Boroomand Nasab, S. 2009. Evan region qualitative soil evaluation for wheat, barley, alfalfa and maize. *Journal of Food, Agriculture & Environment* 7 (2), 843–851.

Boroomand Nasab, S., Albaji, M., Naseri, A.A. 2010. Investigation of different irrigation systems based on the parametric evaluation approach in Boneh Basht plain, Iran. *African Journal of Agricultural Research* 5 (5), 372–379.

Chegah, S., Chehrazi, M., Albaji, M. 2013. Effects of drought stress on growth and development frankenia plant (*Frankenia Leavis*). *Bulgarian Journal of Agricultural Science* 19 (4), 659–666.

Khuzestan Water and Power Authority (KWPA). 2010. Semi-detailed soil study report of Dezful plain, Iran (in Persian).

Landi, A., Boroomand-Nasab, S., Behzad, M., Tondrow, M.R., Albaji, M., Jazaieri, A. 2008. Land suitability evaluation for surface, sprinkle and drip irrigation methods in Fakkeh Plain, Iran. *Journal of Applied Sciences* 8 (20), 3646–3653.

Naseri, A.A., Albaji, M., Boroomand Nasab, S., Landi, A., Papan, P., Bavi, A. 2009a. Land suitability evaluation for principal crops in the Abbas Plain, Southwest Iran. *Journal of Food, Agriculture & Environment* 7 (1), 208–213.

Naseri, A.A., Rezania, A.R., Albaji, M. 2009b. Investigation of soil quality for different irrigation systems in Lali Plain, Iran. *Journal of Food, Agriculture & Environment* 7 (3&4), 955–960.

Rezania, A.R., Naseri, A.A., Albaji, M. 2009. Assessment of soil properties for irrigation methods in North Andimeshk Plain, Iran. *Journal of Food, Agriculture & Environment* 7 (3&4), 728–733.

16 The Best Irrigation System in Shirin Ab Plain, Iran

Mohammad Albaji, Saeed Boroomand Nasab,
Saeid Eslamian, and Abd Ali Naseri

CONTENTS

16.1 INTRODUCTION

The present study was conducted in an area about 9,000 ha in the Shirin Ab Plain in Khuzestan Province, located in the west of Iran, during 2010–2011. The study area is located 30 km west of the city of Shushtar, 3563000–3590000 N and 267000′ to 284000 E. The average annual temperature and precipitation for the period of 1965–2010 were 24°C and 404 mm, respectively. Also, the annual evaporation of the area is 2,550 mm (Khuzestan Water & Power Authority, 2011). The Karun River supplies the bulk of the water demands of the region. The application of irrigated agriculture has been common in the study area. Currently, the irrigation systems used by farmlands in the region are furrow irrigation, basin irrigation, and border irrigation schemes.

The area is composed of two distinct physiographic features, i.e., plateaus and alluvio-colluvial fans. Also, eight different soil series were found in the area. The semi-detailed soil survey report of the Shirin Ab Plain (Khuzestan Water & Power Authority, 2011) was used in order to determine the soil characteristics.

Over much of the Shirin Ab Plain, the use of surface irrigation systems has been applied specifically for field crops to meet the water demand of both summer and winter crops. The major irrigated broad-acre crops grown in this area are wheat, barley, and maize, in addition to fruits, melons, watermelons, and vegetables such as tomatoes and cucumbers. But like other plains in Khuzstan Province, the important major crops are wheat, barley, and maize (Behzad et al., 2009; Naseri et al., 2009; Albaji et al., 2012b; Albaji and Alboshokeh, 2017). There are very few instances of sprinkle and drip irrigation on large-area farms in the Shirin Ab Plain.

Eight soil series and 24 series phases or land units were derived from the semi-detailed soil study of the area. The land units are shown in Figure 16.1 as the basis for further land evaluation practice. The soils of the area are of Entisols and Inceptisols orders. Also, the soil moisture regime is ustic and xeric tempustic, while the soil temperature regime is hyperthermic (Khuzestan Water & Power Authority, 2011).

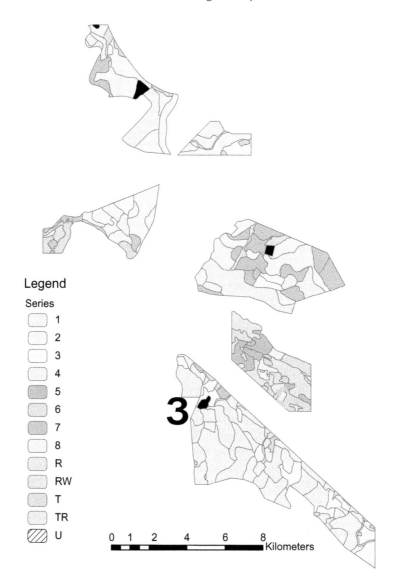

FIGURE 16.1   Soil map of the study area.

16.2   RESULTS AND DISCUSSIONS

As shown in Tables 16.1 and 16.2 for surface irrigation, the soil series coded 1 and 2 (3,724.1 ha – 41.37%) were highly suitable (S_1); soil series coded 3 and 6 (1,001.3 ha – 11.12%) were classified as moderately suitable (S_2); soil series coded 5 and 8 (1,803.2 ha – 20.03%) were found to be marginally suitable (S_3); and soil series coded 4 and 7 (1,497.6 ha – 16.64%) were classified as currently not suitable (N_1) for any surface irrigation practices.

The analysis of the suitability irrigation maps for surface irrigation (Figure 16.2) indicates that the largest portion of the cultivated area in this plain (located in the center and south) is deemed as being highly suitable land due to deep soil, good drainage, texture, salinity, and proper slope of the area. The moderately suitable area is located to the south, center, and north of this area due to gravel soil texture. Other factors such as drainage, salinity, and alkalinity have no influence on the suitability of the area whatsoever. The map also indicates that only some part of the cultivated area

TABLE 16.1

Ci Values and Suitability Classes of Surface, Sprinkle, and Drip Irrigation for Each Land Unit

Codes of Land Units	Surface Irrigation		Sprinkle Irrigation		Drip Irrigation	
	Ci	Suitability Classes	Ci	Suitability Classes	Ci	Suitability Classes
1	87.75	S_1[a]	90	S_1[b]	80	S_1[c]
2	87.75	S_1	90	S_1	80	S_1
3	78.97	$S_{2\,S}$	81	S_1	72	$S_{2\,S}$
4	32.17	$N_{1\,S}$	45.5	$S_{3\,S}$	56.52	$S_{3\,S}$
5	59.94	S_3 sw	75.02	S_2 sw	80	S_1
6	71.07	S_2 sw	81.22	S_1	80	S_1
7	37.46	N_1 s	51.33	S_3 s	53.2	S_3 s
8	47.38	S_3 s	52.65	S_3 s	50.4	S_3 s

[a] Limiting Factors for Surface Irrigation: s: (Gravel Soil Texture & Soil Depth).
[b] Limiting Factors for Sprinkle Irrigation: s: (Soil Depth).
[c] Limiting Factors for Drip Irrigation: s: (Calcium Carbonate & Soil Depth).

TABLE 16.2

Distribution of Surface, Sprinkle, and Drip Irrigation Suitability

Suitability	Surface Irrigation			Sprinkle Irrigation			Drip Irrigation		
	Land Unit	Area (ha)	Ratio (%)	Land Unit	Area (ha)	Ratio (%)	Land Unit	Area (ha)	Ratio (%)
S_1	1, 2	3,724.1	41.37	1, 2, 3, 6	4,725.4	52.49	1, 2, 5, 6	4,260.8	47.33
S_2	3, 6	1,001.3	11.12	5	246.9	2.74	3	711.5	7.9
S_3	5, 8	1,803.2	20.03	4, 7, 8	3,053.9	33.93	4, 7, 8	3,053.9	33.93
N_1	4, 7	1,497.6	16.64	—	—	—	—	—	—
N_2	—	—	—	—	—	—	—	—	—
[a]Mis land		974	10.82		974	10.82		974	10.82
Total		9,000	100		9,000	100		9,000	100

[a] Miscellaneous Land: (Hill, Sand Dune, and Riverbed).

in this plain was evaluated as marginally suitable because of the gravel soil texture and soil depth. The current not suitable land can be observed only in the north of the plain because of severe limitations of gravel soil texture and soil depth. There was no permanently not suitable land in this plain. For almost the total study area, elements such as salinity, drainage, and $CaCO_3$ were not considered as limiting factors.

In order to verify the possible effects of different management practices, the land suitability for sprinkle and drip irrigation was evaluated (Tables 16.1 and 16.2).

For sprinkle irrigation, soil series coded 1, 2, 3, and 6 (4,725.4 ha – 52.49%) were highly suitable (S_1), while only soil series coded 5 (246.9 ha – 2.74%) were classified as moderately suitable (S_2). Further, soil series coded 4, 7, and 8 (3,053.9 ha – 33.93%) were found to be marginally suitable (S_3) for sprinkle irrigation.

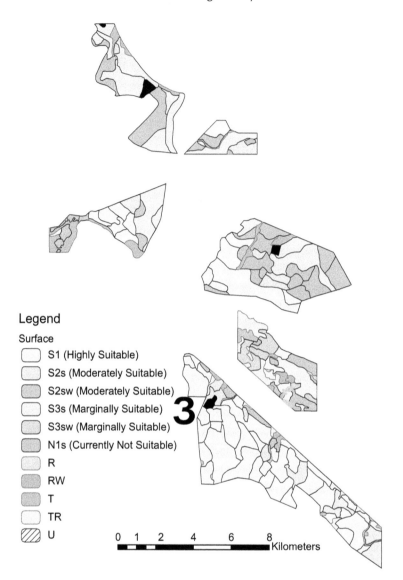

FIGURE 16.2   Land suitability map for surface irrigation.

Regarding sprinkler irrigation, (Figure 16.3) the highly suitable area can be observed in the largest part of the cultivated zone in this plain (located in the center and south) due to deep soil, good drainage, texture, salinity, and proper slope of the area. As seen from the map, some part of the cultivated area in this plain was evaluated as moderately suitable for sprinkle irrigation because of the narrow soil depth. Other factors such as drainage, salinity, and slope never influence the suitability of the area. The marginally suitable lands are located only in the center and northwest of the plain, and their non-suitability of the land is due to the severe limitations of soil depth. The current not suitable lands and permanently not suitable lands did not exist in this plain. Slope, salinity, drainage, and $CaCO_3$ were never taken as limiting factors for almost the entire study area.

For drip irrigation, soil series coded 1, 2, 5, and 6 (4,260.8 ha – 47.33%) were highly suitable (S_1), while only soil series coded 3 (711.5 ha – 7.9%) were classified as moderately suitable (S_2). Further, soil series coded 4, 7, and 8 (3,053.9 ha, 33.93%) were found to be slightly suitable (S_3) for drip irrigation.

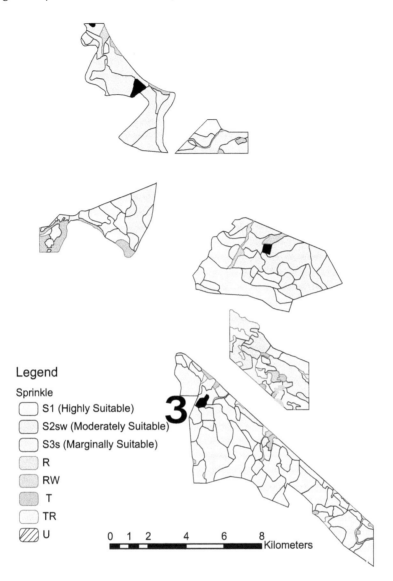

FIGURE 16.3 Land suitability map for sprinkle irrigation.

Regarding drip irrigation (Figure 16.4), the highly suitable area can be observed in the largest part of the cultivated zone in this plain (located in the center and south) due to deep soil, good drainage, texture, salinity, and proper slope of the area. As seen from the map, some part of the cultivated area in this plain was evaluated as moderately suitable for drip irrigation because of the light limitations of calcium carbonate and soil depth. Other factors such as drainage, salinity, and slope never influence the suitability of the area. The marginally suitable lands are located in the center and northwest of the plain, and their non-suitability of the land is due to the severe limitations of calcium carbonate and soil depth. The current not suitable lands and permanently not suitable lands did not exist in this plain. Slope, salinity, and drainage were never taken as limiting factors for almost the entire study area.

The mean capability index (Ci) for surface irrigation was 67.94 (moderately suitable), while for sprinkle irrigation it was 73.55 (moderately suitable). Moreover, for drip irrigation it was 68.78 (moderately suitable). The comparison of the capability indices for surface, sprinkle, and drip irrigation

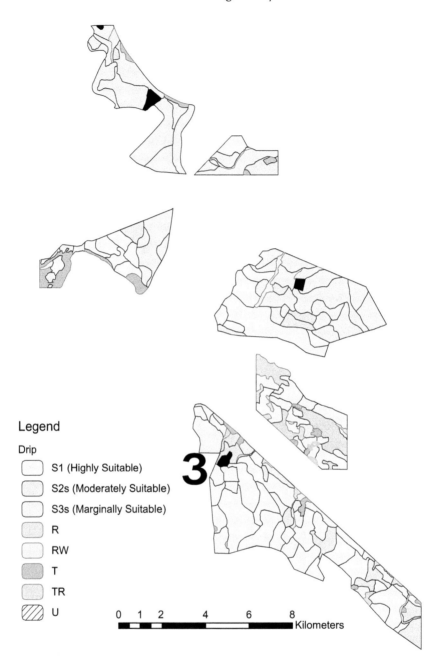

FIGURE 16.4 Land suitability map for drip irrigation.

(Tables 16.1 and 16.3) indicated that in soil series coded 4, 5, and 7, applying drip irrigation systems was the most suitable option as compared to surface and sprinkle irrigation systems. In soil series coded 1, 2, 3, 6, and 8, applying sprinkle irrigation systems was more suitable than surface and drip irrigation systems. Figure 16.5 shows the most suitable map for surface, sprinkle, and drip irrigation systems in the Shirin Ab Plain as per the Ci for different irrigation systems. As seen from this map, the largest part of this plain was suitable for sprinkle irrigation systems, and some parts of this area were suitable for drip irrigation systems.

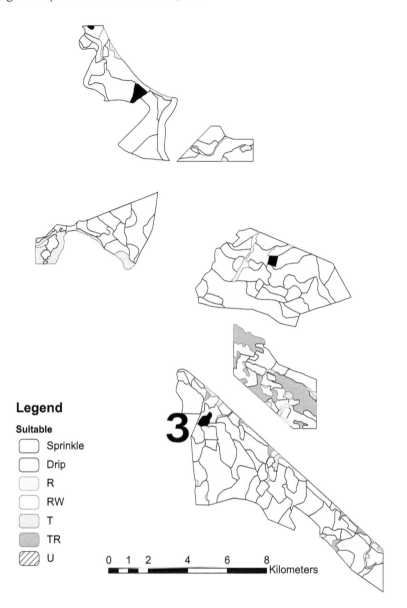

FIGURE 16.5 The most suitable map for different irrigation systems.

16.3 CONCLUSIONS

The results shown in Tables 16.1 and 16.3 indicate that by applying sprinkle irrigation instead of surface and drip irrigation methods, the land suitability of 6281,7 ha (69.78%) of the Shirin Ab Plain's land could be improved substantially. However, by applying drip irrigation instead of surface and sprinkle irrigation methods, the suitability of 1744, 5 ha (19.38%) of this plain's land could be improved. The comparison of the different types of irrigation revealed that sprinkle irrigation was more effective and efficient than the drip and surface irrigation methods, and it improved land suitability for irrigation purposes. The second best option was the application of drip irrigation, which was considered as being more practical than the surface irrigation method. To sum up, the most suitable irrigation systems for the Shirin Ab Plain were sprinkle irrigation, drip irrigation, and surface irrigation, respectively. Other researchers have confirmed these results (Albaji et al., 2008;

TABLE 16.3
The Most Suitable Land Units for Surface, Sprinkle, and Drip Irrigation Systems by Notation to Capability Index (Ci) for Different Irrigation Systems

Codes of Land Units	The Maximum Capability Index for Irrigation (Ci)	Suitability Classes	The Most Suitable Irrigation Systems	Limiting Factors
1	90	S_1	Sprinkle	None exist
2	90	S_1	Sprinkle	None exist
3	81	S_1	Sprinkle	None exist
4	56.52	S_{3s}	Drip	$CaCO_3$ & Soil depth
5	80	S_1	Drip	None exist
6	81.22	S_1	Sprinkle	None exist
7	53.2	S_{3s}	Drip	$CaCO_3$ & Soil depth
8	52.65	S_{3s}	Sprinkle	Soil depth

Landi et al., 2008; Rezania et al., 2009; Albaji et al., 2009; Naseri et al., 2009b, Boroomand Nasab et al., 2010; Albaji, 2010; Albaji and Hemadi, 2011; Jovzi et al., 2012; Albaji et al., 2012a, 2014a, 2014b, 2014c, 2015a, 2015b, 2016).

Moreover, the main limiting factors in using surface irrigation methods in this area were gravel soil texture and soil depth; the main limiting factor in using sprinkle irrigation methods in this area was soil depth; and the main limiting factors in using drip irrigation methods were the calcium carbonate and soil depth.

ACKNOWLEDGMENTS

We are grateful to the Research Council of Shahid Chamran University of Ahvaz for financial support (GN: SCU.WI98.280).

REFERENCES

Albaji, M., Boroomand Nasab, S., Kashkuli, H.A., Naseri, A.A., Sayyad, G., Jafari, S. 2008. Comparison of different irrigation methods based on the parametric evaluation approach in North Molasani Plain, Iran. *Journal of Agronomy* 7 (2), 187–191.

Albaji, M., Boroomand-Nasab, S., Naseri, A.A., Jafari, S. 2009. Comparison of different irrigation methods based on the parametric evaluation approach in Abbas Plain: Iran. *Journal of Irrigation and Drainage Engineering* 136 (2), 131–136.

Albaji, M. 2010. Land suitability evaluation for sprinkler irrigation systems. Khuzestan Water and Power Authority (KWPA), Ahvaz, Iran (in Persian).

Albaji, M., Hemadi, J. 2011. Investigation of different irrigation systems based on the parametric evaluation approach on the Dasht Bozorg Plain. *Transactions of the Royal Society of South Africa* 66 (3), 163–169.

Albaji, M., Boroomand Nasab, S., Hemadi, J. 2012a. Comparison of different irrigation methods based on the parametric evaluation approach in West North Ahvaz Plain. In: *Problems, Perspectives and Challenges of Agricultural Water Management*, M. Kumar, (ed.), InTech, Croatia, pp. 259–274.

Albaji, M., Papan, P., Hosseinzadeh, M., Barani, S. 2012b. Evaluation of land suitability for principal crops in the Hendijan region. *International Journal of Modern Agriculture* 1 (1), 24–32.

Albaji, M., Golabi, M., Boroomand Nasab, S., Jahanshahi. M. 2014a. Land suitability evaluation for surface, sprinkler and drip irrigation systems. *Transactions of the Royal Society of South Africa* 69 (2), 63–73.

Albaji, M., Golabi, M., Piroozfar, V.R., Egdernejad, A., Nazari Zadeh, F. 2014b. Evaluation of agricultural land resources for irrigation in the Ramhormoz Plain by using GIS. *Agriculturae Conspectus Scientificus* 79 (2), 93–102.

Albaji, M., Golabi, M., Egdernejad, A., Nazarizadeh, F. 2014c. Assessment of different irrigation systems in Albaji Plain. *Water Science and Technology: Water Supply* 14 (5), 778–786.

Albaji, M., Golabi, M., Boroomand Nasab, S., Zadeh, F.N. 2015a. Investigation of surface, sprinkler and drip irrigation methods based on the parametric evaluation approach in Jaizan Plain. *Journal of the Saudi Society of Agricultural Sciences* 14 (1), 1–10.

Albaji, M., Boroomand Nasab, S., Golabi, M., Sorkheh Nezhad, M., Ahmadee, M. 2015b. Application possibilities of different irrigation methods in Hofel Plain. *YYÜ TAR BİL DERG (YYU J AGR SCI)* 25 (1), 13–23.

Albaji, M., Golabi, M., Hooshmand, A.R., Ahmadee, M. 2016. Investigation of surface, sprinkler and drip irrigation methods using GIS. *Jordan Journal of Agricultural Sciences* 12 (1), 211–222.

Albaji, M., Alboshokeh, A. 2017. Assessing agricultural land suitability in the Fakkeh region, Iran. *Outlook on Agriculture* 46 (1), 57–65.

Behzad, M., Albaji, M., Papan, P., Boroomand Nasab, S. 2009. Evan region qualitative soil evaluation for wheat, barley, alfalfa and maize. *Journal of Food, Agriculture & Environment* 7 (2), 843–851.

Boroomand Nasab, S., Albaji, M., Naseri, A.A. 2010. Investigation of different irrigation systems based on the parametric evaluation approach in Boneh Basht plain, Iran. *African Journal of Agricultural Research* 5 (5), 372–379.

Jovzi, M., Albaji, M., Gharibzadeh, A. 2012. Investigating the suitability of lands for surface and under-pressure (drip and sprinkler) irrigation in Miheh Plain. *Research Journal of Environmental Sciences* 6 (2), 51–61.

Khuzestan Water and Power Authority (KWPA). 2011. Semi-detailed soil study report of Shirin ab Plain, Iran (in Persian).

Landi, A., Boroomand-Nasab, S., Behzad, M., Tondrow, M.R., Albaji, M., Jazaieri, A. 2008. Land suitability evaluation for surface, sprinkle and drip irrigation methods in Fakkeh Plain, Iran. *Journal of Applied Sciences* 8 (20), 3646–3653.

Naseri, A.A., Albaji, M., Boroomand Nasab, S., Landi, A., Papan, P., Bavi, A. 2009. Land suitability evaluation for principal crops in the Abbas Plain, Southwest Iran. *Journal of Food, Agriculture & Environment* 7 (1), 208–213.

Naseri, A.A., Rezania, A.R., Albaji, M. 2009b. Investigation of soil quality for different irrigation systems in Lali Plain, Iran. *Journal of Food, Agriculture & Environment* 7 (3&4), 955–960.

Rezania, A.R., Naseri, A.A., Albaji, M. 2009. Assessment of soil properties for irrigation methods in North Andimeshk Plain, Iran. *Journal of Food, Agriculture & Environment* 7 (3&4), 728–733.

17 The Best Irrigation System in Mian-Ab Plain, Iran

Mohammad Albaji, Abd Ali Naseri, Mona Golabi, and Saeid Eslamian

CONTENTS

17.1 INTRODUCTION

The scarcity of water in arid and semiarid regions such as Iran is a restrictive element for the agricultural sector; therefore, there is an urgent need to develop initiatives to save water in this particular sector (Chegah et al., 2013; Abyaneh et al., 2017). Among the proposed solutions is the use of pressurized irrigation systems. The present study was conducted in an area about 41,677 hectares in the Mian-Ab Plain in Khuzestan Province, southwest of Iran, during 2010–2011. The study area is located 40 km northwest of the city of Ahvaz, 31° 38′ to 32° 04′ N and 48° 45′ to 49° 01′ E. The average annual temperature and precipitation for the period of 1959–2009 were 25.30°C and 335.70 mm, respectively. Also, the annual evaporation of the area is 1,755.82 mm (Table 17.1) (Khuzestan Water and Power Authority, 2009). The Karun River supplies the bulk of the water demands of the region. The application of irrigated agriculture has been common in the study area. Currently, the irrigation systems used by farmlands in the region are furrow irrigation, basin irrigation, and border irrigation schemes.

The area is composed of three distinct physiographic features, i.e., river alluvial plains, terraces, and plateaus, of which the river alluvial plains' physiographic unit is the dominant feature. Also, 28 different soil series were found in the area (Table 17.2). The semi-detailed soil survey report of the Mian-Ab Plain (Khuzestan Water and Power Authority, 2009) was used in order to determine the soil characteristics. Table 17.3 has shown some of the physicochemical characteristics for reference profiles of different soil series in the plain.

Over much of the Mian-Ab Plain, the use of surface irrigation systems has been applied specifically for field crops to meet the water demand of both summer and winter crops. The major irrigated broad-acre crops grown in this area are wheat, barley, and maize, in addition to fruits, melons, watermelons, and vegetables such as tomatoes and cucumbers. But like other plains in Khuzstan Province, the important major crops are wheat, barley, and maize (Behzad et al., 2009; Naseri et al., 2009a; Albaji et al., 2012b; Albaji and Alboshokeh, 2017). There are very few instances of sprinkle and drip irrigation on large-area farms in the Mian-Ab Plain.

Twenty-eight soil series and 99 series phases or land units were derived from the semi-detailed soil study of the area. The land units are shown in Figure 17.1 as the basis for further land evaluation practice. The soils of the area are of Aridisols and Entisols orders. Also, the soil moisture regime is aridic, while the soil temperature regime is hyperthermic (Khuzestan Water and Power Authority, 2009).

TABLE 17.1
Mean Air Temperature, Relative Humidity and Total Monthly Rainfall and Evaporation (1959–2009) at Mian-Ab Plain

Parameter	Jan	Feb	Mar	Apr	May	Jun	Jul	Aug	Sep	Oct	Nov	Dec	Average
Temperature (°C)	12.80	14.50	18.50	24.20	30.20	34.50	37.20	37.00	33.90	27.90	20.80	15.00	25.30
Relative humidity (%)	72.00	66.20	53.40	42.10	28.30	22.20	23.30	25.10	25.90	34.90	52.70	68.70	42.90
													Total
Rainfall (mm)	69.30	44.00	61.50	28.30	8.70	0	0	0	0	6.50	27.10	90.30	335.70
Evaporation (mm)	28.20	33.80	62.83	110.56	208.73	272.08	272.52	267.81	223.03	143.9	86.42	45.94	1755.82

TABLE 17.2
Soil Series of the Study Area

Characteristics Description	Series No.
Soil texture "Medium: [a]L," slight salinity and alkalinity limitation, Depth 130 cm, gently sloping: 2% to 5%, well drained.	1
Soil texture "Medium: SL," without salinity and alkalinity limitation, Depth 150 cm, level to very gently sloping: 0% to 2%, well drained.	2
Soil texture "Medium: SL," without salinity and alkalinity limitation, Depth 60 cm, gently sloping: 2% to 5%, well drained.	3
Soil texture "Medium: SL," without salinity and alkalinity limitation, Depth 160 cm, gently sloping: 2% to 5%, well drained.	4
Soil texture "Medium: SIL," slight salinity and alkalinity limitation, Depth 135 cm, level to very gently sloping: 0% to 2%, well drained.	5
Soil texture "Heavy: SICL," without salinity and alkalinity limitation, Depth 140 cm, level to very gently sloping: 0% to 2%, well drained.	6
Soil texture "Heavy: SICL," without salinity and alkalinity limitation, Depth 150 cm, level to very gently sloping: 0% to 2%, well drained.	7
Soil texture "Heavy: SCL," without salinity and alkalinity limitation, Depth 150 cm, level to very gently sloping: 0% to 2%, well drained.	8
Soil texture "Medium: SIL," slight salinity and alkalinity limitation, Depth 140 cm, level to very gently sloping: 0% to 2%, well drained.	9
Soil texture "Heavy: SICL," slight salinity and alkalinity limitation, Depth 130 cm, level to very gently sloping: 0% to 2%, well drained.	10
Soil texture "Medium: L," very severe salinity and alkalinity limitation, Depth 140 cm, level to very gently sloping: 0% to 2%, moderately drained.	11
Soil texture "Heavy: SICL," very severe salinity and alkalinity limitation, Depth 140 cm, level to very gently sloping: 0% to 2%, moderately drained.	12
Soil texture "Medium: SIL," without salinity and alkalinity limitation, Depth 150 cm, level to very gently sloping: 0% to 2%, moderately drained.	13
Soil texture "Very Heavy: SIC," without salinity and alkalinity limitation, Depth 140 cm, level to very gently sloping: 0% to 2%, poorly drained.	14
Soil texture "Very Heavy: SIC," without salinity and alkalinity limitation, Depth 140 cm, level to very gently sloping: 0% to 2%, imperfectly drained.	15
Soil texture "Medium: L," without salinity and alkalinity limitation, Depth 150 cm, level to very gently sloping: 0% to 2%, moderately drained.	16
Soil texture "Heavy: CL," without salinity and alkalinity limitation, Depth 150 cm, level to very gently sloping: 0% to 2%, imperfectly drained.	17

(Continued)

TABLE 17.2 (*Continued*)
Soil Series of the Study Area

Characteristics Description	Series No.
Soil texture "Medium: SL," without salinity and alkalinity limitation, Depth 150 cm, level to very gently sloping: 0% to 2%, moderately drained.	18
Soil texture "Heavy: SIC," without salinity and alkalinity limitation, Depth 120 cm, level to very gently sloping: 0% to 2%, imperfectly drained.	19
Soil texture "Very Heavy: C," slight salinity and alkalinity limitation, Depth 125 cm, level to very gently sloping: 0% to 2%, poorly drained.	20
Soil texture "Very Heavy: SIC," very severe salinity and alkalinity limitation, Depth 140 cm, level to very gently sloping: 0% to 2%, very poorly drained.	21
Soil texture "Very Heavy: C," severe salinity and alkalinity limitation, Depth 150 cm, level to very gently sloping: 0% to 2%, very poorly drained.	22
Soil texture "Very Heavy: C," severe salinity and alkalinity limitation, Depth 150 cm, level to very gently sloping: 0% to 2%, very poorly drained.	23
Soil texture "Heavy: SICL," without salinity and alkalinity limitation, Depth 150 cm, level to very gently sloping: 0% to 2%, well drained.	24
Soil texture "Very Heavy: SIC," moderate salinity and alkalinity limitation, Depth 150 cm, level to very gently sloping: 0% to 2%, imperfectly drained.	25
Soil texture "Very Heavy: C," slight salinity and alkalinity limitation, Depth 150 cm, level to very gently sloping: 0% to 2%, imperfectly drained.	26
Soil texture "Heavy: SCL," without salinity and alkalinity limitation, Depth 150 cm, level to very gently sloping: 0% to 2%, well drained.	27
Soil texture "Light: LS," without salinity and alkalinity limitation, Depth 150 cm, gently sloping: 2% to 5%, well drained.	28

[a] Texture symbols: LS: Loamy Sand, SL: Sandy Loam, L: Loam, SIL: Silty Loam, CL: Clay Loam, SICL: Silty Clay Loam, SCL: Sandy Clay Loam, SC: Sandy Clay, SIC: Silty Clay, C: Clay.

TABLE 17.3
Some of Physicochemical Characteristics for Reference Profiles of Different Soil Series

CaCO₃ (%)	CEC (meq/100 g)	OM (%)	pH	ECe (ds m⁻¹)	Soil Texture	Depth (cm)	Soil Series Name	Soil Series No.
43.00	12.36	0.62	7.60	4.10	L	130	Qhaleh Khan	1
42.00	11.92	0.14	7.50	1.50	SL	150	Sheykh Hossein	2
44.00	10.54	0.24	7.90	3.20	SL	60	Bala	3
51.00	11.47	0.08	8.20	0.50	SL	160	Kakoli	4
44.00	12.73	0.39	7.60	5.90	SIL	135	Abbasieh	5
51.00	10.22	0.41	7.60	1.00	SICL	140	Karun	6
49.00	10.81	0.32	7.50	2.90	SICL	150	Deylam	7
49.00	12.05	0.26	7.90	1.50	SCL	150	Qalimeh	8
51.00	11.56	0.38	7.60	4.20	SIL	140	Ghaleh Nasir	9
46.00	10.38	0.57	7.80	7.50	SICL	130	Abdul Amir	10
57.00	10.24	0.14	7.90	39.00	L	140	Kheyrabad	11
44.00	10.87	0.20	7.60	31.00	SICL	140	Khoshmakan	12
50.00	9.26	0.37	7.70	1.90	SIL	150	Karkheh	13
41.00	10.09	0.27	8.10	1.50	SIC	140	Dasht	14

(Continued)

TABLE 17.3 (*Continued*)
Some of Physicochemical Characteristics for Reference Profiles of Different Soil Series

CaCO$_3$ (%)	CEC (meq/100 g)	OM (%)	pH	ECe (ds m^{-1})	Soil Texture	Depth (cm)	Soil Series Name	Soil Series No.
40.00	10.64	0.39	7.60	3.50	SIC	140	Dez	15
44.00	12.07	0.76	7.50	1.22	L	150	Arab Asad	16
48.00	8.54	0.24	7.90	1.50	CL	150	Veyss	17
41.00	8.19	0.39	7.80	1.10	SL	150	Ramin	18
48.00	10.31	0.23	8.50	3.50	SIC	120	Amerabad	19
35.00	15.24	0.52	8.00	4.10	C	125	Band Ghir	20
45.00	11.43	0.37	8.10	52.00	SIC	140	Abu Baghal	21
46.00	13.26	0.56	8.40	17.50	C	150	Sheykh Mussa	22
39.00	12.91	0.36	7.90	21.50	C	150	Molla Sani	23
47.00	9.21	0.25	7.70	2.20	SICL	150	Karun	24
47.00	8.66	0.60	7.90	9.50	SIC	150	Shoteyt	25
42.00	9.37	0.46	7.80	6.50	C	150	Valiabad	26
36.00	5.27	0.29	7.90	1.30	SCL	150	Naghishat	27
33.00	3.60	0.24	8.00	0.60	LS	150	Elleh	28

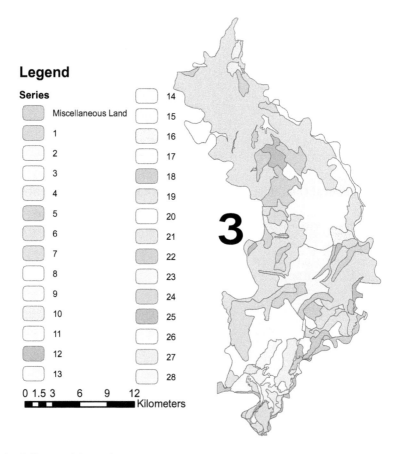

FIGURE 17.1   Soil map of the study area.

17.2   RESULTS AND DISCUSSIONS

As shown in Tables 17.4 and 17.5 for surface irrigation, the soil series coded 5, 7, 8, 16, 24, and 27 (10,454.12 ha – 25.1%) were highly suitable (S_1); soil series coded 1, 2, 6, 9, 10, 13, and 17 (19,587.74 ha – 47 %) were classified as moderately suitable (S_2); and soil series coded 3, 4, 15, 18, 19, 26, and 28 (4,617.81 ha – 11%) were found to be marginally suitable (S_3). Soil series coded 12, 14, 20, and 25 (4,975.06 ha – 11.9%) were classified as currently not suitable (N_1), and soil series coded 11, 21, 22, and 23 (1,150.9 ha – 2.8 %) were classified as permanently not suitable (N_2) for any surface irrigation practices.

The analysis of the suitability irrigation maps for surface irrigation (Figure 17.2) indicates that the highly suitable area is located to the center and south of this area due to deep soil, good drainage, texture, salinity, and proper slope of the area. The major portion of the cultivated area in this

TABLE 17.4

Ci Values and Suitability Classes of Surface, Sprinkle, and Drip Irrigation for Each Land Unit

Codes of Land Units	Surface Irrigation		Sprinkle Irrigation		Drip Irrigation	
	Ci	Suitability Classes	Ci	Suitability Classes	Ci	Suitability Classes
1	71.17	$S_2 s$[a]	75.02	$S_2 s$	68.4	$S_2 s$
2	65.81	$S_2 s$	81	S_1	76	$S_2 s$
3	49.95	$S_3 s$	67.12	$S_2 s$	68.4	$S_2 s$
4	55.5	$S_3 s$	70.2	$S_2 s$	66.5	$S_2 s$
5	83.36	S_1	85.5	S_1	76	$S_2 s$
6	78	$S_2 s$	80	S_1	70	$S_2 s$
7	87.75	S_1	90	S_1	80	S_1
8	83.36	S_1	85.5	S_1	76	$S_2 s$
9	74.1	$S_2 s$	76	$S_2 s$	66.5	$S_2 s$
10	78.97	S_2 sn	85.5	S_1	76	$S_2 s$
11	12.63	N_2 snw	13.68	N_2 sn	12.6	N_2 sn
12	42.12	N_1 snw	52.65	S_3 snw	52	S_3 sn
13	78.97	S_2 sw	85.5	S_1	80	S_1
14	44.75	N_1 sw	49.72	S_3 sw	47.6	S_3 sw
15	52.21	S_3 sw	57.37	S_3 sw	54.4	S_3 sw
16	83.36	S_1	85.5	S_1	76	$S_2 s$
17	70.2	S_2 sw	76.5	S_2 sw	72	S_2 sw
18	59.23	S_3 sw	76.95	S_2 s	76	$S_2 s$
19	52.21	S_3 sw	57.37	S_3 sw	54.4	S_3 sw
20	40.27	N_1 snw	47.23	S_3 sw	45.22	S_3 sw
21	17.90	N_2 snw	22.37	N_2 snw	22.1	N_2 snw
22	20.88	N_2 snw	25.81	N_2 snw	25.5	N_2 snw
23	20.88	N_2 snw	25.81	N_2 snw	25.5	N_2 snw
24	87.75	S_1	90	S_1	80	S_1
25	41.76	N_1 snw	48.76	S_3 snw	46.24	S_3 snw
26	46.99	S_3 snw	54.50	S_3 sw	51.68	S_3 sw
27	83.36	S_1	85.5	S_1	76	$S_2 s$
28	45.78	$S_3 s$	61.42	$S_2 s$	68	$S_2 s$

[a] Limiting Factors for Surface, Sprinkle, and Drip Irrigations: s: (Calcium Carbonate), w: (Drainage) and n: (Salinity & Alkalinity).

TABLE 17.5

Distribution of Surface, Sprinkle, and Drip Irrigation Suitability

	Surface Irrigation			Sprinkle Irrigation			Drip Irrigation		
Suitability	Land Unit	Area (ha)	Ratio (%)	Land Unit	Area (ha)	Ratio (%)	Land Unit	Area (ha)	Ratio (%)
S₁	5, 7, 8, 16, 24, 27	10,454.12	25.1	2, 5, 6, 7, 8, 10, 13, 16, 24, 27	28,910.55	69.3	7, 13, 24	3,329.88	8
S₂	1, 2, 6, 9, 10, 13, 17	19,587.74	47	1, 3, 4, 9, 17, 18, 28	3,111.53	7.5	1, 2, 3, 4, 5, 6, 8, 9, 10, 16, 17, 18, 27, 28	28,692.2	68.8
S₃	3, 4, 15, 18, 19, 26, 28	4,617.81	11	12, 14, 15, 19, 20, 25, 26	7,612.65	18.2	12, 14, 15, 19, 20, 25, 26	7,612.65	18.2
N₁	12, 14, 20, 25	4,975.06	11.9	—	—	—	—	—	—
N₂	11, 21, 22, 23	1,150.9	2.8	11, 21, 22, 23	1,150.9	2.8	11, 21, 22, 23	1,150.9	2.8
ªMis Land		891.64	2.1		891.64	2.1		891.64	2.1
Total		41,677.25	100		41,677.25	100		41,677.25	100

ª Miscellaneous Land: (Hill, Sand Dune, and Riverbed).

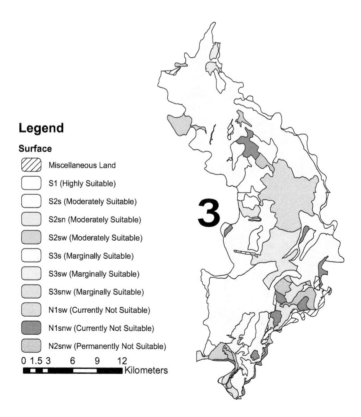

Legend

Surface

- Miscellaneous Land
- S1 (Highly Suitable)
- S2s (Moderately Suitable)
- S2sn (Moderately Suitable)
- S2sw (Moderately Suitable)
- S3s (Marginally Suitable)
- S3sw (Marginally Suitable)
- S3snw (Marginally Suitable)
- N1sw (Currently Not Suitable)
- N1snw (Currently Not Suitable)
- N2snw (Permanently Not Suitable)

0 1.5 3   6   9   12
Kilometers

FIGURE 17.2   Land suitability map for surface irrigation.

plain (located in the north and center) is deemed as being moderately suitable land due to medium limitations of calcium carbonates content and drainage. Other factors such as depth, salinity, and alkalinity have no influence on the suitability of the area whatsoever. The map also indicates that only some parts of the cultivated area in this plain were evaluated as marginally suitable because of severe limitations of drainage and calcium carbonates content. The current not suitable land and permanently not suitable land can be observed only in the center and southeast of the plain. This was because of very severe limitations of salinity, alkalinity, drainage, and calcium carbonates content. For almost the total study area, elements such as soil depth, soil texture, and slope were not considered as limiting factors.

In order to verify the possible effects of different management practices, the land suitability for sprinkle and drip irrigation was evaluated (Tables 17.4 and 17.5).

For sprinkle irrigation, soil series coded 2, 5, 6, 7, 8, 10, 13, 16, 24, and 27 (28,910.55 ha – 69.3%) were highly suitable (S_1), while soil series coded 1, 3, 4, 9, 17, 18, and 28 (3,111.53 ha – 7.5%) were classified as moderately suitable (S_2). Further, soil series coded 12, 14, 15, 19, 20, 25, and 26 (7,612.65 ha – 18.2%) were found to be marginally suitable (S_3), and soil series coded 11, 21, 22, and 23 (1,150.9 ha – 2.8%) were classified as permanently not suitable (N_2) for sprinkle irrigation.

Regarding sprinkler irrigation (Figure 17.3), the highly suitable area can be observed in the largest part of the cultivated zone in this plain (north, south, west, and the east parts) due to deep soil, good drainage, texture, salinity, and proper slope of the area. As seen from the map, some parts of the cultivated area in this plain were evaluated as moderately suitable for sprinkle irrigation because of the medium calcium carbonates content. Other factors such as drainage, depth, salinity, and slope never influence the suitability of the area. The marginally suitable lands are located only in the center and southeast of the plain, and their marginal

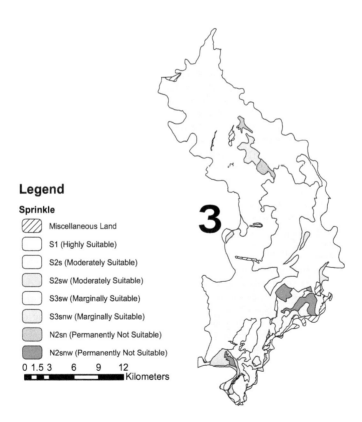

Legend

Sprinkle

- Miscellaneous Land
- S1 (Highly Suitable)
- S2s (Moderately Suitable)
- S2sw (Moderately Suitable)
- S3sw (Marginally Suitable)
- S3snw (Marginally Suitable)
- N2sn (Permanently Not Suitable)
- N2snw (Permanently Not Suitable)

0 1.5 3   6   9   12
Kilometers

FIGURE 17.3 Land suitability map for sprinkle irrigation.

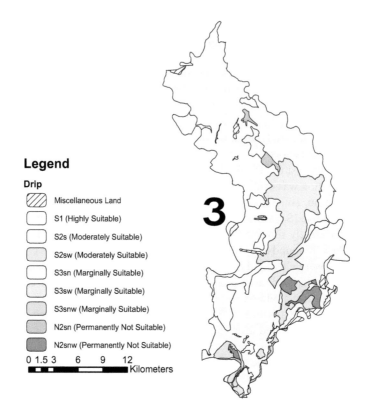

FIGURE 17.4   Land suitability map for drip irrigation.

suitability of the land is due to the severe limitations of calcium carbonates content, drainage, salinity, and alkalinity. The current not suitable lands did not exist in this plain. Finally, the permanently not suitable lands can be observed only in the smallest portion of the plain because of very severe limitations of salinity, alkalinity, drainage, and calcium carbonates content. Slope, soil depth, and soil texture were never taken as limiting factors for almost the entire study area.

For drip irrigation, soil series coded 7, 13, and 24 (3,329.88 ha – 8%) were highly suitable (S_1), while soil series coded 1, 2, 3, 4, 5, 6, 8, 9, 10, 16, 17, 18, 27, and 28 (28,692.2 ha – 68.8%) were classified as moderately suitable (S_2). Further, soil series coded 12, 14, 15, 19, 20, 25, and 26 (7,612.65 ha, 18.2%) were found to be slightly suitable (S_3), and soil series coded 11, 21, 22, and 23 (1,150.9 ha – 2.8 %) were classified as permanently not suitable (N_2) for drip irrigation.

In this case, (Figure 17.4) the highly suitable lands covered some parts of the plain. The slope, soil texture, soil depth, calcium carbonate, salinity, and drainage were all in desirable conditions. The moderately suitable lands could be observed over the largest portion of the plain (west, east, north, and south parts) due to the medium calcium carbonates content. The marginally suitable lands were found only in the center and southeast of the area studied. The limiting factors for these land units were high content of calcium carbonates and severe drainage limitations. The current not suitable lands did not exist in this plain. The permanently not suitable lands can be observed only in the smallest portion of the plain because of very severe limitations of salinity, alkalinity,

TABLE 17.6

The Most Suitable Land Units for Surface, Sprinkle, and Drip Irrigation Systems by Notation to Capability Index (Ci) for Different Irrigation Systems

Codes of Land Units	The Maximum Capability Indexfor Irrigation (Ci)	Suitability Classes	The Most Suitable Irrigation Systems	Limiting Factors
1	75.02	S_{2s}	Sprinkle	$CaCO_3$
2	81	S_1	Sprinkle	None Exist
3	68.4	S_{2s}	Drip	$CaCO_3$
4	70.2	S_{2s}	Sprinkle	$CaCO_3$
5	85.5	S_1	Sprinkle	None Exist
6	80	S_1	Sprinkle	None Exist
7	90	S_1	Sprinkle	None Exist
8	85.5	S_1	Sprinkle	None Exist
9	76	S_{2s}	Sprinkle	$CaCO_3$
10	85.5	S_1	Sprinkle	None Exist
11	13.68	N_2sn	Sprinkle	$CaCO_3$, Salinity, and Alkalinity
12	52.65	$S_3 snw$	Sprinkle	$CaCO_3$, Salinity, Alkalinity, and Drainage
13	85.5	S_1	Sprinkle	No Exist
14	49.72	$S_3 sw$	Sprinkle	$CaCO_3$ and Drainage
15	57.37	$S_3 sw$	Sprinkle	$CaCO_3$ and Drainage
16	85.5	S_1	Sprinkle	None Exist
17	76.5	$S_2 sw$	Sprinkle	$CaCO_3$ and Drainage
18	76.95	S_{2s}	Sprinkle	$CaCO_3$
19	57.37	$S_3 sw$	Sprinkle	$CaCO_3$ and Drainage
20	47.23	$S_3 sw$	Sprinkle	$CaCO_3$ and Drainage
21	22.37	$N_2 snw$	Sprinkle	$CaCO_3$, Salinity, Alkalinity, and Drainage
22	25.81	$N_2 snw$	Sprinkle	$CaCO_3$, Salinity, Alkalinity, and Drainage
23	25.81	$N_2 snw$	Sprinkle	$CaCO_3$, Salinity, Alkalinity, and Drainage
24	90	S_1	Sprinkle	None Exist
25	48.76	$S_3 snw$	Sprinkle	$CaCO_3$, Salinity, Alkalinity, and Drainage
26	54.50	$S_3 sw$	Sprinkle	$CaCO_3$ and Drainage
27	85.5	S_1	Sprinkle	None Exist
28	68	S_{2s}	Drip	$CaCO_3$

drainage, and calcium carbonates content. For almost the entire study area slope, soil depth and soil texture were never taken as limiting factors.

The comparison of the capability indexes for surface, sprinkle, and drip irrigation (Tables 17.4 and 17.6) indicated that in soil series coded 3 and 28, applying drip irrigation systems was the most suitable option as compared to surface and sprinkle irrigation systems. In soil series coded 1, 2, 4, 5, 6, 7, 8, 9, 10, 11, 12, 13, 14, 15, 16, 17, 18, 19, 20, 21, 22, 23, 24, 25, 26, and 27, applying sprinkle irrigation systems was more suitable than surface and drip irrigation systems. Figure 17.5 shows the most suitable map for surface, sprinkle, and drip irrigation systems in the Mian-Ab Plain as per the capability index (Ci) for different irrigation systems. As seen from this map, the largest part of this plain was suitable for sprinkle irrigation systems, and some parts of this area were suitable for drip irrigation systems.

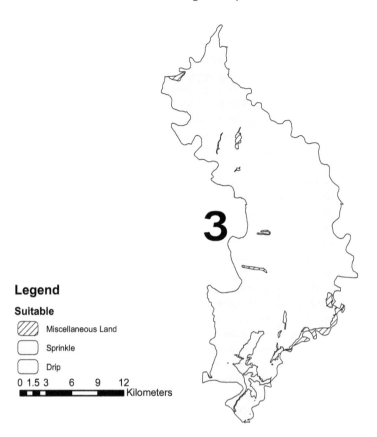

FIGURE 17.5 The most suitable map for different irrigation systems.

17.3 CONCLUSIONS

The results of Tables 17.4 and 17.6 indicated that by applying sprinkle irrigation instead of surface and drip irrigation methods, the land suitability of 39,925, 69 ha (95.7%) of the Mian-Ab Plain's land could be improved substantially. However, by applying drip irrigation instead of surface and sprinkle irrigation methods, the suitability of 859, 94 ha (2.1%) of this plain's land could be improved. The comparison of the different types of irrigation revealed that sprinkle irrigation was more effective and efficient than the drip and surface irrigation methods, and it improved land suitability for irrigation purposes. The second best option was the application of drip irrigation, which was considered as being more practical than the surface irrigation method. To sum up, the most suitable irrigation systems for the Mian-Ab Plain were sprinkle irrigation, drip irrigation, and surface irrigation, respectively. Research carried out by Landi et al. (2008); Naseri et al. (2009b); Albaji et al. (2008); Albaji et al. (2009); Rezania et al. (2009); Boroomand Nasab et al. (2010); Albaji. (2010); Albaji and Hemadi (2011); Albaji et al. (2012a); Albaji et al. (2014a); Albaji et al. (2014b); Albaji et al. (2014c); Albaji et al. (2015a); Albaji et al. (2015b); and Albaji et al. (2016) confirm these results.

Moreover, the main limiting factors in using surface and sprinkle irrigation methods in this area were drainage and calcium carbonates, and the main limiting factor in using drip irrigation methods was calcium carbonates.

ACKNOWLEDGMENTS

We are grateful to the Research Council of Shahid Chamran University of Ahvaz for financial support (GN: SCU.WI98.280).

REFERENCES

Abyaneh, H.Z., Jovzi, M., Albaji, M. 2017. Effect of regulated deficit irrigation, partial root drying and N-fertilizer levels on sugar beet crop (*Beta vulgaris* L.). *Agricultural Water Management* 194, 13–23.

Albaji, M., Boroomand Nasab, S., Kashkuli, H.A., Naseri, A.A., Sayyad, G., Jafari, S. 2008. Comparison of different irrigation methods based on the parametric evaluation approach in North Molasani Plain, Iran. *Journal of Agronomy* 7 (2), 187–191.

Albaji, M., Boroomand-Nasab, S., Naseri, A.A., Jafari, S. 2009. Comparison of different irrigation methods based on the parametric evaluation approach in abbas plain: Iran. *Journal of Irrigation and Drainage Engineering* 136 (2), 131–136.

Albaji, M. 2010. Land suitability evaluation for sprinkler irrigation systems. Khuzestan Water and Power Authority (KWPA), Ahvaz, Iran (in Persian).

Albaji, M., Hemadi, J. 2011. Investigation of different irrigation systems based on the parametric evaluation approach on the Dasht Bozorg Plain. *Transactions of the Royal Society of South Africa* 66 (3), 163–169.

Albaji, M., Boroomand Nasab, S., Hemadi, J. 2012a. Comparison of different irrigation methods based on the parametric evaluation approach in West North Ahvaz plain. In: *Problems, Perspectives and Challenges of Agricultural Water Management*. M. Kumar, (ed.), InTech, Croatia, pp. 259–274.

Albaji, M., Papan, P., Hosseinzadeh, M., Barani, S. 2012b. Evaluation of land suitability for principal crops in the Hendijan region. *International Journal of Modern Agriculture* 1 (1), 24–32.

Albaji, M., Golabi, M., Boroomand Nasab, S., Jahanshahi. M. 2014a. Land suitability evaluation for surface, sprinkler and drip irrigation systems. *Transactions of the Royal Society of South Africa* 69 (2), 63–73.

Albaji, M., Golabi, M., Piroozfar, V.R., Egdernejad, A., Nazari Zadeh, F. 2014b. Evaluation of agricultural land resources for irrigation in the Ramhormoz plain by using GIS. *Agriculturae Conspectus Scientificus* 79 (2), 93–102.

Albaji, M., Golabi, M., Egdernejad, A., Nazarizadeh, F. 2014c. Assessment of different irrigation systems in Albaji Plain. *Water Science and Technology: Water Supply* 14 (5), 778–786.

Albaji, M., Golabi, M., Boroomand Nasab, S., Zadeh, F.N. 2015a. Investigation of surface, sprinkler and drip irrigation methods based on the parametric evaluation approach in Jaizan Plain. *Journal of the Saudi Society of Agricultural Sciences* 14 (1), 1–10.

Albaji, M., Boroomand Nasab, S., Golabi, M., Sorkheh Nezhad, M., Ahmadee, M. 2015b. Application possibilities of different irrigation methods in Hofel Plain. *Journal of Agricultural Science* 25 (1), 13–23.

Albaji, M., Golabi, M., Hooshmand, A.R., Ahmadee, M. 2016. Investigation of surface, sprinkler and drip irrigation methods using GIS. *Jordan Journal of Agricultural Sciences* 12 (1), 211–222.

Albaji, M., Alboshokeh, A. 2017. Assessing agricultural land suitability in the Fakkeh region, Iran. *Outlook on Agriculture* 46 (1), 57–65.

Behzad, M., Albaji, M., Papan, P., Boroomand Nasab, S. 2009. Evan region qualitative soil evaluation for wheat, barley, alfalfa and maize. *Journal of Food, Agriculture & Environment* 7 (2), 843–851.

Boroomand Nasab, S., Albaji, M., Naseri, A.A. 2010. Investigation of different irrigation systems based on the parametric evaluation approach in Boneh Basht plain, Iran. *African Journal of Agricultural Research* 5 (5), 372–379.

Chegah, S., Chehrazi, M., Albaji, M. 2013. Effects of drought stress on growth and development frankenia plant (*Frankenia Leavis*). *Bulgarian Journal of Agricultural Science* 19 (4), 659–666.

Khuzestan Water and Power Authority (KWPA). 2009. Semi-detailed soil study report of Mian Ab Plain, Ahvaz, Iran (in Persian).

Landi, A., Boroomand-Nasab, S., Behzad, M., Tondrow, M.R., Albaji, M., Jazaieri, A. 2008. Land suitability evaluation for surface, sprinkle and drip irrigation methods in Fakkeh plain, Iran. *Journal of Applied Sciences* 8 (20), 3646–3653.

Naseri, A.A., Albaji, M., Boroomand Nasab, S., Landi, A., Papan, P., Bavi, A. 2009a. Land suitability evaluation for principal crops in the Abbas Plain, Southwest Iran. *Journal of Food, Agriculture & Environment* 7 (1), 208–213.

Naseri, A.A., Rezania, A.R., Albaji, M. 2009b. Investigation of soil quality for different irrigation systems in Lali Plain, Iran. *Journal of Food, Agriculture & Environment* 7 (3&4), 955–960.

Rezania, A.R., Naseri, A.A., Albaji, M. 2009. Assessment of soil properties for irrigation methods in North Andimeshk Plain, Iran. *Journal of Food, Agriculture & Environment* 7 (3&4), 728–733.

18 The Best Irrigation System in East North Ahvaz Plain, Iran

Mohammad Albaji, Abd Ali Naseri, and Saeid Eslamian

CONTENTS

18.1 INTRODUCTION

The scarcity of water in arid and semiarid regions such as Iran is a restrictive element for the agricultural sector; therefore, there is an urgent need to develop initiatives to save water in this particular sector (Chegah et al., 2013; Abyaneh et al., 2017). The present study was conducted in an area about 36,502 hectares in the East North Ahvaz Plain in Khuzestan Province, located in the southwest of Iran, during 2011–2012. The study area is located 5 km northeast of the city of Ahvaz, 31° 20′ to 31° 40′ N and 48° 47′ to 48° 58′ E. The average annual temperature and precipitation for the period of 1966–2011 were 25.4°C and 230 mm, respectively. Also, the annual evaporation of the area is 2,035 mm (Table 18.1) (Khuzestan Water and Power Authority, 2010). The Karun River supplies the bulk of the water demands of the region. The application of irrigated agriculture has been common in the study area. Currently, the irrigation systems used by farmlands in the region are furrow irrigation, basin irrigation, and border irrigation schemes.

The area is composed of two distinct physiographic features, i.e., river alluvial plains and terraces, of which the river alluvial plains' physiographic unit is the dominant feature. Also, 16 different soil series were found in the area (Table 18.2). The semi-detailed soil survey report of the East North Ahvaz Plain (Khuzestan Water and Power Authority, 2010) was used in order to determine the soil characteristics. Table 18.3 shows some of the physicochemical characteristics for reference profiles of different soil series in the plain.

Over much of the East North Ahvaz Plain, the use of surface irrigation systems has been applied specifically for field crops to meet the water demand of both summer and winter crops. The major irrigated broad-acre crops grown in this area are wheat, barley, and maize, in addition to fruits, melons, watermelons, and vegetables such as tomatoes and cucumbers. But like other plains in Khuzstan Province, the important major crops are wheat, barley, and maize (Behzad et al., 2009a, 2009b; Naseri et al., 2009a; Albaji et al., 2012b; Albaji and Alboshokeh, 2017). There are very few instances of sprinkle and drip irrigation on large-area farms in the East North Ahvaz Plain.

Sixteen soil series and 67 series phases or land units were derived from the semi-detailed soil study of the area. The land units are shown in Figure 18.1 as the basis for further land evaluation practice. The soils of the area are of Aridisols and Entisols orders. Also, the soil moisture regime is aridic and aquic, while the soil temperature regime is hyperthermic (Khuzestan Water and Power Authority, 2010). Therefore, the organic matter content is low due to high soil temperature and high decomposition ratio of this matter (Mahjoobi et al., 2010).

TABLE 18.1

Mean Air Temperature, Relative Humidity, and Total Monthly Rainfall and Evaporation (1966–2011) at Ahvaz

Parameter	Jan	Feb	Mar	Apr	May	Jun	Jul	Aug	Sep	Oct	Nov	Dec	Average
Temperature (°C)	12.2	14.3	18.6	24.5	30.6	35.0	37.2	36.7	33.2	27.7	20.2	14.2	25.4
Relative humidity (%)	74.1	66.5	57.1	47.7	36.5	29.4	31.2	34.5	36.6	44.2	57.3	72	48.9
													Total
Rainfall (mm)	48.4	37.2	26	22.6	10.3	0.2	0.0	0.1	0.2	2.5	30.2	52.6	230.3
Evaporation (mm)	53	71.96	124	174.9	249	301	300	274.4	209.2	143	82	53	2,035.5

TABLE 18.2

Soil Series of the Study Area

Characteristics Description	Series No
Soil texture "Heavy: [a]CL," without salinity and alkalinity limitation, Depth 150 cm, level to very gently sloping: 0% to 2%, imperfectly drained.	1
Soil texture "Heavy: CL," very severe salinity and alkalinity limitation, Depth 100 cm, level to very gently sloping: 0% to 2%, poorly drained.	2
Soil texture "Medium: SL," without salinity and alkalinity limitation, Depth 150 cm, level to very gently sloping: 0% to 2%, moderately drained.	3
Soil texture "Heavy: SIC," without salinity and alkalinity limitation, Depth 120 cm, level to very gently sloping: 0% to 2%, imperfectly drained.	4
Soil texture "Medium: SL," without salinity and alkalinity limitation, Depth 150 cm, level to very gently sloping: 0% to 2%, moderately drained.	5
Soil texture "Very Heavy: C," slight salinity and alkalinity limitation, Depth 125 cm, level to very gently sloping: 0% to 2%, poorly drained.	6
Soil texture "Very Heavy: SIC," very severe salinity and alkalinity limitation, Depth 140 cm, level to very gently sloping: 0% to 2%, very poorly drained.	7
Soil texture "Very Heavy: C," severe salinity and alkalinity limitation, Depth 150 cm, level to very gently sloping: 0% to 2%, very poorly drained.	8
Soil texture "Very Heavy: C," severe salinity and alkalinity limitation, Depth 150 cm, level to very gently sloping: 0% to 2%, very poorly drained.	9
Soil texture "Very Heavy: C," very severe salinity and alkalinity limitation, Depth 110 cm, level to very gently sloping: 0% to 2%, very poorly drained.	10
Soil texture "Medium: L," without salinity and alkalinity limitation, Depth 170 cm, level to very gently sloping: 0% to 2%, moderately drained.	11
Soil texture "Heavy: SICL," without salinity and alkalinity limitation, Depth 150 cm, level to very gently sloping: 0% to 2%, well drained.	12
Soil texture "Very Heavy: SIC," moderate salinity and alkalinity limitation, Depth 150 cm, level to very gently sloping: 0% to 2%, imperfectly drained.	13
Soil texture "Light: LS," without salinity and alkalinity limitation, Depth 150 cm, gently sloping: 2% to 5%, well drained.	14
Soil texture "Medium: SL," without salinity and alkalinity limitation, Depth 160 cm, gently sloping: 2% to 5%, well drained.	15
Soil texture "Medium: SIL," slight salinity and alkalinity limitation, Depth 140 cm, level to very gently sloping: 0% to 2%, well drained.	16

[a] Texture symbols: LS: Loamy Sand, SL: Sandy Loam, L: Loam, SIL: Silty Loam, CL: Clay Loam, SICL: Silty Clay Loam, SCL: Sandy Clay Loam, SC: Sandy Clay, SIC: Silty Clay, C: Clay.

TABLE 18.3
Some of Physicochemical Characteristics for Reference Profiles of Different Soil Series

CaCO$_3$ (%)	CEC (meq/100 g)	OM (%)	pH	ECe (ds.m^{-1})	Soil Texture	Depth (cm)	Soil Series Name	Soil Series No.
48.00	8.54	0.24	7.90	1.50	CL	150	Veyss	1
49.00	5.61	0.46	7.70	48.00	CL	100	Omel Gharib	2
41.00	8.19	0.39	7.80	1.10	SL	150	Ramin	3
48.00	10.31	0.23	8.50	3.50	SIC	120	Amerabad	4
34.00	5.57	0.29	7.90	3.40	SL	150	Solieh	5
35.00	15.24	0.52	8.00	4.10	C	125	Band Ghir	6
45.00	11.43	0.37	8.10	52.00	SIC	140	Abu Baghal	7
46.00	13.26	0.56	8.40	17.50	C	150	Sheykh Mussa	8
39.00	12.91	0.36	7.90	21.50	C	150	Molla Sani	9
49.00	9.85	0.68	7.90	55.00	C	110	Teal Bomeh	10
46.00	6.49	0.29	7.70	2.70	L	170	Karkheh	11
47.00	9.21	0.25	7.70	2.20	SICL	150	Karun	12
47.00	8.66	0.60	7.90	9.50	SIC	150	Shoteyt	13
33.00	3.60	0.24	8.00	0.60	LS	150	Elleh	14
51.00	11.47	0.08	8.20	0.50	SL	160	Kakoli	15
51.00	11.56	0.38	7.60	4.20	SIL	140	Ghaleh Nasir	16

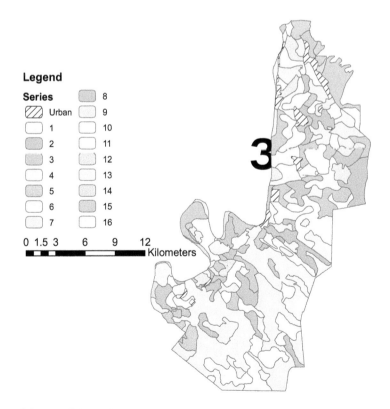

FIGURE 18.1 Soil map of the study area.

18.2 RESULTS AND DISCUSSIONS

As shown in Tables 18.4 and 18.5 for surface irrigation, only soil series coded 12 (253.83 ha – 0.7%) were highly suitable (S_1); soil series coded 1, 11, and 16 (6,903.15 ha – 18.92 %) were classified as moderately suitable (S_2), soil series coded 3, 4, 5, 14, and 15 (1,4671.08 ha – 40.18%) were found to be marginally suitable (S_3). Soil series coded 6 and 13 (665.89 ha – 1.82%) were classified as currently not suitable (N_1), and soil series coded 2, 7, 8, 9, and 10 (12,944.7 ha – 35.46%) were classified as permanently not suitable (N_2) for any surface irrigation practices.

The analysis of the suitability irrigation maps for surface irrigation (Figure 18.2) indicate that the smallest portion of the cultivated area in this plain (located in the southwest) is deemed as being highly suitable land due to deep soil, good drainage, texture, salinity, and proper slope of the area. The moderately suitable area is located in the south, center, and north of this area due to light limitations of drainage and heavy soil texture. Other factors such as slope, depth, salinity, and alkalinity have no influence on the suitability of the area whatsoever. The map also indicates that the largest part of the cultivated area in this plain was evaluated as marginally suitable because of high limitations of heavy soil texture and drainage. The current not suitable land can be observed only in the north of the plain because of severe limitations of heavy soil texture, drainage, salinity, and alkalinity. The map also indicates that some part of the cultivated area in this plain was evaluated as permanently not suitable land because of very severe limitations of heavy soil texture, drainage, salinity, and alkalinity. For almost the total study area, elements such as soil depth and slope were not considered as limiting factors.

TABLE 18.4
Ci Values and Suitability Classes of Surface, Sprinkle, and Drip Irrigation for Each Land Unit

Codes of Land Units	Surface Irrigation		Sprinkle Irrigation		Drip Irrigation	
	Ci	Suitability Classes	Ci	Suitability Classes	Ci	Suitability Classes
1	70.2	S_2 sw[a]	76.5	S_2 sw[b]	72	S_2 sw[c]
2	11.40	N_2 snw	12.6	N_2 snw	12.8	N_2 snw
3	59.23	S_3 sw	76.95	S_2 s	76	S_2 s
4	52.21	S_3 sw	57.37	S_3 sw	54.4	S_3 sw
5	59.23	S_3 sw	76.95	S_2 s	76	S_2 s
6	40.27	N_1 snw	47.23	S_3 sw	45.22	S_3 sw
7	17.90	N_2 snw	22.37	N_2 snw	22.1	N_2 snw
8	20.88	N_2 snw	25.81	N_2 snw	25.5	N_2 snw
9	20.88	N_2 snw	25.81	N_2 snw	25.5	N_2 snw
10	17.90	N_2 snw	22.37	N_2 snw	22.1	N_2 snw
11	71..07	S_2 sw	76.95	S_2 s	72	S_2 s
12	87.75	S_1	90	S_1	80	S_1
13	41.76	N_1 snw	48.76	S_3 snw	46.24	S_3 snw
14	45.78	S_3 s	61.42	S_2 s	68	S_2 s
15	55.5	S_3 s	70.2	S_2 s	66.5	S_2 s
16	74.1	S_{2S}	76	S_2 s	66.5	S_2 s

[a&b] Limiting Factors for Surface and Sprinkle Irrigations: n: (Salinity & Alkalinity), w: (Drainage), and s: (Heavy Soil Texture).

[c] Limiting Factors for Drip Irrigation: s: (Calcium Carbonate & Heavy Soil Texture), w: (Drainage), and n: (Salinity & Alkalinity).

TABLE 18.5

Distribution of Surface, Sprinkle, and Drip Irrigation Suitability

	Surface Irrigation			Sprinkle Irrigation			Drip Irrigation		
Suitability	Land Unit	Area (ha)	Ratio (%)	Land Unit	Area (ha)	Ratio (%)	Land Unit	Area (ha)	Ratio (%)
S_1	12	253.83	0.7	12	253.83	0.7	12	253.83	0.7
S_2	1, 11, 16	6,903.15	18.92	1, 3, 5, 11, 14, 15, 16,	12,475.99	34.18	1, 3, 5, 11, 14, 15, 16,	12,475.99	34.18
S_3	3, 4, 5, 14, 15,	14,671.08	40.18	4, 6, 13	9,764.13	26.74	4, 6, 13	9,764.13	26.74
N_1	6, 13	665.89	1.82	—	—	—	—	—	—
N_2	2, 7, 8, 9, 10	12,944.7	35.46	2, 7, 8, 9, 10	12,944.7	35.46	2, 7, 8, 9, 10	12,944.7	35.46
[a]Mis Land		1,063.08	2.91		1,063.08	2.91		1,063.08	2.91
Total		36,501.74	100		36,501.74	100		36,501.74	100

[a] Miscellaneous Land: (Hill, Sand Dune, and Riverbed).

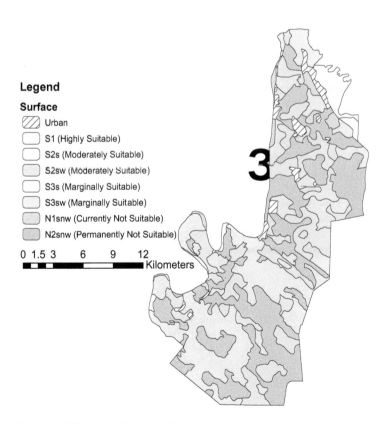

Legend

Surface

- ▨ Urban
- ▢ S1 (Highly Suitable)
- ▢ S2s (Moderately Suitable)
- ▢ S2sw (Moderately Suitable)
- ▢ S3s (Marginally Suitable)
- ▢ S3sw (Marginally Suitable)
- ▨ N1snw (Currently Not Suitable)
- ▨ N2snw (Permanently Not Suitable)

0 1.5 3   6   9   12
━━━━━━━━━━━━Kilometers

FIGURE 18.2 Land suitability map for surface irrigation.

In order to verify the possible effects of different management practices, the land suitability for sprinkle and drip irrigation was evaluated (Tables 18.4 and 18.5).

For sprinkle irrigation, only soil series coded 12 (253.83 ha – 0.7%) were highly suitable (S_1), while soil series coded 1, 3, 5, 11, 14, 15, and 16 (12,475.99 ha – 34.18%) were classified as moderately suitable (S_2). Further, soil series coded 4, 6, and 13 (9,764.13 ha – 26.74%) were found to be marginally suitable (S_3), and soil series coded 2, 7, 8, 9, and 10 (12,944.7 ha – 35.46%) were classified as permanently not suitable (N_2) for sprinkle irrigation.

The analysis of the suitability irrigation maps for sprinkle irrigation (Figure 18.3) indicates that the smallest portion of the cultivated area in this plain (located in the southwest) is deemed as being highly suitable land due to deep soil, good drainage, texture, salinity, and proper slope of the area. The moderately suitable area is located in the south, center, and north of this plain due to light limitations of drainage and heavy soil texture. Other factors such as slope, depth, salinity, and alkalinity have no influence on the suitability of the area whatsoever. The map also indicates that the some part of the cultivated area in this plain was evaluated as marginally suitable because of high limitations of heavy soil texture and drainage. The currently not suitable land did not exist in this plain. The map also indicates that the largest part of the cultivated area in this plain was evaluated as permanently not suitable land because of very severe limitations of heavy soil texture, drainage, salinity, and alkalinity. For almost the total study area, elements such as soil depth and slope were not considered as limiting factors.

For drip irrigation, only soil series coded 12 (253.83 ha – 0.7%) were highly suitable (S_1), while soil series coded 1, 3, 5, 11, 14, 15, and 16 (12,475.99 ha – 34.18%) were classified as moderately suitable (S_2). Further, soil series coded 4, 6, and 13 (9,764.13 ha, 26.74%) were found to be slightly

FIGURE 18.3 Land suitability map for sprinkle irrigation.

Legend

Drip

- ⬚ Urban
- ⬚ S1 (Highly Suitable)
- ⬚ S2s (Moderately Suitable)
- ⬚ S2sw (Moderately Suitable)
- ⬚ S3sw (Marginally Suitable)
- ⬚ S3snw (Marginally Suitable)
- ⬚ N2snw (Permanently Not Suitable)

0 1.5 3   6   9   12
Kilometers

FIGURE 18.4 Land suitability map for drip irrigation.

suitable (S₃), and soil series coded 2, 7, 8, 9, and 10 (12,944.7 ha – 35.46%) were classified as permanently not suitable (N₂) for drip irrigation.

Regarding drip irrigation (Figure 18.4), the smallest portion of the cultivated area in this plain (located in the southwest) is deemed as being highly suitable land due to deep soil, good drainage, texture, salinity, and proper slope of the area. The moderately suitable area is located in the south, center, and north of this plain due to light limitations of calcium carbonate, drainage, and heavy soil texture. Other factors such as slope, depth, salinity, and alkalinity have no influence on the suitability of the area whatsoever. The map also indicates that some part of the cultivated area in this plain was evaluated as marginally suitable because of high limitations of calcium carbonate, heavy soil texture, and drainage. The current not suitable land did not exist in this plain. The map also indicates that the largest part of the cultivated area in this plain was evaluated as permanently not suitable land because of very severe limitations of calcium carbonate, heavy soil texture, drainage, salinity, and alkalinity. For almost the total study area, elements such as soil depth and slope were not considered as limiting factors.

The comparison of the capability indexes for surface, sprinkle, and drip irrigation (Tables 18.4 and 18.6) indicated that in soil series coded 2 and 14, applying drip irrigation systems was the most suitable option as compared to surface and sprinkle irrigation systems. In soil series coded 1, 3, 4, 5, 6, 7, 8, 9, 10, 11, 12, 13, 15, and 16, applying sprinkle irrigation systems was more suitable than surface and drip irrigation systems. Figure 18.5 shows the most suitable map for surface, sprinkle, and drip irrigation systems in the East North Ahvaz Plain as per the capability index (Ci) for different irrigation systems. As seen from this map, the largest part of this plain was suitable for sprinkle irrigation systems, and some parts of this area were suitable for drip irrigation systems.

TABLE 18.6

The Most Suitable Land Units for Surface, Sprinkle, and Drip Irrigation Systems by Notation to Capability Index (Ci) for Different Irrigation Systems

Codes of Land Units	The Maximum Capability Index for Irrigation (Ci)	Suitability Classes	The Most Suitable Irrigation Systems	Limiting Factors
1	76.5	S_2 sw	Sprinkle	Soil Texture and Drainage
2	12.8	N_2 snw	Drip	$CaCO_3$ & Soil Texture, Salinity & Alkalinity, and Drainage
3	76.95	S_2 s	Sprinkle	Soil Texture
4	57.37	S_3 sw	Sprinkle	Soil Texture and Drainage
5	76.95	S_2 s	Sprinkle	Soil Texture
6	47.23	S_3 sw	Sprinkle	Soil Texture and Drainage
7	22.37	N_2 snw	Sprinkle	Soil Texture, Salinity & Alkalinity, and Drainage
8	25.81	N_2 snw	Sprinkle	Soil Texture, Salinity & Alkalinity, and Drainage
9	25.81	N_2 snw	Sprinkle	Soil Texture, Salinity & Alkalinity, and Drainage
10	22.37	N_2 snw	Sprinkle	Soil Texture, Salinity & Alkalinity, and Drainage
11	76.95	S_2 s	Sprinkle	Soil Texture
12	90	S_1	Sprinkle	None Exist
13	48.76	S_3 snw	Sprinkle	Soil Texture, Salinity & Alkalinity, and Drainage
14	68	S_{2S}	Drip	$CaCO_3$ & Soil Texture
15	70.2	S_{2S}	Sprinkle	Soil Texture
16	76	S_{2S}	Sprinkle	Soil Texture

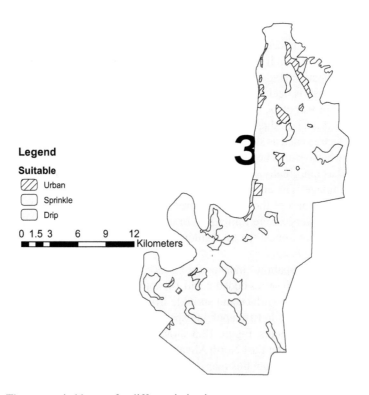

Legend

Suitable

- Urban
- Sprinkle
- Drip

0 1.5 3   6   9   12
Kilometers

FIGURE 18.5 The most suitable map for different irrigation systems.

18.3 CONCLUSIONS

The results shown in Tables 18.4 and 18.6 indicate that by applying sprinkle irrigation instead of surface and drip irrigation methods, the land suitability of 32819.05 ha (89.91%) of the East North Ahvaz Plain land could be improved substantially. However, by applying drip irrigation instead of surface and sprinkle irrigation methods, the suitability of 2619.6 ha (7.17%) of this plain's land could be improved. The comparison of the different types of irrigation revealed that sprinkle irrigation was more effective and efficient than the drip and surface irrigation methods, and it improved land suitability for irrigation purposes. The second best option was the application of drip irrigation, which was considered as being more practical than the surface irrigation method. To sum up, the most suitable irrigation systems for the East North Ahvaz Plain were sprinkle irrigation, drip irrigation, and surface irrigation, respectively. Other researchers have confirmed these results (Landi et al., 2008; Albaji et al., 2008, 2009, 2012a, 2014a, 2014b, 2014c, 2015a, 2015b, 2016; Rezania et al., 2009; Naseri et al., 2009b; Albaji, 2010; Boroomand Nasab et al., 2010; Albaji and Hemadi, 2011; Jovzi et al., 2012).

Moreover, the main limiting factor in using surface and sprinkle irrigation methods in this area were salinity, alkalinity, drainage, and heavy soil texture, and the main limiting factors in using drip irrigation methods were the salinity, alkalinity, drainage, heavy soil texture, and calcium carbonate.

ACKNOWLEDGMENTS

We are grateful to the Research Council of Shahid Chamran University of Ahvaz for financial support (GN: SCU.WI98.280).

REFERENCES

Abyaneh, H.Z., Jovzi, M., Albaji, M. 2017. Effect of regulated deficit irrigation, partial root drying and N-fertilizer levels on sugar beet crop (*Beta vulgaris* L.). *Agricultural Water Management* 194, 13–23.

Albaji, M., Boroomand Nasab, S., Kashkuli, H.A., Naseri, A.A., Sayyad, G., Jafari, S. 2008. Comparison of different irrigation methods based on the parametric evaluation approach in North Molasani Plain, Iran. *Journal of Agronomy* 7 (2), 187–191.

Albaji, M., Boroomand-Nasab, S., Naseri, A.A., Jafari, S. 2009. Comparison of different irrigation methods based on the parametric evaluation approach in Abbas Plain: Iran. *Journal of Irrigation and Drainage Engineering* 136 (2), 131–136.

Albaji, M. 2010. Land suitability evaluation for sprinkler irrigation systems. Khuzestan Water and Power Authority (KWPA), Ahvaz, Iran (in Persian).

Albaji, M., Hemadi, J. 2011. Investigation of different irrigation systems based on the parametric evaluation approach on the Dasht Bozorg Plain. *Transactions of the Royal Society of South Africa* 66 (3), 163–169.

Albaji, M., Boroomand Nasab, S., Hemadi, J. 2012a. Comparison of different irrigation methods based on the parametric evaluation approach in West North Ahvaz Plain. In: *Problems, Perspectives and Challenges of Agricultural Water Management*, M. Kumar, (ed.), InTech, Croatia, pp. 259–274.

Albaji, M., Papan, P., Hosseinzadeh, M., Barani, S. 2012b. Evaluation of land suitability for principal crops in the Hendijan region. *International Journal of Modern Agriculture* 1 (1), 24–32.

Albaji, M., Golabi, M., Boroomand Nasab, S., Jahanshahi, M. 2014a. Land suitability evaluation for surface, sprinkler and drip irrigation systems. *Transactions of the Royal Society of South Africa* 69 (2), 63–73.

Albaji, M., Golabi, M., Piroozfar, V.R., Egdernejad, A., Nazari Zadeh, F. 2014b. Evaluation of agricultural land resources for irrigation in the Ramhormoz Plain by using GIS. *Agriculturae Conspectus Scientificus* 79 (2), 93–102.

Albaji, M., Golabi, M., Egdernejad, A., Nazarizadeh, F. 2014c. Assessment of different irrigation systems in Albaji Plain. *Water Science and Technology: Water Supply* 14 (5), 778–786.

Albaji, M., Golabi, M., Boroomand Nasab, S., Zadeh, F.N. 2015a. Investigation of surface, sprinkler and drip irrigation methods based on the parametric evaluation approach in Jaizan Plain. *Journal of the Saudi Society of Agricultural Sciences* 14 (1), 1–10.

Albaji, M., Boroomand Nasab, S., Golabi, M., Sorkheh Nezhad, M., Ahmadee, M. 2015b. Application possibilities of different irrigation methods in Hofel Plain. *Journal of Agricultural Science* 25 (1), 13–23.

Albaji, M., Golabi, M., Hooshmand, A.R., Ahmadee, M. 2016. Investigation of surface, sprinkler and drip irrigation methods using GIS. *Jordan Journal of Agricultural Sciences* 12 (1), 211–222.

Albaji, M., Alboshokeh, A. 2017. Assessing agricultural land suitability in the Fakkeh region, Iran. *Outlook on Agriculture* 46 (1), 57–65.

Behzad, M., Albaji, M., Papan, P., Boroomand Nasab, S., Naseri, A.A., Bavi, A. 2009a. Qualitative evaluation of land suitability for principal crops in the Gargar Region, Khuzestan Province, Southwest Iran. *Asian Journal of Plant Sciences* 8 (1), 28.

Behzad, M., Albaji, M., Papan, P., Boroomand Nasab, S. 2009b. Evan region qualitative soil evaluation for wheat, barley, alfalfa and maize. *Journal of Food, Agriculture & Environment* 7 (2), 843–851.

Boroomand Nasab, S., Albaji, M., Naseri, A.A. 2010. Investigation of different irrigation systems based on the parametric evaluation approach in Boneh Basht plain, Iran. *African Journal of Agricultural Research* 5 (5), 372–379.

Chegah, S., Chehrazi, M., Albaji, M. 2013. Effects of drought stress on growth and development frankenia plant (*Frankenia Leavis*). *Bulgarian Journal of Agricultural Science* 19 (4), 659–666.

Jovzi, M., Albaji, M., Gharibzadeh, A. 2012. Investigating the suitability of lands for surface and under-pressure (drip and sprinkler) irrigation in Miheh Plain. *Research Journal of Environmental Sciences* 6 (2), 51–61.

Khuzestan Water and Power Authority (KWPA). 2010. Semi-detailed soil study report of East North Ahvaz Plain, Iran (in Persian).

Landi, A., Boroomand-Nasab, S., Behzad, M., Tondrow, M.R., Albaji, M., Jazaieri, A. 2008. Land suitability evaluation for surface, sprinkle and drip irrigation methods in Fakkeh Plain, Iran. *Journal of Applied Sciences* 8 (20), 3646–3653.

Mahjoobi, A., Albaji, M., Torfi, K. 2010. Determination of heavy metal levels of Kondok soills-haftgel. *Research Journal of Environmental Sciences* 4 (3), 294–299.

Naseri, A.A., Albaji, M., Boroomand Nasab, S., Landi, A., Papan, P., Bavi, A. 2009a. Land suitability evaluation for principal crops in the Abbas Plain, Southwest Iran. *Journal of Food, Agriculture & Environment* 7 (1), 208–213.

Naseri, A.A., Rezania, A.R., Albaji, M. 2009b. Investigation of soil quality for different irrigation systems in Lali Plain, Iran. *Journal of Food, Agriculture & Environment* 7 (3&4), 955–960.

Rezania, A.R., Naseri, A.A., Albaji, M. 2009.Assessment of soil properties for irrigation methods in North Andimeshk Plain, Iran. *Journal of Food, Agriculture & Environment* 7 (3&4), 728–733.

19 Determining the Best Irrigation System in Malashih Plain, Iran

Mohammad Albaji, Saeed Boroomand Nasab, and Saeid Eslamian

CONTENTS

19.1 INTRODUCTION

The present study was conducted in an area about 18,250 hectares in the Malashih Plain in Khuzestan Province, located in the southwest of Iran, during 2011–2012. The study area is located 5 km southwest of the city of Ahvaz, 31° 00′ to 31° 20′ N and 48° 20′ to 48° 40′ E. The average annual temperature and precipitation for the period of 1965–2011 were 25.4°C and 230 mm, respectively. Also, the annual evaporation of the area is 2,035 mm (Table 19.1) (Khuzestan Water and Power Authority, 2011). The Kharkhe River supplies the bulk of the water demands of the region. The application of irrigated agriculture has been common in the study area. Currently, the irrigation systems used by farmlands in the region are furrow irrigation, basin irrigation, and border irrigation schemes.

The area is composed of two distinct physiographic features, i.e., river alluvial plains and sand dunes, of which the river alluvial plains' physiographic unit is the dominant feature. Also, five different soil series were found in the area. The semi-detailed soil survey report of the Malashih Plain (Khuzestan Water and Power Authority, 2011) was used in order to determine the soil characteristics.

Over much of the Malashih Plain, the use of surface irrigation systems has been applied specifically for field crops to meet the water demand of both summer and winter crops. The major irrigated broad-acre crops grown in this area are wheat, barley, and maize, in addition to fruits, melons, watermelons, and vegetables such as tomatoes and cucumbers. But like other plains in Khuzstan Province, the important major crops are wheat, barley, and maize (Behzad et al., 2009a, 2009b; Naseri et al., 2009a; Albaji et al., 2012b; Albaji and Alboshokeh, 2017). There are very few instances of sprinkle and drip irrigation on large-area farms in the Malashih Plain.

Five soil series and 16 series phases or land units were derived from the semi-detailed soil study of the area. The land units are shown in Figure 19.1 as the basis for further land evaluation practice. The soils of the area are of Entisols and Aridisols orders. Also, the soil moisture regime is aridic and torric, while the soil temperature regime is hyperthermic (Khuzestan Water and Power Authority, 2011). Therefore, the organic matter content is low due to high soil temperature and high decomposition ratio of this matter (Mahjoobi et al., 2010).

TABLE 19.1

Mean Air Temperature, Relative Humidity, and Total Monthly Rainfall and Evaporation (1965–2011) at Ahvaz

Parameter	Jan	Feb	Mar	Apr	May	Jun	Jul	Aug	Sep	Oct	Nov	Dec	Average
Temperature (°C)	12.2	14.3	18.6	24.5	30.6	35.0	37.2	36.7	33.2	27.7	20.2	14.2	25.4
Relative humidity (%)	74.1	66.5	57.1	47.7	36.5	29.4	31.2	34.5	36.6	44.2	57.3	72	48.9
													Total
Rainfall (mm)	48.4	37.2	26	22.6	10.3	0.2	0.0	0.1	0.2	2.5	30.2	52.6	230.3
Evaporation (mm)	53	71.96	124	174.9	249	301	300	274.4	209.2	143	82	53	2035.5

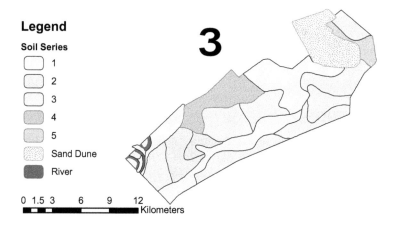

FIGURE 19.1   Soil map of the study area.

19.2   RESULTS AND DISCUSSIONS

As shown in Tables 19.2 and 19.3 for surface irrigation, soil series coded 1, 2, and 5 (8,750 ha – 47.94%) were classified as moderately suitable (S_2); only soil series coded 3 (7,400 ha – 40.55%) were found to be marginally suitable (S_3); and only soil series coded 4 (2,050 ha – 11.23%) were classified as permanently not suitable (N_2) for any surface irrigation practices.

The analyses of the suitability irrigation map for surface irrigation (Figure 19.2) indicate that there was no highly suitable land in this area. The major portion of the cultivated area in this plain (located in the west, center, and south) is deemed as being moderately suitable land due to light limitations of calcium carbonate and drainage. The marginally suitable area is located to the center and east of this area because of high limitations of drainage and calcium carbonate. Other factors such as slope, depth, salinity, and alkalinity have no influence on the suitability of the area whatsoever. The currently not suitable land did not exist in this plain. The map also indicates that only part of the cultivated area in this plain was evaluated as permanently not suitable due to very severe limitations of drainage, calcium carbonate, salinity, and alkalinity.

In order to verify the possible effects of different management practices, the land suitability for sprinkle and drip irrigation was evaluated (Tables 19.2 and 19.3).

For sprinkle irrigation, only soil series coded 2 (2,675 ha – 14.66%) were highly suitable (S_1), while soil series coded 1 and 5 (6,075 ha – 33.28%) were classified as moderately suitable (S_2). Further, only soil series coded 3 (7,400 ha – 40.55%) were found to be marginally suitable (S_3), and only soil series coded 4 (2,050 ha – 11.23%) were classified as permanently not suitable (N_2) for sprinkle irrigation.

TABLE 19.2
The Ci Values and Suitability Classes of Surface, Sprinkle, and Drip Irrigation for Each Land Unit

Codes of Land Units	Surface Irrigation		Sprinkle Irrigation		Drip Irrigation	
	Ci	Suitability Classes	Ci	Suitability Classes	Ci	Suitability Classes
1	68.4	$S_{2\ sw}$[a]	72.2	$S_{2\ s}$[b]	66.5	$S_{2\ s}$[c]
2	72	S_2 sw	81	S_1	80	S_1
3	54	S_3 sw	58.5	S_3 sw	56	S_3 sw
4	16.2	N_2 snw	17.1	N_2 sn	16	N_2 sn
5	60	$S_{2\ S}$	72	$S_{2\ S}$	66.5	$S_{2\ S}$

[a, b, c] The Limiting Factors for Surface, Sprinkle, and Drip Irrigations: w: (Drainage), n: (Salinity & Alkalinity), and s: (Calcium Carbonate).

TABLE 19.3
Distribution of Surface, Sprinkle, and Drip Irrigation Suitability

Suitability	Surface Irrigation			Sprinkle Irrigation			Drip Irrigation		
	Land Unit	Area (ha)	Ratio (%)	Land Unit	Area (ha)	Ratio (%)	Land Unit	Area (ha)	Ratio (%)
S_1	—	—	—	2	2,675	14.66	2	2,675	14.66
S_2	1, 2, 5	8,750	47.94	1, 5	6,075	33.28	1, 5	6,075	33.28
S_3	3	7,400	40.55	3	7,400	40.55	3	7,400	40.55
N_1	—	—	—	—	—	—	—	—	—
N_2	4	2,050	11.23	4	2,050	11.23	4	2,050	11.23
[a]Mis Land		50	0.27		50	0.27		50	0.27
Total		18,250	100		18,250	100		18,250	100

[a] Miscellaneous Land: (Sand Dune and Riverbed).

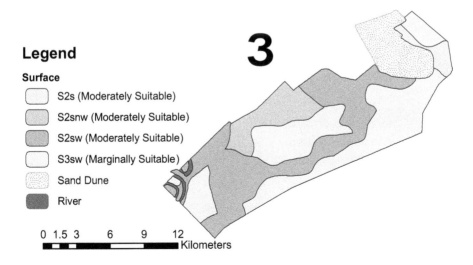

FIGURE 19.2 Land suitability map for surface irrigation.

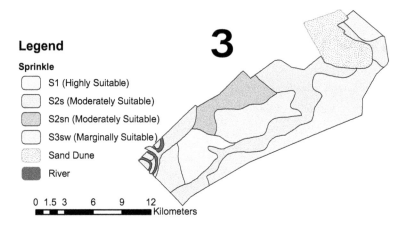

FIGURE 19.3 Land suitability map for sprinkle irrigation.

Regarding sprinkler irrigation (Figure 19.3), the highly suitable area can be observed in part of the cultivated zone in this plain (located in the south) due to deep soil, good drainage, texture, salinity, and proper slope of the area. As seen from the map, some portion of the cultivated area in this plain was evaluated as moderately suitable for sprinkle irrigation because of light limitations of calcium carbonate. Other factors such as drainage, depth, salinity, and slope never influence the suitability of the area. The major portion of the cultivated area in this plain (located in the east and center) is deemed as being marginally suitable land due to high limitations of calcium carbonate and drainage. The currently not suitable area did not exist in this plain. The map also indicates that only part of the cultivated area in this plain was evaluated as permanently not suitable because of very severe limitations of calcium carbonate, salinity, and alkalinity. For almost the total study area, elements such as soil depth, slope, and soil texture were not considered as limiting factors.

For drip irrigation, only soil series coded 2 (2,675 ha – 14.66%) were highly suitable (S_1), while soil series coded 1 and 5 (6,075 ha – 33.28%) were classified as moderately suitable (S_2). Further, only soil series coded 3 (7,400 ha – 40.55%) were found to be slightly suitable (S_3), and only soil series coded 4 (2,050 ha – 11.23%) were classified as permanently not suitable (N_2) for drip irrigation.

Regarding drip irrigation (Figure 19.4), the highly suitable area can be observed in part of the cultivated zone in this plain (located in the south) due to deep soil, good drainage, texture, salinity,

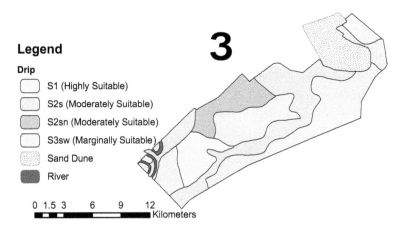

FIGURE 19.4 Land suitability map for drip irrigation.

and proper slope of the area. As seen from the map, a portion of the cultivated area in this plain was evaluated as moderately suitable for drip irrigation because of light limitations of calcium carbonate. Other factors such as drainage, depth, salinity, and slope never influence the suitability of the area. The major portion of the cultivated area in this plain (located in the east and center) is deemed as being marginally suitable land due to high limitations of calcium carbonate and drainage. The currently not suitable area did not exist in this plain. The map also indicates that only part of the cultivated area in this plain was evaluated as permanently not suitable because of very severe limitations of calcium carbonate, salinity, and alkalinity. For almost the total study area, elements such as soil depth, slope, and soil texture were not considered as limiting factors.

The comparison of the capability indices for surface, sprinkle, and drip irrigation (Tables 19.2 and 19.4) indicated that in soil series coded 1, 2, 3, 4, and 5, applying sprinkle irrigation systems was more suitable than surface and drip irrigation systems. Figure 19.5 shows the most suitable map for surface, sprinkle, and drip irrigation systems in the Malashih Plain as per the capability index (Ci) for different irrigation systems. As seen from this map, all of this plain was suitable for a sprinkle irrigation system.

TABLE 19.4

The Suitable Land Units for Surface, Sprinkle, and Drip Irrigation Systems by Notation to Capability Index (Ci) for Different Irrigation Systems

Codes of Land Units	The Maximum Capability Index for Irrigation (Ci)	Suitability Classes	The Most Suitable Irrigation Systems	Limiting Factors
1	72.2	S_2 s	Sprinkle	$CaCO_3$
2	81	S_1	Sprinkle	None Exist
3	58.5	S_3 sw	Sprinkle	$CaCO_3$ and Drainage
4	17.1	N_2 sn	Sprinkle	$CaCO_3$, Salinity and Alkalinity
5	72	S_2 s	Sprinkle	$CaCO_3$

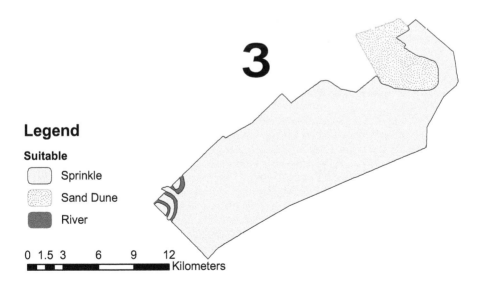

FIGURE 19.5 The most suitable map for different irrigation systems.

19.3　CONCLUSIONS

The results shown in Tables 19.2 and 19.4 indicate that by applying sprinkle irrigation instead of surface and drip irrigation methods, the land suitability of 18,200 ha (99.72%) of the Malashih Plain's land could be improved substantially. The comparison of the different types of irrigation revealed that sprinkle irrigation was more effective and efficient than the drip and surface irrigation methods, and it improved the land suitability for irrigation purposes. The second best option was the application of drip irrigation, which was considered as being more practical than the surface irrigation method. To sum up, the most suitable irrigation systems for the Malashih Plain were sprinkle irrigation, drip irrigation, and surface irrigation, respectively. Rezania et al. (2009); Naseri et al. (2009b); Landi et al. (2008); Albaji et al. (2008, 2009, 2012a, 2014a, 2014b, 2014c, 2015a, 2015b, 2016); Boroomand Nasab et al. (2010); Albaji (2010); Albaji and Hemadi (2011); and Jovzi et al. (2012) found similar results.

Moreover, the main limiting factors in using surface, sprinkle, and drip irrigation methods in this area were calcium carbonate, drainage, salinity, and alkalinity.

ACKNOWLEDGMENTS

We are grateful to the Research Council of Shahid Chamran University of Ahvaz for financial support (GN: SCU.WI98.280).

REFERENCES

Albaji, M., Boroomand Nasab, S., Kashkuli, H.A., Naseri, A.A., Sayyad, G., Jafari, S. 2008. Comparison of different irrigation methods based on the parametric evaluation approach in North Molasani Plain, Iran. *Journal of Agronomy* 7 (2), 187–191.

Albaji, M,. Boroomand-Nasab, S., Naseri, A.A., Jafari, S. 2009. Comparison of different irrigation methods based on the parametric evaluation approach in Abbas Plain: Iran. *Journal of Irrigation and Drainage Engineering* 136 (2), 131–136.

Albaji, M. 2010. Land suitability evaluation for sprinkler irrigation systems. Khuzestan Water and Power Authority (KWPA), Ahvaz, Iran (in Persian).

Albaji, M., Hemadi, J. 2011. Investigation of different irrigation systems based on the parametric evaluation approach on the Dasht Bozorg Plain. *Transactions of the Royal Society of South Africa* 66 (3), 163–169.

Albaji, M., Boroomand Nasab, S., Hemadi, J. 2012a. Comparison of different irrigation methods based on the parametric evaluation approach in West North Ahvaz Plain. In: *Problems, Perspectives and Challenges of Agricultural Water Management*. M. Kumar, (ed.), InTech, Croatia, pp. 259–274.

Albaji, M., Papan, P., Hosseinzadeh, M., Barani, S. 2012b. Evaluation of land suitability for principal crops in the Hendijan region. *International Journal of Modern Agriculture* 1 (1), 24–32.

Albaji, M., Golabi, M., Boroomand Nasab, S., Jahanshahi. M. 2014a. Land suitability evaluation for surface, sprinkler and drip irrigation systems. *Transactions of the Royal Society of South Africa* 69 (2), 63–73.

Albaji, M., Golabi, M., Piroozfar, V.R., Egdernejad, A., Nazari Zadeh, F. 2014b. Evaluation of agricultural land resources for irrigation in the Ramhormoz Plain by using GIS. *Agriculturae Conspectus Scientificus* 79 (2), 93–102.

Albaji, M., Golabi, M., Egdernejad, A., Nazarizadeh, F. 2014c. Assessment of different irrigation systems in Albaji Plain. *Water Science and Technology: Water Supply* 14 (5), 778–786.

Albaji, M., Golabi, M., Boroomand Nasab, S., Zadeh, F.N. 2015a. Investigation of surface, sprinkler and drip irrigation methods based on the parametric evaluation approach in Jaizan Plain. *Journal of the Saudi Society of Agricultural Sciences* 14 (1), 1–10.

Albaji, M., Boroomand Nasab, S., Golabi, M., Sorkheh Nezhad, M., Ahmadee, M. 2015b. Application possibilities of different irrigation methods in Hofel Plain. *Journal of Agricultural Science* 25 (1), 13–23.

Albaji, M., Golabi, M., Hooshmand, A.R., Ahmadee, M. 2016. Investigation of surface, sprinkler and drip irrigation methods using GIS. *Jordan Journal of Agricultural Sciences* 12 (1), 211–222.

Albaji, M., Alboshokeh, A. 2017. Assessing agricultural land suitability in the Fakkeh region, Iran. *Outlook on Agriculture* 46 (1), 57–65.

Behzad, M., Albaji, M., Papan, P., Boroomand Nasab, S., Naseri, A.A., Bavi, A. 2009a. Qualitative evaluation of land suitability for principal crops in the Gargar Region, Khuzestan Province, Southwest Iran. *Asian Journal of Plant Sciences* 8 (1), 28.

Behzad, M., Albaji, M., Papan, P., Boroomand Nasab, S. 2009b. Evan region qualitative soil evaluation for wheat, barley, alfalfa and maize. *Journal of Food, Agriculture & Environment* 7 (2), 843–851.

Boroomand Nasab, S., Albaji, M., Naseri, A.A. 2010. Investigation of different irrigation systems based on the parametric evaluation approach in Boneh Basht plain, Iran. *African Journal of Agricultural Research* 5 (5), 372–379.

Jovzi, M., Albaji, M., Gharibzadeh, A. 2012. Investigating the suitability of lands for surface and under-pressure (drip and sprinkler) irrigation in Miheh Plain. *Research Journal of Environmental Sciences* 6 (2), 51–61.

Khuzestan Water and Power Authority (KWPA). 2011. Semi-detailed soil study report of Malashih Plain, Ahvaz, Iran (in Persian).

Landi, A., Boroomand-Nasab, S., Behzad, M., Tondrow, M.R., Albaji, M., Jazaieri, A. 2008. Land suitability evaluation for surface, sprinkle and drip irrigation methods in Fakkeh Plain, Iran. *Journal of Applied Sciences* 8 (20), 3646–3653.

Mahjoobi, A., Albaji, M., Torfi, K. 2010. Determination of heavy metal levels of Kondok soills-haftgel. *Research Journal of Environmental Sciences* 4 (3), 294–299.

Naseri, A.A., Albaji, M., Boroomand Nasab, S., Landi, A., Papan, P., Bavi, A. 2009a. Land suitability evaluation for principal crops in the Abbas Plain, Southwest Iran. *Journal of Food, Agriculture & Environment* 7 (1), 208–213.

Naseri, A.A., Rezania, A.R., Albaji, M. 2009b. Investigation of soil quality for different irrigation systems in Lali Plain, Iran. *Journal of Food,Agriculture & Environment* 7 (3&4), 955–960.

Rezania, A.R., Naseri, A.A., Albaji, M. 2009. Assessment of soil properties for irrigation methods in North Andimeshk Plain, Iran. *Journal of Food, Agriculture & Environment* 7 (3&4), 728–733.

20 Best Irrigation System Determination in Ramshir Plain, Iran

Mohammad Albaji, Abd Ali Naseri, Mona Golabi, and Saeid Eslamian

CONTENTS

20.1 INTRODUCTION

The scarcity of water in arid and semiarid regions such as Iran is a restrictive element for the agricultural sector; therefore, there is an urgent need to develop initiatives to save water in this particular sector (Chegah et al., 2013; Abyaneh et al., 2017). Among the proposed solutions is the use of pressurized irrigation systems.

The present study was conducted in an area about 16,000 hectares in the Ramshir Plain, in Khuzestan Province, located in the southwest of Iran, during 2011–2012. The study area is located 100 km west of the city of Ahvaz, 30° 45′ to 30° 56′ N and 49° 15′ to 49° 28′ E. The average annual temperature and precipitation for the period of 1965–2011 were 25.2°C and 218 mm, respectively. Also, the annual evaporation of the area is 2,550 mm (Khuzestan Water and Power Authority, 2011). The Jarahi River supplies the bulk of the water demands of the region. The application of irrigated agriculture has been common in the study area. Currently, the irrigation systems used by farmlands in the region are furrow irrigation, basin irrigation, and border irrigation schemes.

The area is composed of two distinct physiographic features, i.e., river alluvial plains and river terraces, of which the river alluvial plains' physiographic unit is the dominant feature. Also, five different soil series were found in the area. The semi-detailed soil survey report of the Ramshir Plain (Khuzestan Water and Power Authority, 2011) was used in order to determine the soil characteristics.

Over much of the Ramshir Plain, the use of surface irrigation systems has been applied specifically for field crops to meet the water demand of both summer and winter crops. The major irrigated broad-acre crops grown in this area are wheat, barley, and maize, in addition to fruits, melons, watermelons, and vegetables such as tomatoes and cucumbers. But like other plains in Khuzstan Province, the important major crops are wheat and barley (Behzad et al., 2009a, 2009b; Naseri et al., 2009a; Albaji et al., 2012b; Albaji and Alboshokeh, 2017). There are very few instances of sprinkle and drip irrigation on large-area farms in the Ramshir Plain.

Five soil series and 29 series phases or land units were derived from the semi-detailed soil study of the area. The land units are shown in Figure 20.1 as the basis for further land evaluation practice. The soils of the area are of Aridisols orders. Also, the soil moisture regime is aridic and ustic, while the soil temperature regime is hyperthermic (Khuzestan Water and Power Authority, 2011).

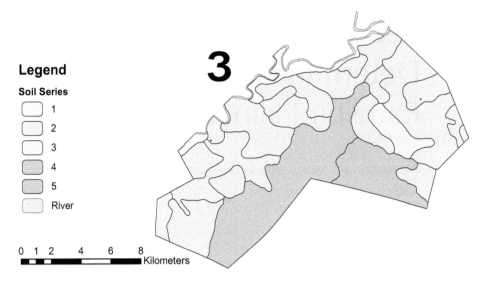

FIGURE 20.1   Soil map of the study area.

Therefore, the organic matter content is low due to high soil temperature and high decomposition ratio of this matter (Mahjoobi et al., 2010).

20.2   RESULTS AND DISCUSSIONS

As shown in Tables 20.1 and 20.2 for surface irrigation, only soil series coded 2 (9,900 ha – 61.88%) were classified as moderately suitable (S_2); only soil series coded 1 (625 ha – 3.91%) were found to be marginally suitable (S_3). Soil series coded 3 and 4 (3,875 ha – 24.21%) were classified as currently not suitable (N_1), and only soil series coded 5(1,400 ha – 8.75%) were classified as permanently not suitable (N_2) for any surface irrigation practices.

The analyses of the suitability irrigation maps for surface irrigation (Figure 20.2) indicate that there was no highly suitable land in this plain. The major portion of the cultivated area in this plain (located in the north, south, and center) is deemed as being moderately suitable due to light limitations of calcium carbonate and heavy soil texture. Other factors such as drainage, depth, salinity, and alkalinity have no influence on the suitability of the area whatsoever. The marginally suitable area is located in

TABLE 20.1
Ci Values and Suitability Classes of Surface, Sprinkle, and Drip irrigation for Each Land Unit

Codes of Land Units	Surface Irrigation		Sprinkle Irrigation		Drip Irrigation	
	Ci	Suitability Classes	Ci	Suitability Classes	Ci	Suitability Classes
1	58.5	S_{3S}[a]	72	$S_{2}S$[b]	66.5	$S_{2}S$[c]
2	70.2	S_{2S}	72	$S_{2}S$	63	$S_{2}S$
3	37.44	N_1 snw	44.2	N_1 snw	41.65	N_1 snw
4	31.32	N_1 snw	37.29	N_1 snw	35.7	N_1 snw
5	26.85	N_2 snw	32.32	N_1 snw	30.94	N_1 snw

[a,b & c] Limiting Factors for Surface and Sprinkle and Drip Irrigations: s: (Calcium Carbonate & Heavy Soil Texture), n: (Salinity & Alkalinity), and w: (Drainage).

TABLE 20.2

Distribution of Surface, Sprinkle, and Drip Irrigation Suitability

	Surface Irrigation			Sprinkle Irrigation			Drip Irrigation		
Suitability	Land Unit	Area (ha)	Ratio (%)	Land Unit	Area (ha)	Ratio (%)	Land Unit	Area (ha)	Ratio (%)
S_1	—	—	—	—	—	—	—	—	—
S_2	2	9,900	61.88	1, 2	10,525	65.79	1, 2	10,525	65.79
S_3	1	625	3.91	—	—	—	—	—	—
N_1	3, 4	3,875	24.21	3, 4, 5	5,275	32.96	3, 4, 5	5,275	32.96
N_2	5	1,400	8.75	—	—	—	—	—	—
[a]Mis Land		200	1.25		200	1.25		200	1.25
Total		16,000	100		16,000	100		16,000	100

[a] Miscellaneous Land: (Hill, Sand Dune, and Riverbed).

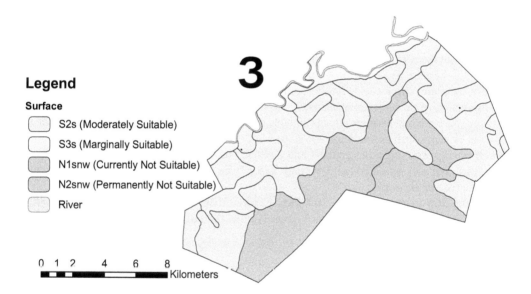

FIGURE 20.2 Land suitability map for surface irrigation.

the west of this area due to high limitations of calcium carbonate and heavy soil texture. The currently not suitable land can be observed in the east of the plain because of severe limitations of calcium carbonate, heavy soil texture, drainage, salinity, and alkalinity. The map also indicates that some part of the cultivated area in this plain was evaluated as permanently not suitable land because of very severe limitations of calcium carbonate, heavy soil texture, drainage, salinity, and alkalinity. For almost the total study area, elements such as soil depth and slope were not considered as limiting factors.

In order to verify the possible effects of different management practices, the land suitability for sprinkle and drip irrigation was evaluated (Tables 20.1 and 20.2).

For sprinkle irrigation, soil series coded 1 and 2 (10,525 ha – 65.79%) were classified as moderately suitable (S_2), and soil series coded 3, 4, and 5 (5,275 ha – 32.96%) were classified as currently not suitable (N_1) for sprinkle irrigation.

Regarding sprinkler irrigation, Figure 20.3 indicates that there was no highly suitable land in this plain. The moderately suitable area can be observed in the largest part of the cultivated zone in this

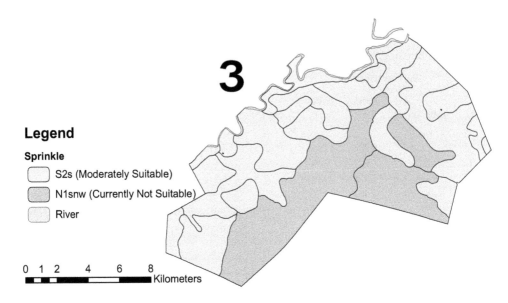

FIGURE 20.3 Land suitability map for sprinkle irrigation.

plain (located in the north, south, and center) due to light limitations of calcium carbonate and heavy soil texture of the area. As seen from the map, part of the cultivated area in this plain (located only in the east of the plain) was evaluated as currently not suitable lands for sprinkle irrigation because of severe limitations of calcium carbonate, heavy soil texture, drainage, salinity, and alkalinity. The marginally suitable lands and permanently not suitable lands did not exist in this plain. For almost the total study area, elements such as soil depth and slope were not considered as limiting factors.

For drip irrigation, soil series coded 1 and 2 (10,525 ha – 65.79%) were classified as moderately suitable (S_2), and soil series coded 3, 4, and 5 (5,275 ha – 32.96%) were classified as currently not suitable (N_1) for drip irrigation.

Regarding drip irrigation, Figure 20.4 indicates that there was no highly suitable land in this plain. The moderately suitable area can be observed in the largest part of the cultivated zone in

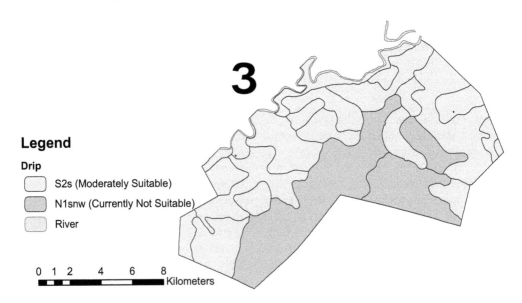

FIGURE 20.4 Land suitability map for drip irrigation.

this plain (located in the north, south, and center) due to light limitations of calcium carbonate and heavy soil texture of the area. As seen from the map, part of the cultivated area in this plain (located only in the east of the plain) was evaluated as currently not suitable lands for drip irrigation because of severe limitations of calcium carbonate, heavy soil texture, drainage, salinity, and alkalinity. The marginally suitable lands and permanently not suitable lands did not exist in this plain. For almost the total study area, elements such as soil depth and slope were not considered as limiting factors.

The comparison of the capability indices for surface, sprinkle, and drip irrigation (Tables 20.1 and 20.3) indicated that in soil series coded 1, 2, 3, 4 and 5, applying sprinkle irrigation systems was more suitable than surface and drip irrigation systems. Figure 20.5 shows the most suitable map for surface, sprinkle, and drip irrigation systems in the Ramshir Plain as per the capability index (Ci) for different irrigation systems. As seen from this map, all of this plain was suitable for sprinkle irrigation systems.

TABLE 20.3

The Most Suitable Land Units for Surface, Sprinkle, and Drip Irrigation Systems by Notation to Capability Index (Ci) for Different Irrigation Systems

Codes of Land Units	The Maximum Capability Index for Irrigation (Ci)	Suitability Classes	The Most Suitable Irrigation Systems	Limiting Factors
1	72	S_{2s}	Sprinkle	$CaCO_3$ and Heavy Soil Texture
2	72	S_{2s}	Sprinkle	$CaCO_3$ and Heavy Soil Texture
3	44.2	N_1 snw	Sprinkle	$CaCO_3$, Heavy Soil Texture, Salinity, Alkalinity, and Drainage
4	37.29	N_1 snw	Sprinkle	$CaCO_3$, Heavy Soil Texture, Salinity, Alkalinity, and Drainage
5	32.32	N_1 snw	Sprinkle	$CaCO_3$, Heavy Soil Texture, Salinity, Alkalinity, and Drainage

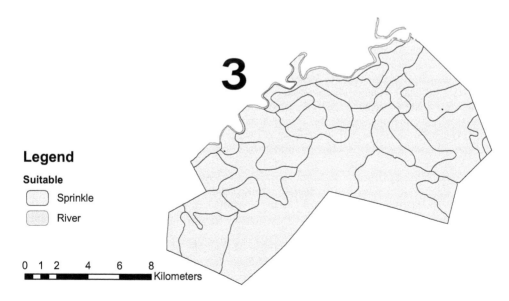

FIGURE 20.5   The most suitable map for different irrigation systems.

20.3   CONCLUSIONS

The results of Tables 20.1 and 20.3 indicated that by applying sprinkle irrigation instead of surface and drip irrigation methods, the land suitability of 15,800 ha (98.75%) of the Ramshir Plain's land could be improved substantially. The comparison of the different types of irrigation revealed that sprinkle irrigation was more effective and efficient than the drip and surface irrigation methods, and it improved land suitability for irrigation purposes. The second best option was the application of drip irrigation, which was considered as being more practical than the surface irrigation method. To sum up, the most suitable irrigation systems for the Ramshir Plain were sprinkle irrigation, drip irrigation, and surface irrigation, respectively. Other researchers have confirmed these results (Albaji et al. 2008, 2009, 2012a, 2014a, 2014b, 2014c, 2015a, 2015b, 2016; Landi et al., 2008; Rezania et al., 2009; Naseri et al., 2009b; Albaji, 2010; Boroomand Nasab et al., 2010; Albaji and Hemadi, 2011).

Moreover, the main limiting factors in using surface sprinkle and drip irrigation methods in this area were calcium carbonate, heavy soil texture, drainage, salinity, and alkalinity.

ACKNOWLEDGMENTS

We are grateful to the Research Council of Shahid Chamran University of Ahvaz for financial support (GN: SCU.WI98.280).

REFERENCES

Abyaneh, H.Z., Jovzi, M., Albaji, M. 2017. Effect of regulated deficit irrigation, partial root drying and N-fertilizer levels on sugar beet crop (*Beta vulgaris* L.). *Agricultural Water Management* 194, 13–23.

Albaji, M., Boroomand Nasab, S., Kashkuli, H.A., Naseri, A.A., Sayyad, G., Jafari, S. 2008. Comparison of different irrigation methods based on the parametric evaluation approach in North Molasani Plain, Iran. *Journal of Agronomy* 7 (2), 187–191.

Albaji, M,. Boroomand-Nasab, S., Naseri, A.A., Jafari, S. 2009. Comparison of different irrigation methods based on the parametric evaluation approach in Abbas Plain: Iran. *Journal of Irrigation and Drainage Engineering* 136 (2), 131–136.

Albaji, M. 2010. Land suitability evaluation for sprinkler irrigation systems. Khuzestan Water and Power Authority (KWPA), Ahvaz, Iran (in Persian).

Albaji, M., Hemadi, J. 2011. Investigation of different irrigation systems based on the parametric evaluation approach on the Dasht Bozorg Plain. *Transactions of the Royal Society of South Africa* 66 (3), 163–169.

Albaji, M., Boroomand Nasab, S., Hemadi, J. 2012a. Comparison of different irrigation methods based on the parametric evaluation approach in West North Ahvaz Plain. In: *Problems, Perspectives and Challenges of Agricultural Water Management*. M. Kumar, (ed.), InTech, Croatia, pp. 259–274.

Albaji, M., Papan, P., Hosseinzadeh, M., Barani, S. 2012b. Evaluation of land suitability for principal crops in the Hendijan region. *International Journal of Modern Agriculture* 1 (1), 24–32.

Albaji, M., Golabi, M., Boroomand Nasab, S., Jahanshahi. M. 2014a. Land suitability evaluation for surface, sprinkler and drip irrigation systems. *Transactions of the Royal Society of South Africa* 69 (2), 63–73.

Albaji, M., Golabi, M., Piroozfar, V.R., Egdernejad, A., Nazari Zadeh, F. 2014b. Evaluation of agricultural land resources for irrigation in the Ramhormoz Plain by using GIS. *Agriculturae Conspectus Scientificus* 79 (2), 93–102.

Albaji, M., Golabi, M., Egdernejad, A., Nazarizadeh, F. 2014c. Assessment of different irrigation systems in Albaji Plain. *Water Science and Technology: Water Supply* 14 (5), 778–786.

Albaji, M., Golabi, M., Boroomand Nasab, S., Zadeh, F.N. 2015a. Investigation of surface, sprinkler and drip irrigation methods based on the parametric evaluation approach in Jaizan Plain. *Journal of the Saudi Society of Agricultural Sciences* 14 (1), 1–10.

Albaji, M., Boroomand Nasab, S., Golabi, M., Sorkheh Nezhad, M., Ahmadee, M. 2015b. Application possibilities of different irrigation methods in Hofel Plain. *Journal of Agricultural Science* 25 (1), 13–23.

Albaji, M., Golabi, M., Hooshmand, A.R., Ahmadee, M. 2016. Investigation of surface, sprinkler and drip irrigation methods using GIS. *Jordan Journal of Agricultural Sciences* 12 (1), 211–222.

Albaji, M., Alboshokeh, A. 2017. Assessing agricultural land suitability in the Fakkeh region, Iran. *Outlook on Agriculture* 46 (1), 57–65.

Behzad, M., Albaji, M., Papan, P., Boroomand Nasab, S., Naseri, A.A., Bavi, A. 2009a. Qualitative evaluation of land suitability for principal crops in the Gargar Region, Khuzestan Province, Southwest Iran. *Asian Journal of Plant Sciences* 8 (1), 28.

Behzad, M., Albaji, M., Papan, P., Boroomand Nasab, S. 2009b. Evan region qualitative soil evaluation for wheat, barley, alfalfa and maize. *Journal of Food, Agriculture & Environment* 7 (2), 843–851.

Boroomand Nasab, S., Albaji, M., Naseri, A.A. 2010. Investigation of different irrigation systems based on the parametric evaluation approach in Boneh Basht plain, Iran. *African Journal of Agricultural Research* 5 (5), 372–379.

Chegah, S., Chehrazi, M., Albaji, M. 2013. Effects of drought stress on growth and development frankenia plant (Frankenia Leavis). *Bulgarian Journal of Agricultural Science* 19 (4), 659–666.

Khuzestan Water and Power Authority (KWPA). 2011. Semi-detailed soil study report of Ramshir Plain, Ahvaz, Iran (in Persian).

Landi, A., Boroomand-Nasab, S., Behzad, M., Tondrow, M.R., Albaji, M., Jazaieri, A. 2008. Land suitability evaluation for surface, sprinkle and drip irrigation methods in Fakkeh Plain, Iran. *Journal of Applied Sciences* 8 (20), 3646–3653.

Mahjoobi, A., Albaji, M., Torfi, K. 2010. Determination of heavy metal levels of Kondok soills-haftgel. *Research Journal of Environmental Sciences* 4 (3), 294–299.

Naseri, A.A., Albaji, M., Boroomand Nasab, S., Landi, A., Papan, P., Bavi, A. 2009a. Land suitability evaluation for principal crops in the Abbas Plain, Southwest Iran. *Journal of Food, Agriculture & Environment* 7 (1), 208–213.

Naseri, A.A., Rezania, A.R., Albaji, M. 2009b. Investigation of soil quality for different irrigation systems in Lali Plain, Iran. *Journal of Food, Agriculture & Environment* 7 (3&4), 955–960.

Rezania, A.R., Naseri, A.A., Albaji, M. 2009. Assessment of soil properties for irrigation methods in North Andimeshk Plain, Iran. *Journal of Food, Agriculture & Environment* 7 (3&4), 728–733.

21 Irrigation System in Khalf Abad Plain, Iran

Mohammad Albaji, Abd Ali Naseri, and Saeid Eslamian

CONTENTS

21.1 INTRODUCTION

The scarcity of water in arid and semiarid regions such as Iran is a restrictive element for the agricultural sector; therefore, there is an urgent need to develop initiatives to save water in this particular sector (Chegah et al., 2013; Abyaneh et al., 2017). The present study was conducted in an area about 63,525 hectares in the Khalf Abad Plain in Khuzestan Province, located in the southwest of Iran, during 2011–2012. The study area is located 80 km west of the city of Ahvaz, 30° 40′ to 31° 00′ N and 49° 00′ to 49° 30′ E. The average annual temperature and precipitation for the period of 1965–2011 were 24.42°C and 218.9 mm, respectively. Also, the annual evaporation of the area is 2550 mm (Khuzestan Water and Power Authority, 2010). The Jarahi River supplies the bulk of the water demands of the region. The application of irrigated agriculture has been common in the study area. Currently, the irrigation systems used by farmlands in the region are furrow irrigation, basin irrigation, and border irrigation schemes.

The area is composed of two distinct physiographic features, i.e., river alluvial plains and plateaus, of which the river alluvial plains' physiographic unit is the dominant feature. Also, eight different soil scrics wcrc found in the area. The semi-detailed soil survey report of the Khalf Abad Plain (Khuzestan Water and Power Authority, 2010) was used in order to determine the soil characteristics.

Over much of the Khalf Abad Plain, the use of surface irrigation systems has been applied specifically for field crops to meet the water demand of both summer and winter crops. The major irrigated broad-acre crops grown in this area are wheat, barley, and maize, in addition to fruits, melons, watermelons, and vegetables such as tomatoes and cucumbers. But like other plains in Khuzstan Province, the important major crops are wheat and barley (Behzad et al., 2009a, 2009b; Naseri et al., 2009; Albaji et al., 2012b; Albaji and Alboshokeh, 2017). There are very few instances of sprinkle and drip irrigation on large-area farms in the Khalf Abad Plain.

Eight soil series and 38 series phases or land units were derived from the semi-detailed soil study of the area. The land units are shown in Figure 21.1 as the basis for further land evaluation practice. The soils of the area are of Aridisols and Entisols orders. Also, the soil moisture regime is aridic, while the soil temperature regime is hyperthermic (Khuzestan Water and Power Authority, 2010). Therefore, the organic matter content is low due to high soil temperature and high decomposition ratio of this matter (Mahjoobi et al., 2010).

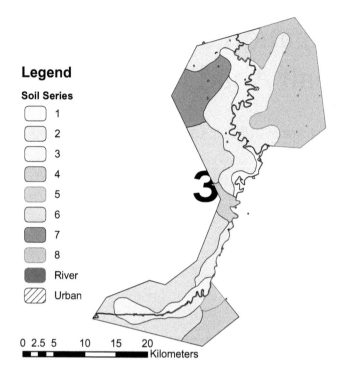

FIGURE 21.1   Soil map of the study area.

21.2   RESULTS AND DISCUSSIONS

As shown in Tables 21.1 and 21.2 for surface irrigation, only soil series coded 2 (18,025 ha – 28.38%) were classified as moderately suitable (S_2); soil series coded 1 and 8 (15,900 ha – 25.03%) were found to be marginally suitable (S_3). Only soil series coded 3 (4,525 ha – 7.12%) were classified as

---

TABLE 21.1

Ci Values and Suitability Classes of Surface, Sprinkle, and Drip irrigation for Each Land Unit

Codes of Land Units	Surface Irrigation		Sprinkle Irrigation		Drip Irrigation	
	Ci	Suitability Classes	Ci	Suitability Classes	Ci	Suitability Classes
1	55.57	$S_{3\,S}$[a]	68.4	$S_{2\,S}$[b]	63.17	$S_{2\,S}$[c]
2	60.02	S_2 sw	64.98	$S_{2\,S}$	59.15	$S_{3\,S}$
3	35.1	N_1 snw	38	N_1 sn	35	N_1 sn
4	12.63	N_2 snw	13.68	N_2 sn	12.6	N_2 sn
5	22.11	N_2 snw	23.94	N_2 sn	22.05	N_2 sn
6	9.36	N_2 snw	11.2	N_2 snw	11.2	N_2 snw
7	9.36	N_2 snw	11.2	N_2 snw	11.2	N_2 snw
8	49.14	S_3 snw	60.75	S_2 snw	60	S_2 sn

[a, b, c] Limiting Factors for Surface and Sprinkle and Drip Irrigations: n: (Salinity & Alkalinity), s: (Calcium Carbonate), and w: (Drainage).

TABLE 21.2

Distribution of Surface, Sprinkle, and Drip Irrigation Suitability

Suitability	Surface Irrigation			Sprinkle Irrigation			Drip Irrigation		
	Land Unit	Area (ha)	Ratio (%)	Land unit	Area (ha)	Ratio (%)	Land Unit	Area (ha)	Ratio (%)
S_1	—	—	—	—	—	—	—	—	—
S_2	2	18,025	28.38	1, 2, 8	33,925	53.41	1, 8	15,900	25.03
S_3	1, 8	15,900	25.03	—	—	—	2	18,025	28.38
N_1	3	4,525	7.12	3	4,525	7.12	3	4,525	7.12
N_2	4, 5, 6, 7	23,975	37.74	4, 5, 6, 7	23,975	37.74	4, 5, 6, 7	23,975	37.74
[a]Mis Land		1,100	1.73		1,100	1.73		1,100	1.73
Total		63,525	100		63,525	100		63,525	100

[a] Miscellaneous Land: (Hill, Sand Dune, and Riverbed).

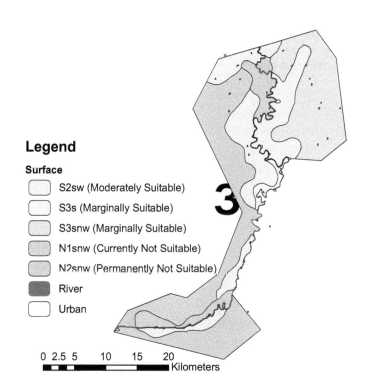

Legend

Surface

- S2sw (Moderately Suitable)
- S3s (Marginally Suitable)
- S3snw (Marginally Suitable)
- N1snw (Currently Not Suitable)
- N2snw (Permanently Not Suitable)
- River
- Urban

0 2.5 5   10   15   20
Kilometers

FIGURE 21.2 Land suitability map for surface irrigation.

currently not suitable (N_1), and soil series coded 4, 5, 6, and 7 (23,975 ha – 37.74%) were classified as permanently not suitable (N_2) for any surface irrigation practices.

The analyses of the suitability irrigation maps for surface irrigation (Figure 21.2) indicate that there was no highly suitable land in this plain. The moderately suitable area is located to the north and east of this area due to light limitations of drainage and calcium carbonate. Other factors such as slope, depth, salinity, and alkalinity have no influence on the suitability of the area whatsoever. The map also indicates that only part of the cultivated area in this plain was evaluated as marginally suitable because of the high limitations of calcium carbonate, drainage, salinity, and alkalinity. The currently not suitable land can be observed only in the north of the plain due to severe limitations of drainage, salinity, and alkalinity. The major portion of the cultivated area in this plain (located in the south, west, and center) is deemed as being permanently not suitable land due to very severe limitations of calcium carbonate, drainage, salinity, and alkalinity. For almost the total study area, elements such as soil depth, soil texture, and slope were not considered as limiting factors.

In order to verify the possible effects of different management practices, the land suitability for sprinkle and drip irrigation was evaluated (Tables 21.1 and 21.2).

For sprinkle irrigation, soil series coded 1, 2 and 8 (33,925 ha – 53.41%) were classified as moderately suitable (S_2). Only soil series coded 3 (4,525 ha – 7.12%) were classified as currently not suitable (N_1), and soil series coded 4, 5, 6, and 7 (23,975 ha – 37.74%) were classified as permanently not suitable (N_2) for sprinkle irrigation.

Regarding sprinkle irrigation (Figure 21.3), there was no highly suitable land in this plain. The moderately suitable area can be observed in the largest part of the cultivated zone in this plain (located in the north and east) due to light limitations of calcium carbonate, drainage, salinity, and alkalinity. Other factors such as soil texture, depth, and slope never influence the suitability of the area. The marginally suitable lands did not exist in this plain. As seen from the map, part of the cultivated area in this plain was evaluated as currently not suitable lands for sprinkle irrigation because of the severe limitations of calcium carbonate, drainage, salinity, and alkalinity. Some portions of the cultivated area in this plain (located in the south, west, and center) are deemed as being permanently not suitable land due to very severe limitations of calcium carbonate, drainage,

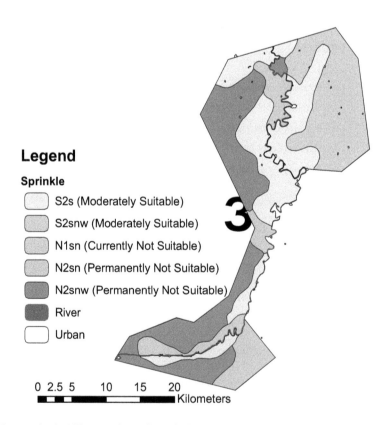

Legend

Sprinkle

- S2s (Moderately Suitable)
- S2snw (Moderately Suitable)
- N1sn (Currently Not Suitable)
- N2sn (Permanently Not Suitable)
- N2snw (Permanently Not Suitable)
- River
- Urban

0 2.5 5   10   15   20
Kilometers

FIGURE 21.3   Land suitability map for sprinkle irrigation.

salinity, and alkalinity. For almost the total study area, elements such as soil depth, soil texture, and slope were not considered as limiting factors.

For drip irrigation, soil series coded 1 and 8 (15,900 ha – 25.03%) were classified as moderately suitable (S_2). Further, only soil series coded 2 (18,025 ha – 28.38%) were found to be slightly suitable (S_3). Only soil series coded 3 (4,525 ha – 7.12%) were classified as currently not suitable (N_1), and soil series coded 4, 5, 6, and 7 (23,975 ha – 37.74%) were classified as permanently not suitable (N_2) for drip irrigation.

Regarding drip irrigation (Figure 21.4), there was no highly suitable land in this plain. The moderately suitable lands could be observed over some portion of the plain (north and northeast parts) due to the medium content of calcium carbonate and light limitations of salinity and alkalinity. The marginally suitable lands were found in the north and east of the area. The limiting factor for this land unit was high content of calcium carbonate. The currently not suitable land can be observed in the north and south of the plain due to severe limitations of salinity, alkalinity, and high content of calcium. The major portion of the cultivated area in this plain (located in the south, west, and center) is deemed as being permanently not suitable land due to very severe limitations of calcium carbonate, drainage, salinity, and alkalinity. For almost the total study area, elements such as soil depth, soil texture, and slope were not considered as limiting factors.

The comparison of the capability indices for surface, sprinkle, and drip irrigation (Tables 21.1 and 21.3) indicated that in soil series coded 1 and 8, applying drip and sprinkle irrigation systems was more suitable than surface irrigation systems. In soil series coded 2, applying sprinkle and

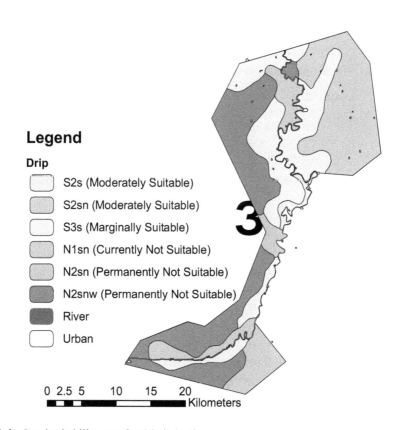

Legend

Drip

☐ S2s (Moderately Suitable)

☐ S2sn (Moderately Suitable)

☐ S3s (Marginally Suitable)

☐ N1sn (Currently Not Suitable)

☐ N2sn (Permanently Not Suitable)

☐ N2snw (Permanently Not Suitable)

☐ River

☐ Urban

0 2.5 5   10   15   20
Kilometers

FIGURE 21.4   Land suitability map for drip irrigation.

TABLE 21.3

The Most Suitable Land Units for Surface, Sprinkle, and Drip Irrigation Systems by Notation to Capability Index (Ci) for Different Irrigation Systems

Codes of Land Units	Suitability Classes	The Most Suitable Irrigation Systems	Limiting Factors
1	S_{2s}	Sprinkle and Drip	$CaCO_3$
2	S_{2s}	Sprinkle and Surface	$CaCO_3$ and Drainage
3	N_1 sn	Sprinkle, Drip, and Surface	$CaCO_3$, Salinity, Alkalinity and Drainage
4	N_2 sn	Sprinkle, Drip, and Surface	$CaCO_3$, Salinity, Alkalinity and Drainage
5	N_2 sn	Sprinkle, Drip, and Surface	$CaCO_3$, Salinity, Alkalinity and Drainage
6	N_2 snw	Sprinkle, Drip, and Surface	$CaCO_3$, Salinity, Alkalinity and Drainage
7	N_2 snw	Sprinkle, Drip, and Surface	$CaCO_3$, Salinity, Alkalinity and Drainage
8	S_2 snw	Sprinkle and Drip	$CaCO_3$, Salinity, Alkalinity and Drainage

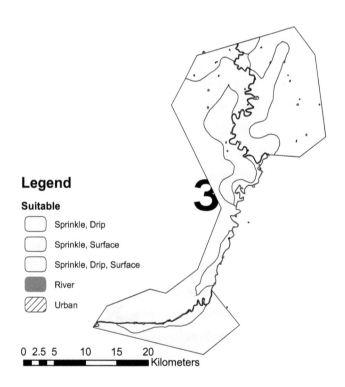

FIGURE 21.5 The most suitable map for different irrigation systems.

surface irrigation systems was the most suitable option as compared to drip irrigation systems. In soil series coded 3, 4, 5, 6, and 7, applying different irrigation systems was the same. Figure 21.5 shows the most suitable map for surface, sprinkle, and drip irrigation systems in the Khalf Abad Plain as per the capability index (Ci) for different irrigation systems. As seen from this map, the largest part of this plain was suitable for sprinkle irrigation systems, and some parts of this area were suitable for surface and drip irrigation systems.

21.3 CONCLUSIONS

The comparison of the different types of irrigation revealed that sprinkle irrigation was more effective and efficient than the drip and surface irrigation methods, and it improved land suitability for irrigation purposes. The second best option was the application of drip and surface irrigation. Research carried out by Albaji et al. (2008, 2009, 2014a, 2014b, 2014c, 2015a, 2015b and 2016); Landi et al. (2008); Rezania et al. (2009); Boroomand Nasab et al. (2010); Albaji (2010); Albaji and Hemadi (2011); Albaji et al. (2012a); and Jovzi et al. (2012) confirm these results.

Moreover, the main limiting factors in using surface, sprinkle, and drip irrigation methods in this area were salinity, alkalinity, drainage, and carbonate content.

ACKNOWLEDGMENTS

We are grateful to the Research Council of Shahid Chamran University of Ahvaz for financial support (GN: SCU.WI98.280).

REFERENCES

Abyaneh, H.Z., Jovzi, M., Albaji, M. 2017. Effect of regulated deficit irrigation, partial root drying and N-fertilizer levels on sugar beet crop (*Beta vulgaris* L.). *Agricultural Water Management* 194, 13–23.

Albaji, M., Boroomand Nasab, S., Kashkuli, H.A., Naseri, A.A., Sayyad, G., Jafari, S. 2008. Comparison of different irrigation methods based on the parametric evaluation approach in North Molasani Plain, Iran. *Journal of Agronomy* 7 (2), 187–191.

Albaji, M., Boroomand-Nasab, S., Naseri, A.A., Jafari, S. 2009. Comparison of different irrigation methods based on the parametric evaluation approach in Abbas Plain: Iran. *Journal of Irrigation and Drainage Engineering* 136 (2), 131–136.

Albaji, M. 2010. Land suitability evaluation for sprinkler irrigation systems. Khuzestan Water and Power Authority (KWPA), Ahvaz, Iran (in Persian).

Albaji, M., Hemadi, J. 2011. Investigation of different irrigation systems based on the parametric evaluation approach on the Dasht Bozorg Plain. *Transactions of the Royal Society of South Africa* 66 (3), 163–169.

Albaji, M., Boroomand Nasab, S., Hemadi, J. 2012a. Comparison of different irrigation methods based on the parametric evaluation approach in West North Ahvaz Plain. In: *Problems, Perspectives and Challenges of Agricultural Water Management*. M. Kumar, (ed.), InTech, Croatia, pp. 259–274.

Albaji, M., Papan, P., Hosseinzadeh, M., Barani, S. 2012b. Evaluation of land suitability for principal crops in the Hendijan region. *International Journal of Modern Agriculture* 1 (1), 24–32.

Albaji, M., Golabi, M., Boroomand Nasab, S., Jahanshahi, M. 2014a. Land suitability evaluation for surface, sprinkler and drip irrigation systems. *Transactions of the Royal Society of South Africa* 69 (2), 63–73.

Albaji, M., Golabi, M., Piroozfar, V.R., Egdernejad, A., Nazari Zadeh, F. 2014b. Evaluation of agricultural land resources for irrigation in the Ramhormoz plain by using GIS. *Agriculturae Conspectus Scientificus* 79 (2), 93–102.

Albaji, M., Golabi, M., Egdernejad, A., Nazarizadeh, F. 2014c. Assessment of different irrigation systems in Albaji Plain.*Water Science and Technology: Water Supply* 14 (5), 778–786.

Albaji, M., Golabi, M., Boroomand Nasab, S., Zadeh, F.N. 2015a. Investigation of surface, sprinkler and drip irrigation methods based on the parametric evaluation approach in Jaizan Plain. *Journal of the Saudi Society of Agricultural Sciences* 14 (1), 1–10.

Albaji, M., Boroomand Nasab, S., Golabi, M., Sorkheh Nezhad, M., Ahmadee, M. 2015b. Application possibilities of different irrigation methods in Hofel Plain. *Journal of Agricultural Science* 25 (1), 13–23.

Albaji, M., Golabi, M., Hooshmand, A.R., Ahmadee, M. 2016. Investigation of surface, sprinkler and drip irrigation methods using GIS. *Jordan Journal of Agricultural Sciences* 12 (1), 211–222.

Albaji, M., Alboshokeh, A. 2017. Assessing agricultural land suitability in the Fakkeh region, Iran. *Outlook on Agriculture* 46 (1), 57–65.

Behzad, M., Albaji, M., Papan, P., Boroomand Nasab, S., Naseri, A.A., Bavi, A. 2009a. Qualitative evaluation of land suitability for principal crops in the Gargar Region, Khuzestan Province, Southwest Iran. *Asian Journal of Plant Sciences* 8 (1), 28.

Behzad, M., Albaji, M., Papan, P., Boroomand Nasab, S. 2009b. Evan region qualitative soil evaluation for wheat, barley, alfalfa and maize. *Journal of Food, Agriculture & Environment* 7 (2), 843–851.

Boroomand Nasab, S., Albaji, M., Naseri, A.A. 2010. Investigation of different irrigation systems based on the parametric evaluation approach in Boneh Basht plain, Iran. *African Journal of Agricultural Research* 5 (5), 372–379.

Chegah, S., Chehrazi, M., Albaji, M. 2013. Effects of drought stress on growth and development frankenia plant (Frankenia Leavis). *Bulgarian Journal of Agricultural Science* 19 (4), 659–666.

Jovzi, M., Albaji, M., Gharibzadeh, A. 2012. Investigating the suitability of lands for surface and under-pressure (drip and sprinkler) irrigation in Miheh Plain. *Research Journal of Environmental Sciences* 6 (2), 51–61.

Khuzestan Water and Power Authority (KWPA). 2010. Semi-detailed soil study report of Khalf Abad Plain, Ahvaz, Iran (in Persian).

Landi, A., Boroomand-Nasab, S., Behzad, M., Tondrow, M.R., Albaji, M., Jazaieri, A. 2008. Land suitability evaluation for surface, sprinkle and drip irrigation methods in Fakkeh Plain, Iran. *Journal of Applied Sciences* 8 (20), 3646–3653.

Mahjoobi, A., Albaji, M., Torfi, K. 2010. Determination of heavy metal levels of Kondok soills-haftgel. *Research Journal of Environmental Sciences* 4 (3), 294–299.

Naseri, A.A., Albaji, M., Boroomand Nasab, S., Landi, A., Papan, P., Bavi, A. 2009. Land suitability evaluation for principal crops in the Abbas Plain, Southwest Iran. *Journal of Food, Agriculture & Environment* 7 (1), 208–213.

Rezania, A.R., Naseri, A.A., Albaji, M. 2009. Assessment of soil properties for irrigation methods in North Andimeshk Plain, Iran. *Journal of Food, Agriculture & Environment* 7 (3&4), 728–733.

22 Irrigation System Option for Shahid Rajaie Plain, Iran

Mohammad Albaji, Saeed Boroomand Nasab, and Saeid Eslamian

CONTENTS

22.1 INTRODUCTION

The scarcity of water in arid and semiarid regions such as Iran is a restrictive element for the agricultural sector; therefore, there is an urgent need to develop initiatives to save water in this particular sector (Chegah et al., 2013; Abyaneh et al., 2017). The present study was conducted in an area about 2,250 hectares in the Shahid Rajaie Plain in Khuzestan Province, located in the southwest of Iran, during 2010–2011. The study area is located 5 km west of the city of Omidieh, 30° 35′ to 30° 39′ N and 49° 44′ to 49° 47′ E. The average annual temperature and precipitation for the period of 1965–2010 were 23.6°C and 204 mm, respectively. Also, the annual evaporation of the area is 3,230 mm (Table 22.1) (Khuzestan Water & Power Authority, 2010). The Zohreh River supplies the bulk of the water demands of the region. The application of irrigated agriculture has been common in the study area. Currently, the irrigation systems used by farmlands in the region are furrow irrigation, basin irrigation, and border irrigation schemes. The area is composed of one distinct physiographic feature, i.e., river alluvial plains. Also, five different soil series were found in the area. The semi-detailed soil survey report of the Shahid Rajaie Plain (Khuzestan Water & Power Authority, 2010) was used in order to determine the soil characteristics.

Over much of the Shahid Rajaie Plain, the use of surface irrigation systems has been applied specifically for field crops to meet the water demand of both summer and winter crops. The major irrigated broad-acre crops grown in this area are wheat, barley, and maize, in addition to fruits, melons, watermelons and vegetables such as tomatoes and cucumbers. But like other plains in Khuzstan province, the important major crops are wheat and barley (Behzad et al., 2009; Naseri et al., 2009a; Albaji et al., 2012b; Albaji and Alboshokeh, 2017). There are very few instances of sprinkle and drip irrigation on large-area farms in the Shahid Rajaie Plain.

Five soil series and 14 series/phases or land units were derived from the semi-detailed soil study of the area. The land units are shown in Figure 22.1 as the basis for further land evaluation practice. The soils of the area are of Aridisols and Entisols orders. Also, the soil moisture regime is aridic, while the soil temperature regime is hyperthermic (Khuzestan Water & Power Authority, 2010). Therefore, the organic matter content is low due to high soil temperature and high decomposition ratio of this matter (Mahjoobi et al., 2010).

TABLE 22.1

Ci Values and Suitability Classes of Surface, Sprinkle, and Drip Irrigation for Each Land Unit

Codes of Land Units	Surface Irrigation		Sprinkle Irrigation		Drip Irrigation	
	Ci	Suitability Classes	Ci	Suitability Classes	Ci	Suitability Classes
1	49.14	S_3 nw[a]	60.75	S_2 nw[b]	60	S_2 sn[c]
2	50.87	S_3	68.25	S_2	80.75	S_1
3	36.85	N_1 nw	43.87	N_1 nw	41.6	N_1 snw
4	27.64	N_2 nw	29.92	N_2 n	28	N_2 sn
5	47.38	S_3 nw	55.57	S_3 w	53.2	S_3 sw

[a,b] Limiting Factors for Surface and Sprinkle Irrigations: n: (Salinity & Alkalinity) and w: (Drainage).
[c] Limiting Factors for Drip Irrigation: n: (Salinity & Alkalinity), w: (Drainage), and s: (Calcium Carbonate).

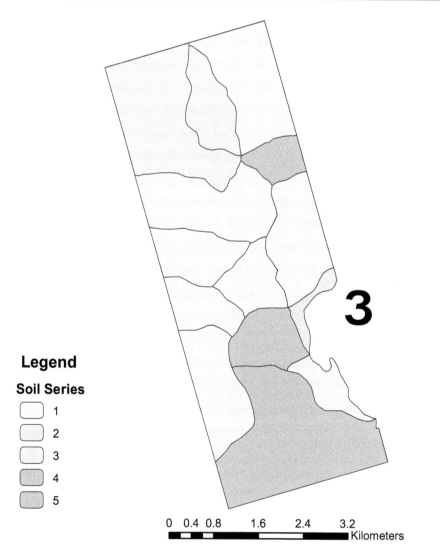

Legend

Soil Series

☐ 1
☐ 2
☐ 3
▨ 4
▨ 5

0 0.4 0.8   1.6   2.4   3.2
                                    Kilometers

FIGURE 22.1   Soil map of the study area.

22.2 RESULTS AND DISCUSSIONS

As shown in Tables 22.1 and 22.2 for surface irrigation, soil series coded 1, 2, and 5 (704 ha – 31.29%) were found to be marginally suitable (S_3). Only soil series coded 3 (866 ha – 38.49%) were classified as currently not suitable (N_1), and only soil series coded 4 (680 ha – 30.22 %) were classified as permanently not suitable (N_2) for any surface irrigation practices.

The analysis of the suitability irrigation maps for surface irrigation (Figure 22.2) indicates that there was no highly suitable land, and moderately suitable land in this plain. The major portion of the cultivated area in this plain (located in the center) is deemed as being marginally suitable due to severe limitations of drainage, salinity, and alkalinity. The currently not suitable land and permanently not suitable land can be observed in the northwest and south of the plain due to very severe limitations of drainage, salinity, and alkalinity. For almost the total study area, elements such as soil depth, slope, and $CaCO_3$ were not considered as limiting factors.

In order to verify the possible effects of different management practices, the land suitability for sprinkle and drip irrigation was evaluated (Tables 22.1 and 22.2).

For sprinkle irrigation, while soil series coded 1 and 2 (652 ha – 28.98%) were classified as moderately suitable (S_2). Further, only soil series coded 5 (52 ha – 2.31%) were found to be marginally suitable (S_3). Soil series coded 3 (866 ha – 38.49%) were classified as currently not suitable (N_1), and only soil series coded 4 (680 ha – 30.22%) were classified as permanently not suitable (N_2) for sprinkle irrigation.

Regarding sprinkler irrigation, (Figure 22.3) there was no highly suitable land in this plain. The moderately suitable area can be observed in part of the cultivated zone in this plain (located in the center and the north) due to light limitations of drainage, salinity, and alkalinity of the area. As seen from the map, only a part of the cultivated area in this plain was evaluated as marginally suitable for sprinkle irrigation because of severe limitations of drainage, salinity, and alkalinity. Other factors such as depth, $CaCO_3$, and slope never influence the suitability of the area. The currently not suitable lands and permanently not suitable lands are located in the northwest and south of the plain, and their non-suitability of the land is due to very severe limitations of drainage, salinity, and alkalinity. Slope, soil depth, and $CaCO_3$ were never taken as limiting factors for almost the entire study area.

For drip irrigation, only soil series coded 2 (8 ha – 0.36%) were highly suitable (S_1), while only soil series coded 1 (644 ha – 28.62%) were classified as moderately suitable (S_2). Further, only soil series coded 5 (52 ha – 2.31%) were found to be slightly suitable (S_3). Only soil series coded 3 (866 ha – 38.49%) were classified as currently not suitable (N_1), and only soil series coded 4 (680 ha – 30.22%) were classified as permanently not suitable (N_2) for drip irrigation.

TABLE 22.2
Distribution of Surface, Sprinkle, and Drip Irrigation Suitability

Suitability	Surface Irrigation			Sprinkle Irrigation			Drip Irrigation		
	Land Unit	Area (ha)	Ratio (%)	Land Unit	Area (ha)	Ratio (%)	Land unit	Area (ha)	Ratio (%)
S_1	—	—	—	—	—	—	2	8	0.36
S_2	—	—	—	1, 2	652	28.98	1	644	28.62
S_3	1, 2, 5	704	31.29	5	52	2.31	5	52	2.31
N_1	3	866	38.49	3	866	38.49	3	866	38.49
N_2	4	680	30.22	4	680	30.22	4	680	30.22
[a]Mis Land	—	—	—	—	—	—	—	—	—
Total		2,250	100		2,250	100		2,250	100

[a] Miscellaneous Land: (Hill, Sand Dune, and Riverbed).

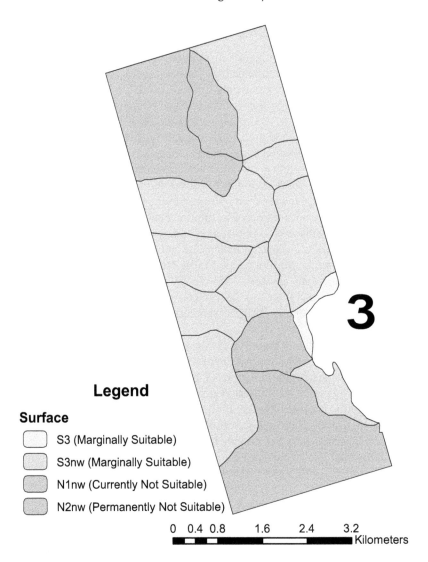

FIGURE 22.2 Land suitability map for surface irrigation.

Regarding drip irrigation (Figure 22.4), the highly suitable lands covered the smallest part of the plain. The slope, soil texture, soil depth, calcium carbonate, salinity, and drainage were in good conditions. The moderately suitable lands could be observed over some portion of the plain (center and north parts) due to the medium content of calcium carbonate. The marginally suitable lands were found only in the east of the area. The limiting factors for this land unit were severe limitations of drainage, salinity, and alkalinity, and the medium content of calcium carbonate. The currently not suitable lands and permanently not suitable lands are located in the north and south of the plain, and their non-suitability of the land is due to very severe limitations of drainage, salinity, and alkalinity, and high content of calcium carbonate. Slope, soil depth, and soil texture were never considered as limiting factors for the entire study area.

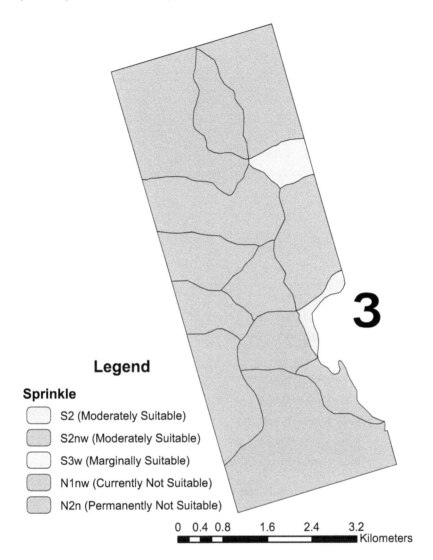

FIGURE 22.3   Land suitability map for sprinkle irrigation.

The mean capability index (Ci) for surface irrigation was 37.87 (currently not suitable), while for sprinkle irrigation it was 46.84 (marginally suitable). Moreover, for drip irrigation it was 44.96 (marginally suitable). For the comparison of the capability indices for surface, sprinkle, and drip irrigation, Tables 22.1 and 22.3 show that in soil series coded 2, applying drip irrigation systems was the most suitable option as compared to surface and sprinkle irrigation systems. In soil series coded 1, 3, 4, and 5, applying sprinkle irrigation systems was more suitable than surface and drip irrigation systems. Figure 22.5 shows the most suitable map for surface, sprinkle, and drip irrigation systems in the Shahid Rajaie Plain as per the Ci for different irrigation systems. As seen from this map, the largest part of this plain was suitable for sprinkle irrigation systems, and some parts of this area were suitable for drip irrigation systems.

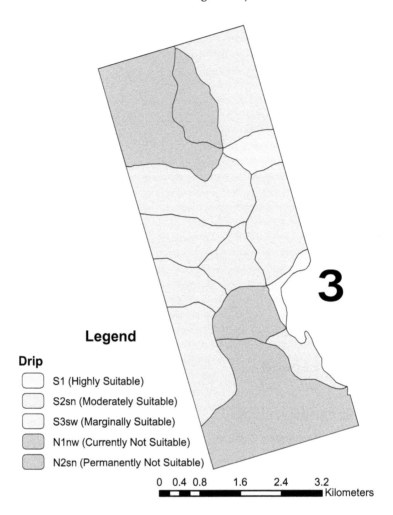

Legend

Drip

☐ S1 (Highly Suitable)

☐ S2sn (Moderately Suitable)

☐ S3sw (Marginally Suitable)

☐ N1nw (Currently Not Suitable)

☐ N2sn (Permanently Not Suitable)

0   0.4 0.8   1.6   2.4   3.2
████▪███▪████████████████▪ Kilometers

FIGURE 22.4   Land suitability map for drip irrigation.

TABLE 22.3
The Most Suitable Land Units for Surface, Sprinkle, and Drip Irrigation Systems by Notation to Capability Index (Ci) for Different Irrigation Systems

Codes of Land Units	The Maximum Capability Index for Irrigation (Ci)	Suitability Classes	The Most Suitable Irrigation Systems	Limiting Factors
1	60.75	S_2 nw	Sprinkle	Salinity, Alkalinity, and Drainage
2	80.75	S_1	Drip	None Exist
3	43.87	N_1 nw	Sprinkle	Salinity, Alkalinity, and Drainage
4	29.92	N_2 n	Sprinkle	Salinity and Alkalinity
5	55.57	S_3 w	Sprinkle	Drainage

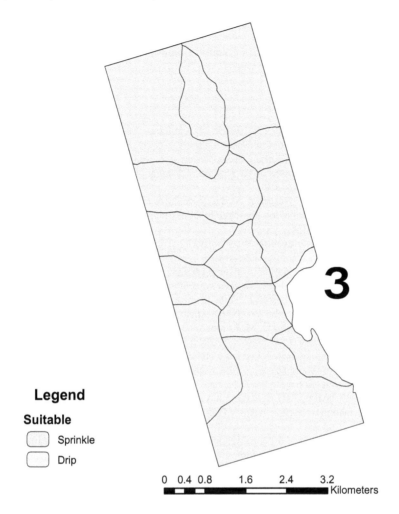

FIGURE 22.5 The most suitable map for different irrigation systems.

22.3 CONCLUSIONS

The results shown in Tables 22.1 and 22.3 indicate that by applying sprinkle irrigation instead of surface and drip irrigation methods, the land suitability of 2,242 ha (99.64%) of the Shahid Rajaie Plain's land could be improved substantially. However, by applying drip irrigation instead of surface and sprinkle irrigation methods, the suitability of 8 ha (0.36%) of this plain's land could be improved. The comparison of the different types of irrigation revealed that sprinkle irrigation was more effective and efficient than the drip and surface irrigation methods, and it improved land suitability for irrigation purposes. The second best option was the application of drip irrigation, which was considered as being more practical than the surface irrigation method. To sum up, the most suitable irrigation systems for the Shahid Rajaie Plain were sprinkle irrigation, drip irrigation, and surface irrigation, respectively. Other researchers have confirmed these results (Landi et al., 2008; Albaji et al., 2008, 2009, 2012a, 2014a, 2014b, 2014c, 2015a, 2015b, 2016; Rezania et al., 2009; Naseri et al., 2009b; Albaji., 2010; Boroomand Nasab et al., 2010; Albaji and Hemadi, 2011; Jovzi et al., 2012).

Moreover, the main limiting factors in using surface and sprinkle irrigation methods in this area were salinity, alkalinity, and drainage, and the main limiting factors in using drip irrigation methods were salinity, alkalinity, drainage, and calcium carbonate content.

ACKNOWLEDGMENTS

We are grateful to the Research Council of Shahid Chamran University of Ahvaz for financial support (GN: SCU.WI98.280).

REFERENCES

Abyaneh, H.Z., Jovzi, M., Albaji, M. 2017. Effect of regulated deficit irrigation, partial root drying and N-fertilizer levels on sugar beet crop (*Beta vulgaris* L.). *Agricultural Water Management* 194, 13–23.

Albaji, M., Boroomand Nasab, S., Kashkuli, H.A., Naseri, A.A., Sayyad, G., Jafari, S. 2008. Comparison of different irrigation methods based on the parametric evaluation approach in North Molasani Plain, Iran. *Journal of Agronomy* 7 (2), 187–191.

Albaji, M,. Boroomand-Nasab, S., Naseri, A.A., Jafari, S. 2009. Comparison of different irrigation methods based on the parametric evaluation approach in Abbas Plain: Iran. *Journal of Irrigation and Drainage Engineering* 136 (2), 131–136.

Albaji, M. 2010. Land suitability evaluation for sprinkler irrigation systems. Khuzestan Water and Power Authority (KWPA), Ahvaz, Iran (in Persian).

Albaji, M., Hemadi, J. 2011. Investigation of different irrigation systems based on the parametric evaluation approach on the Dasht Bozorg Plain. *Transactions of the Royal Society of South Africa* 66 (3), 163–169.

Albaji, M., Boroomand Nasab, S., Hemadi, J. 2012a. Comparison of Different irrigation methods based on the parametric evaluation approach in West North Ahvaz Plain. In: *Problems, Perspectives and Challenges of Agricultural Water Management*. M. Kumar, (ed.), InTech, Croatia, pp. 259–274.

Albaji, M., Papan, P., Hosseinzadeh, M., Barani, S. 2012b. Evaluation of land suitability for principal crops in the Hendijan region. *International Journal of Modern Agriculture* 1 (1), 24–32.

Albaji, M., Golabi, M., Boroomand Nasab, S., Jahanshahi. M. 2014a. Land suitability evaluation for surface, sprinkler and drip irrigation systems. *Transactions of the Royal Society of South Africa* 69 (2), 63–73.

Albaji, M., Golabi, M., Piroozfar, V.R., Egdernejad, A., Nazari Zadeh, F.2014b. Evaluation of agricultural land resources for irrigation in the Ramhormoz Plain by using GIS. *Agriculturae Conspectus Scientificus* 79 (2), 93–102.

Albaji, M., Golabi, M., Egdernejad, A., Nazarizadeh, F. 2014c. Assessment of different irrigation systems in Albaji Plain. *Water Science and Technology: Water Supply* 14 (5), 778–786.

Albaji, M., Golabi, M., Boroomand Nasab, S., Zadeh, F.N. 2015a. Investigation of surface, sprinkler and drip irrigation methods based on the parametric evaluation approach in Jaizan Plain. *Journal of the Saudi Society of Agricultural Sciences* 14 (1), 1–10.

Albaji, M., Boroomand Nasab, S., Golabi, M., Sorkheh Nezhad, M., Ahmadee, M. 2015b. Application possibilities of different irrigation methods in Hofel Plain. *Journal of Agricultural Science* 25 (1), 13–23.

Albaji, M., Golabi, M., Hooshmand, A.R., Ahmadee, M. 2016. Investigation of surface, sprinkler and drip irrigation methods using GIS. *Jordan Journal of Agricultural Sciences* 12 (1), 211–222.

Albaji, M., Alboshokeh, A. 2017. Assessing agricultural land suitability in the Fakkeh region, Iran. *Outlook on Agriculture* 46 (1), 57–65.

Behzad, M., Albaji, M., Papan, P., Boroomand Nasab, S. 2009. Evan region qualitative soil evaluation for wheat, barley, alfalfa and maize. *Journal of Food, Agriculture & Environment* 7 (2), 843–851.

Boroomand Nasab, S., Albaji, M., Naseri, A.A. 2010. Investigation of different irrigation systems based on the parametric evaluation approach in Boneh Basht plain, Iran. *African Journal of Agricultural Research* 5 (5), 372–379.

Chegah, S., Chehrazi, M., Albaji, M. 2013. Effects of drought stress on growth and development frankenia plant (Frankenia Leavis). *Bulgarian Journal of Agricultural Science* 19 (4), 659–666.

Jovzi, M., Albaji, M., Gharibzadeh, A. 2012. Investigating the suitability of lands for surface and under-pressure (drip and sprinkler) irrigation in Miheh Plain. *Research Journal of Environmental Sciences* 6 (2), 51–61.

Khuzestan Water and Power Authority (KWPA). 2010. Semi-detailed soil study report of Shahid Rajaie Plain, Ahvaz, Iran (in Persian).

Landi, A., Boroomand-Nasab, S., Behzad, M., Tondrow, M.R., Albaji, M., Jazaieri, A. 2008. Land suitability evaluation for surface, sprinkle and drip irrigation methods in Fakkeh Plain, Iran. *Journal of Applied Sciences* 8 (20), 3646–3653.

Mahjoobi, A., Albaji, M., Torfi, K. 2010. Determination of heavy metal levels of Kondok soills-haftgel. *Research Journal of Environmental Sciences* 4 (3), 294–299.

Naseri, A.A., Albaji, M., Boroomand Nasab, S., Landi, A., Papan, P., Bavi, A. 2009a. Land suitability evaluation for principal crops in the Abbas Plain, Southwest Iran. *Journal of Food, Agriculture & Environment* 7 (1), 208–213.

Naseri, A.A., Rezania, A.R., Albaji, M. 2009b. Investigation of soil quality for different irrigation systems in Lali Plain, Iran. *Journal of Food, Agriculture & Environment* 7 (3&4), 955–960.

Rezania, A.R., Naseri, A.A., Albaji, M. 2009. Assessment of soil properties for irrigation methods in North Andimeshk Plain, Iran. *Journal of Food, Agriculture & Environment* 7 (3&4), 728–733.

23 Irrigation System Preference in Dehmola Plain, Iran

Mohammad Albaji, Mona Golabi, and Saeid Eslamian

CONTENTS

23.1 INTRODUCTION

The scarcity of water in arid and semiarid regions such as Iran is a restrictive element for the agricultural sector; therefore, there is an urgent need to develop initiatives to save water in this particular sector (Abyaneh et al., 2017; Chegah et al., 2013). Among the proposed solutions is the use of pressurized irrigation systems. The present study was conducted in an area about 20043 hectares in the Dehmola Plain in Khuzestan Province, located in southwest Iran, during 2011–2012. The study area is located 50 km northwest of the city of Handijan, 30° 15′ to 30° 30′ N and 49° 35′ to 49° 45′ E. The average annual temperature and precipitation for the period of 1985–2011 were 23.6 °C and 204 mm, respectively. Also, the annual evaporation of the area is 2,550 mm (Table 23.1) (Khuzestan Water and Power Authority, 2011). The Zohreh River supplies the bulk of the water demands of the region. The application of irrigated agriculture has been common in the study area. Currently, the irrigation systems used by farmlands in the region are furrow irrigation, basin irrigation, and border irrigation schemes.

The area is composed of two distinct physiographic features, i.e., river alluvial plains and terraces, of which the river alluvial plains' physiographic unit is the dominant feature. Also, six different soil series were found in the area. The semi-detailed soil survey report of the Dehmola Plain (Khuzestan Water and Power Authority, 2011) was used in order to determine the soil characteristics.

Over much of the Dehmola Plain, the use of surface irrigation systems has been applied specifically for field crops to meet the water demand of both summer and winter crops. The major irrigated broad-acre crops grown in this area are wheat, barley, and maize, in addition to fruits, melons, watermelons, and vegetables such as tomatoes and cucumbers. But like other plains in Khuzstan Province, the important major crops are wheat and barley (Behzad et al., 2009a, 2009b; Naseri et al., 2009; Albaji et al., 2012b; Albaji and Alboshokeh, 2017). There are very few instances of sprinkle and drip irrigation on large-area farms in the Dehmola Plain.

Six soil series and 28 series phases or land units were derived from the semi-detailed soil study of the area. The land units are shown in Figure 23.1 as the basis for further land evaluation practice. The soils of the area are of aridisols and entisols orders. Also, the soil moisture regime is aridic and aquic, while the soil temperature regime is hyperthermic (Khuzestan Water and Power Authority, 2011). Therefore, the organic matter content is low due to high soil temperature and high decomposition ratio of this matter (Mahjoobi et al., 2010).

TABLE 23.1

The Ci Values and Suitability Classes of Surface, Sprinkle, and Drip Irrigation for Each Land Unit

Codes of Land Units	Surface Irrigation		Sprinkle Irrigation		Drip Irrigation	
	Ci	Suitability Classes	Ci	Suitability Classes	Ci	Suitability Classes
1	35.53	N_1 nw[a]	38.47	N_1 n[b]	36	N_1 sn[c]
2	28.51	N_2 nw	31.5	N_1 nw	32	N_1 snw
3	26.25	N_2 nw	28.42	N_2 n	26.6	N_2 sn
4	31.32	N_1 nw	37.29	N_1 nw	35.7	N_1 snw
5	27.64	N_2 nw	29.92	N_2 n	28	N_2 sn
6	15.79	N_2 nw	17.1	N_2 n	16	N_2 sn

[a, b] The Limiting Factors for Surface and Sprinkle Irrigations: n: (Salinity & Alkalinity) and w: (Drainage)

[c] The Limiting Factors for Drip Irrigation: n: (Salinity & Alkalinity), w: (Drainage), and s: (Calcium Carbonate).

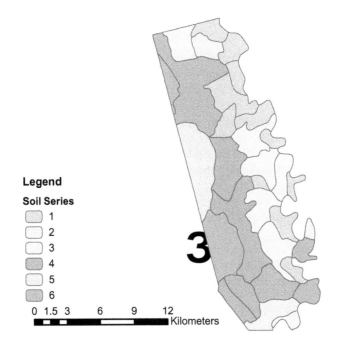

Legend

Soil Series

1
2
3
4
5
6

0 1.5 3　6　9　12
Kilometers

FIGURE 23.1　Soil map of the study area.

23.2　RESULTS AND DISCUSSIONS

As shown in Tables 23.1 and 23.2 for surface irrigation, soil series coded 1 and 4 (9,384 ha – 46.82%) were classified as currently not suitable (N_1), and soil series coded 2, 3, 5, and 6 (10,659 ha – 53.18%) were classified as permanently not suitable (N_2) for any surface irrigation practices.

The analyses of the suitability irrigation map for surface irrigation (Figure 23.2) indicate that there was no highly suitable, moderately suitable, and marginally suitable land in this plain. The map also indicates that some part of the cultivated area in this plain was evaluated as currently not suitable land due to severe limitations of drainage, salinity, and alkalinity. The major portion of this plain (located in the center, west, south, and north) is deemed as being permanently not suitable land

TABLE 23.2

Distribution of Surface, Sprinkle, and Drip Irrigation Suitability

Suitability	Surface Irrigation			Sprinkle Irrigation			Drip Irrigation		
	Land Unit	Area (ha)	Ratio (%)	Land Unit	Area (ha)	Ratio (%)	Land Unit	Area (ha)	Ratio (%)
S_1	—	—	—	—	—	—	—	—	—
S_2	—	—	—	—	—	—	—	—	—
S_3	—	—	—	—	—	—	—	—	—
N_1	1, 4	9,384	46.82	1, 2, 4	11,918	59.46	1, 2, 4	11,918	59.46
N_2	2, 3, 5, 6	10,659	53.18	3, 5, 6	8,125	40.54	3, 5, 6	8,125	40.54
Total		20,043	100		20,043	100		20,043	100

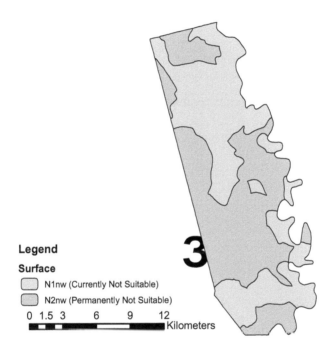

Legend

Surface

N1nw (Currently Not Suitable)

N2nw (Permanently Not Suitable)

0 1.5 3   6   9   12
Kilometers

FIGURE 23.2 Land suitability map for surface irrigation.

due to very severe limitations of drainage, salinity, and alkalinity. For almost the total study area, elements such as soil depth, slope, and calcium carbonate were not considered as limiting factors.

In order to verify the possible effects of different management practices, the land suitability for sprinkle and drip irrigation was evaluated (Tables 23.1 and 23.2).

For sprinkle irrigation, soil series coded 1, 2, and 4 (11,918 ha – 59.46%) were classified as currently not suitable (N_1), and soil series coded 3, 5, and 6 (8,125 ha – 40.54%) were classified as permanently not suitable (N_2) for sprinkle irrigation.

Regarding sprinkler irrigation (Figure 23.3), there was no highly suitable, moderately suitable, and marginally suitable land in this plain. The map also indicates that the largest part of the cultivated area in this plain (located in the center, south, and north) was evaluated as currently not suitable land due to severe limitations of drainage, salinity, and alkalinity. Some portion of this plain (located in the west and east) is deemed as being permanently not suitable land due to very severe limitations of salinity and alkalinity. For almost the total study area, elements such as soil depth, slope, and calcium carbonate were not considered as limiting factors.

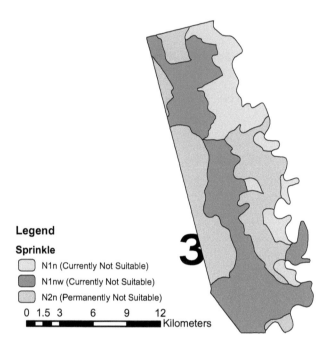

FIGURE 23.3   Land suitability map for sprinkle irrigation.

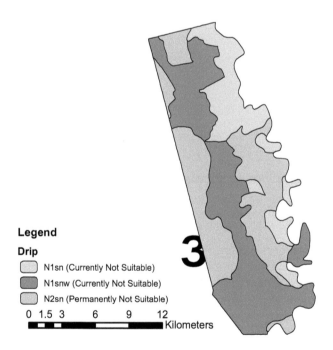

FIGURE 23.4   Land suitability map for drip irrigation.

For drip irrigation, soil series coded 1, 2, and 4 (11,918 ha – 59.46%) were classified as currently not suitable (N_1), and soil series coded 3, 5, and 6 (8,125 ha – 40.54%) were classified as permanently not suitable (N_2) for drip irrigation.

Regarding drip irrigation (Figure 23.4), there was no highly suitable, moderately suitable, and marginally suitable land in this plain. The map also indicates that the largest part of the cultivated area in this plain (located in the center, south, and north) was evaluated as currently not suitable land due to severe limitations of calcium carbonate, drainage, salinity, and alkalinity. Some portion of this plain (located in the west and east) is deemed as being permanently not suitable land due to very severe limitations of calcium carbonate, salinity, and alkalinity. For almost the total study area, elements such as soil depth, slope, and soil texture were not considered as limiting factors.

The mean capability index (Ci) for surface irrigation was 28.72 (permanently not suitable), while for sprinkle irrigation it was 32.29 (currently not suitable). Moreover, for drip irrigation it was 30.81 (currently not suitable). For the comparison of the capability indices for surface, sprinkle, and drip irrigation, Tables 23.1 and 23.3 indicate that in soil series coded 2, applying drip irrigation systems was the most suitable option as compared to surface and sprinkle irrigation systems. In soil series coded 1, 3, 4, 5, and 6, applying sprinkle irrigation systems was more suitable than surface and drip irrigation systems. Figure 23.5 shows the most suitable map for surface, sprinkle, and drip irrigation systems in the Dehmola Plain as per the Ci for different irrigation systems. As seen from this map,

TABLE 23.3

The Suitable Land Units for Surface, Sprinkle, and Drip Irrigation Systems by Notation to Capability Index (Ci) for Different Irrigation Systems

Codes of Land Units	The Maximum Capability Index for Irrigation (Ci)	Suitability Classes	The Most Suitable Irrigation Systems	Limiting Factors
1	38.47	N_1 n	Sprinkle	Salinity and alkalinity
2	32	N_1 snw	Drip	$CaCO_3$, salinity, alkalinity, and drainage
3	28.42	N_2 n	Sprinkle	Salinity and alkalinity
4	37.29	N_1 nw	Sprinkle	Salinity, alkalinity, and drainage
5	29.92	N_2 n	Sprinkle	Salinity and alkalinity
6	17.1	N_2 n	Sprinkle	Salinity and alkalinity

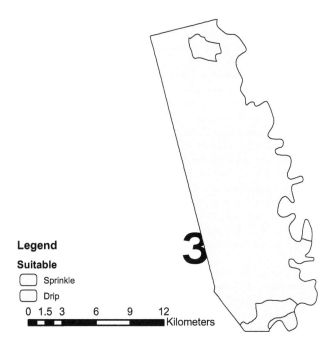

Legend

Suitable

☐ Sprinkle

☐ Drip

0 1.5 3 6 9 12 Kilometers

FIGURE 23.5 The most suitable map for different irrigation systems.

the largest part of this plain was suitable for sprinkle irrigation systems, and some parts of this area were suitable for drip irrigation systems.

23.3 CONCLUSIONS

The results shown in Tables 23.1 and 23.3 indicate that by applying sprinkle irrigation instead of surface and drip irrigation methods, the land suitability of 17,509 ha (87.36%) of the Dehmola Plain's land could be improved substantially. However, by applying drip irrigation instead of surface and sprinkle irrigation methods, the suitability of 2,534 ha (12.64%) of this plain's land could be improved. The comparison of the different types of irrigation revealed that sprinkle irrigation was more effective and efficient than the drip and surface irrigation methods, and it improved the land suitability for irrigation purposes. The second best option was the application of drip irrigation, which was considered as being more practical than the surface irrigation method. To sum up, the most suitable irrigation systems for the Dehmola Plain were sprinkle irrigation, drip irrigation, and surface irrigation, respectively. Landi et al. (2008), Rezania et al. (2009), Boroomand Nasab et al. (2010), Albaji (2010), Albaji and Hemadi (2011), Jovzi et al. (2012), and Albaji et al. (2008, 2012a, 2014a, 2014b, 2014c, 2015a, 2015b, 2016) found similar results.

Moreover, the main limiting factors in using surface and sprinkle irrigation methods in this area were drainage, salinity, and alkalinity, and the main limiting factors in using drip irrigation methods were calcium carbonate, drainage, salinity, and alkalinity.

ACKNOWLEDGMENTS

We are grateful to the Research Council of Shahid Chamran University of Ahvaz for financial support (GN: SCU.WI98.280).

REFERENCES

Abyaneh, H.Z., Jovzi, M., Albaji, M. 2017. Effect of regulated deficit irrigation, partial root drying and *N*-fertilizer levels on sugar beet crop (Beta vulgaris L.). *Agricultural Water Management* 194, 13–23.

Albaji, M., Boroomand Nasab, S., Kashkuli, H.A., Naseri, A.A., Sayyad, G., Jafari, S. 2008. Comparison of different irrigation methods based on the parametric evaluation approach in North Molasani Plain, Iran. *Journal of Agronomy* 7 (2), 187–191.

Albaji, M. 2010. Land suitability evaluation for sprinkler irrigation systems. *Khuzestan Water and Power Authority (KWPA)*, Ahvaz, Iran (in Persian).

Albaji, M., Hemadi, J. 2011. Investigation of different irrigation systems based on the parametric evaluation approach on the Dasht Bozorg Plain. *Transactions of the Royal Society of South Africa* 66 (3), 163–169.

Albaji, M., Boroomand Nasab, S., Hemadi, J. 2012a. Comparison of different irrigation methods based on the parametric evaluation approach in West North Ahvaz Plain. In: *Problems, Perspectives and Challenges of Agricultural Water Management*. M. Kumar, (ed.), InTech, Croatia, pp. 259–274.

Albaji, M., Papan, P., Hosseinzadeh, M., Barani, S. 2012b. Evaluation of land suitability for principal crops in the Hendijan region. *International Journal of Modern Agriculture* 1 (1), 24–32.

Albaji, M., Golabi, M., Boroomand Nasab, S., Jahanshahi. M. 2014a. Land suitability evaluation for surface, sprinkler and drip irrigation systems. *Transactions of the Royal Society of South Africa* 69 (2), 63–73.

Albaji, M., Golabi, M., Piroozfar, V.R., Egdernejad, A., Nazari Zadeh, F. 2014b. Evaluation of agricultural land resources for irrigation in the Ramhormoz Plain by using GIS. *Agriculturae Conspectus Scientificus* 79 (2), 93–102.

Albaji, M., Golabi, M., Egdernejad, A., Nazarizadeh, F. 2014c. Assessment of different irrigation systems in Albaji Plain.*Water Science and Technology: Water Supply* 14 (5), 778–786.

Albaji, M., Golabi, M., Boroomand Nasab, S., Zadeh, F.N. 2015a. Investigation of surface, sprinkler and drip irrigation methods based on the parametric evaluation approach in Jaizan Plain. *Journal of the Saudi Society of Agricultural Sciences* 14 (1), 1–10.

Albaji, M., Boroomand Nasab, S., Golabi, M., Sorkheh Nezhad, M., Ahmadee, M. 2015b. Application possibilities of different irrigation methods in Hofel Plain. *YYÜ TAR BİL DERG (YYU J AGR SCI)* 25 (1), 13–23.

Albaji, M., Golabi, M., Hooshmand, A.R., Ahmadee, M. 2016. Investigation of surface, sprinkler and drip irrigation methods using GIS. *Jordan Journal of Agricultural Sciences* 12 (1), 211–222.

Albaji, M., Alboshokeh, A. 2017. Assessing agricultural land suitability in the Fakkeh region, Iran. *Outlook on Agriculture* 46 (1), 57–65.

Behzad, M., Albaji, M., Papan, P., Boroomand Nasab, S. 2009a. Evan region qualitative soil evaluation for wheat, barley, alfalfa and maize. *Journal of Food, Agriculture & Environment* 7 (2), 843–851.

Behzad, M., Albaji, M., Papan, P., Boroomand Nasab, S., Naseri, A.A., Bavi, A. 2009b. Qualitative evaluation of land suitability for principal crops in the Gargar Region, Khuzestan Province, Southwest Iran. *Asian Journal of Plant Sciences* 8 (1), 28.

Boroomand Nasab, S., Albaji, M., Naseri, A.A. 2010. Investigation of different irrigation systems based on the parametric evaluation approach in Boneh Basht plain, Iran. *African Journal of Agricultural Research* 5 (5), 372–379.

Chegah, S., Chehrazi, M., Albaji, M. 2013. Effects of drought stress on growth and development frankenia plant (Frankenia Leavis). *Bulgarian Journal of Agricultural Science* 19 (4), 659–666.

Jovzi, M., Albaji, M., Gharibzadeh, A. 2012. Investigating the suitability of lands for surface and under-pressure (drip and sprinkler) irrigation in Miheh Plain. *Research Journal of Environmental Sciences* 6 (2), 51–61.

Khuzestan Water and Power Authority (KWPA). 2011. Semi-detailed soil study report of Dehmola Plain, Iran (in Persian).

Landi, A., Boroomand-Nasab, S., Behzad, M., Tondrow, M.R., Albaji, M., Jazaieri, A. 2008. Land suitability evaluation for surface, sprinkle and drip irrigation methods in Fakkeh Plain, Iran. *Journal of Applied Sciences* 8 (20), 3646–3653.

Mahjoobi, A., Albaji, M., Torfi, K. 2010. Determination of heavy metal levels of Kondok soills-haftgel. *Research Journal of Environmental Sciences* 4 (3), 294–299.

Naseri, A.A., Albaji, M., Boroomand Nasab, S., Landi, A., Papan, P., Bavi, A. 2009. Land suitability evaluation for principal crops in the Abbas Plain, Southwest Iran. *Journal of Food, Agriculture & Environment* 7 (1), 208–213.

Rezania, A.R., Naseri, A.A., Albaji, M. 2009. Assessment of soil properties for irrigation methods in North Andimeshk Plain, Iran. *Journal of Food, Agriculture & Environment* 7 (3&4), 728–733.

24 Irrigation System Suggestions in Sudjan Plain, Iran

Mehdi Jovzi, Hamid Zare Abyaneh, Niaz Ali
Ebrahimi Pak, Majid Heydari, Shahrokh Fatehi, Amin
Behmanesh, Mohammad Albaji, and Saeid Eslamian

CONTENTS

24.1   INTRODUCTION

Security and stability of food in the world rests on the management of natural resources. Due to the reduction of water resources and an increase in population, the extent of irrigated area per capita is declining, and irrigated lands now produce 40% of the food supply (Hargreaves and Merkley, 1998).

Land evaluation has traditionally been based primarily on soil surveys. If the land-use requirements are known, they can be matched with the group of land characteristics or the land quality (Ikawa, 1992). Therefore, soil characterization and soil classification—by using soil maps—serve as efficient tools in land evaluation. McBratney et al. (2000) also stated that soil surveys may be thought of as base data involving field description and laboratory analysis, and subsequent classification and mapping of the study area for land evaluation. They also stated that effective soil management requires an understanding of soil distribution patterns within the landscape and that soil survey mapping is the most enabled data for wise decisions regarding land use and land evaluation by planners and policymakers.

According to FAO (1976) methodology, land suitability is strongly related to "land qualities" including erosion resistance, water availability, and flood hazards, which are derived from slope angle and length, rainfall, and soil texture. Sys et al. (1991) suggested a parametric evaluation system for irrigation methods that was primarily based on physical and chemical soil properties. Factors affecting the soil suitability for irrigation are physical properties such as permeability and available water content, chemical properties such as soil salinity, alkalinity and acidity, drainage properties, such as depth of groundwater, and environmental factors such as slope. The capability index for irrigation (CI) is calculated based on the multiplication of the above factors.

Bienvenue et al. (2002) evaluated the land suitability for surface (gravity) and drip (localized) irrigation in the Thies, Senegal, by using the parametric evaluation systems. Regarding surface irrigation, there was no area classified as highly suitable (S1). Only 20.24% of the study area proved suitable (S2, 7.73%) or slightly suitable (S3, 12.51%). Most of the study area (57.66%) was classified as unsuitable (N2). The limiting factor to this kind of land use was mainly the soil drainage status and texture that was mostly sandy, while surface irrigation generally requires heavier soils. For drip irrigation, a good portion (45.25%) of the area was suitable (S2), while 25.03% was classified as highly suitable (S1), and only a small portion was relatively suitable (N1, 5.83%) or unsuitable (N2, 5.83%). The researchers showed the limiting factors for surface irrigation as drainage and soil texture of the region. Also, they specified the soil depth, soil texture with coarse gravel and/or poor drainage. Barberis and Minelli (2005) provided land suitability classification for both surface and drip irrigation methods in Shouyang county, Shanxi Province, China, where the study was carried out by a modified parametric system. The results indicated that due to the unusual morphology, the area suitability for the surface irrigation (34%) is smaller than the surface used for the drip irrigation (62%). The most limiting factors were physical parameters including slope and soil depth. Dengiz (2006) investigated the different methods of irrigation (surface, drip, and local) based on the parametric evaluation in the pilot fields of Ikizce Central Research Institute, located south of Ankara, Turkey, and concluded that the drip irrigation resulted in the increase of suitability of the lands by 38% compared to the surface irrigation. Naseri et al. (2009b) investigated different irrigation methods based upon a parametric evaluation system in an area of 7,000 ha in the Baghe region located in Khuzestan Province, southwest Iran. The comparison of the different types of irrigation techniques revealed that the sprinkler and drip irrigation were more effective and efficient than the surface irrigation methods for improving the suitability to the irrigation purposes. Additionally, the main limiting factors in using surface and sprinkler irrigation methods in this area were soil texture, drainage, salinity, and alkalinity, and the main limiting factors in using drip irrigation methods were soil calcium carbonate content, drainage, salinity, and alkalinity. Fatapour and Eslami (2014) studied the land suitability for drip and sprinkler irrigation methods based on the parametric method in Kouhdasht Plain, located in Lorestan Province, in the west of Iran. Their results showed that all of the arable lands were considered suitable for the drip irrigation method and classified as class S1. Bagherzadeh and Paymard (2015) investigated land capability for different types of irrigation systems including surface, drip, and sprinkler practices by parametric and fuzzy approaches to evaluate the capability of cultivated lands on 6,131 km^2 of the Mashhad Plain, northeast Iran. Based on parametric approach, some 1116.5 ha of the study area were classified as highly suitable (S1) for surface irrigation method, while the corresponding values by fuzzy approach accounted for 6099.7 ha of the region. The moderately suitable class (S2) assessed by parametric and fuzzy approaches included 5014.5 and 31.3 ha of the plain, respectively. It was revealed that the land capability indices were in higher classes (S1 to S2) by drip and sprinkler irrigation compared to the surface irrigation method, and the soil texture was detected as the most limiting factor for using the surface irrigation method. With respect to current soil and climate conditions in the study area, the most efficient irrigation systems are drip and sprinkler methods. Masoudi et al. (2018) studied different irrigation methods based on a parametric evaluation system in an area of 100 ha in the Fars Province, in the south of Iran. The results showed that land suitability of 71.9% of the case study was classified as permanently not suitable (N2) and 28.1% currently not suitable (N1) for surface irrigation. On the other hand, land suitability of 47.3% of the case study was classified as permanently not suitable (N2), 28.5% currently not suitable (N1), and 24.3% marginally suitable (S3) for drip irrigation. The limiting factor for drip irrigation was slope, and the limiting factors for surface irrigation were slope and drainage.

The main purpose of this chapter was to select suitable irrigation systems (surface, sprinkler, and drip irrigation) based on the parametric evaluation methods for the Sudjan Plain.

24.2   MATERIALS AND METHODS

24.2.1   Ecological Conditions of the Study Area

The present study was conducted in an area of about 3306.98 ha in the Sudjan Plain, in the Zayandeh Rood watershed on the Chaharmahal and Bakhtiari Province, located in the center of Iran. The study area is located 70 km northwest of the city of Shahr e Kord, 32° 28′ 37.8″ to 32° 34′ 40.8″ N and 50° 21′ 21″ to 50° 26″ 43.8″ E. The maximum and minimum elevations for the area are 2,300 and 2,200 m for the northeastern and northwestern parts, respectively. A review of the existing statistics for a 36-year period (1970–2006) shows that the average annual temperature and rainfall are 9.4°C and 470 mm, respectively. The annual evaporation of the area is 1,165 mm. The main water supply to this area is Zayandeh Rood River. Currently, the irrigation systems used by farmlands in the region are furrow, basin, and border irrigation systems (Agricultural Jihad Organization of Chaharmahal and Bakhtiari Province, 2008). In this plain, like other plains in Iran, the important major crops are wheat and barley (Behzad et al., 2009; Naseri et al., 2009a; Albaji et al., 2012b; Albaji and Alboshokeh., 2017).

24.2.2   Irrigation System Selection

The land evaluation was determined based on topography and soil characteristics of the region. The topographic characteristics included slope and soil properties such as soil texture, depth, salinity, drainage, and calcium carbonate content, which were taken into account. Soil properties such as cation exchange capacity (CEC), percentage of basic saturation (PBS), organic matter (OM), and pH were considered in terms of soil fertility. Sys et al. (1991) suggested that soil characteristics such as PBS and OM did not require any evaluation in arid regions, whereas clay CEC rate usually exceeds the plant requirement without further limitation—thus, fertility properties can be excluded from land evaluation if it is done for the purpose of irrigation.

To determine the soil characteristics, the semi-detailed soil survey report of the Sudjan Plain, Shahr-e Kord, was used (Agricultural Jihad Organization of Chaharmahal and Bakhtiari Province, 2008). Based on the profile description and laboratory analysis, the groups of soils that had similar properties and were located in the same physiographic unit were categorized as soil series; taxonomy was used to form a soil family as per the Soil Survey Staff (2000). Finally, eight soil series and 48 series phases were selected for the surface, sprinkler, and drip irrigation land suitability.

To obtain the average soil texture, salinity, and $CaCO_3$ for the upper 150 cm of soil surface, the profile was subdivided into six equal sections and weighting factors of 2, 1.5, 1, 0.75, 0.50, and 0.25 were used for each section, respectively. The weight factors are used in order to give more importance to the upper part of the soil profile where the root is developing more (Sys et al., 1991).

To evaluate the suitability of lands for surface, drip, and sprinkler irrigation methods, the parametric evaluation method is applied. The system is based on morphology, physical, and chemical properties of soil (Sys et al., 1991). The parametric evaluation system consists of six parameters including slope, drainage properties, electrical conductivity of soil solution, calcium carbonate status, depth soil, and soil texture. Each of the six aforementioned parameters is scaled according to the related tables, and the capability index for irrigation (Ci) is calculated using them as the following equation:

$$Ci = A \; \frac{B}{100} \times \frac{C}{100} \times \frac{D}{100} \times \frac{E}{100} \times \frac{F}{100} \tag{24.1}$$

where A, B, C, D, E, and F are soil texture, soil depth, calcium carbonate content, electrical conductivity, drainage, and slope rating, respectively. In Table 24.1, the ranges of capability index and the corresponding suitability classes are shown.

TABLE 24.1
Suitability Classes for the Irrigation Capability Indices (Ci) Classes

Capability Index	Definition	Symbol
>80	Highly suitable	S1
60–80	Moderately suitable	S2
45–59	Marginally suitable	S3
30–44	Currently not suitable	N1
<29	Permanently not suitable	N2

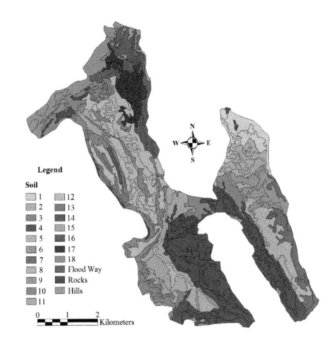

FIGURE 24.1   Soil map of the study area.

The semi-detailed soil map of the Sudjan Plain is shown in Figure 24.1. This soil map was used to develop land suitability maps for different irrigation methods. All the data for soil characteristics were incorporated in the map using Arc GIS 9.2 software. The digital soil map base preparation was the first step toward the presentation of a GIS module for land suitability maps for different irrigation systems. A total of 18 different land mapping units (LMUs) were determined in the base map. Soil characteristics were also given for each LMU. These values were used to generate the land suitability maps for different irrigation systems (surface, sprinkler, and drip irrigation) using geographic information systems (GIS).

24.3   RESULTS AND DISCUSSION

The irrigation system applied in most lands of the Sudjan Plain is surface irrigation. The major plants in agricultural lands and gardens are wheat, barley, alfalfa, walnut, and almond (Agricultural Jihad Organization of Chaharmahal and Bakhtiari Province, 2008). The semi-detailed soil study resulted in the recognition of 18 series of soil and altogether 48 series phases in the studied region.

The series of soil are introduced in Figure 24.1 as the basis of evaluation. The soils of the region are rated as Alfi soil, Molli soil, and Incepti soil. The soil moisture regime is of ceric and aquic (very small part) nature. The soil temperature regime is mesic. Miscellaneous lands are 143.71 ha (Agricultural Jihad Organization of Chaharmahal and Bakhtiari Province, 2008).

24.3.1 LAND SUITABILITY FOR SURFACE IRRIGATION

As shown in Tables 24.2 and 24.3 for surface irrigation, only the soil series coded 3 (43.71 ha – 1.32%) were highly suitable (S1); soil series coded 2, 10, 11, 14, 15, 16, 17, and 18 with the area of 1213.3 ha (36.69%) were classified as moderately suitable (S2). Soil series coded 1, 4, 5, 6, 7, 8, and 12 with the area of 1836.9 ha (55.55%) were found to be marginally suitable (S3). Soil series coded 9 and 13 with the area of 69.34 ha (2.10%) were classified as currently not suitable (N1).

As shown in Figure 24.2, analyzing the maps of land suitability for surface irrigation shows that a small area (located in the east) is highly suitable for surface irrigation. This was due to deep soil, suitable texture, good drainage, salinity, and proper slope of the area; however, the calcium carbonate content was high (31.03%). Moderately suitable lands were identified mostly in northern, southwestern, and central locations. In general, the soil texture of these lands is silty clay, clay loam, and clay, and gently sloped. Other factors such as salinity, drainage, and soil depth (except the lands of 11 soil series) have no effect on the land suitability of the region. Also, this map shows that most lands in the eastern, southern, and central study areas are marginally suitable because of gravels (40% in the lands of 7 soil series), clay and silty clay texture of the soil,

---

TABLE 24.2
Capability Index for Irrigation (Ci) Values and Suitability Classes of Surface, Sprinkler, and Drip Irrigation for Each Land Unit

Codes of Land Units	Surface Irrigation		Sprinkler Irrigation		Drip Irrigation	
	Ci	Suitability Classes	Ci	Suitability Classes	Ci	Suitability Classes
1	53.47	S3 ts[a]	61.64	S2 ts[a]	63.53	S2 ts[a]
2	63.15	S2 t	72.09	S2 s	76.15	S2 s
3	82.49	S1	76.00	S2 s	80.00	S1
4	50.72	S3 ts	55.58	S3 s	57.65	S3 s
5	45.00	S3 ts	57.42	S3 ts	59.40	S2 s
6	56.83	S3 ts	60.71	S2 s	64.12	S2 s
7	47.58	S3 ts	60.52	S2 s	63.84	S2 s
8	45.54	S3 ws	46.51	S3 ws	56.53	S3 ws
9	34.00	N1 ws	32.30	N1 ws	40.38	N1 ws
10	59.67	S2 t	65.38	S2 s	67.82	S2 s
11	61.80	S2 ts	68.40	S2 s	72.00	S2 s
12	53.55	S3 ts	57.80	S3 s	61.20	S2 s
13	42.57	N1 tws	43.32	N1 ws	52.22	S3 ws
14	63.11	S2 s	69.04	S2 s	80.75	S1
15	59.09	S2 s	58.14	S3 s	68.00	S2 s
16	65.66	S2 ts	64.60	S2 s	68.00	S2 s
17	70.12	S2 s	64.60	S2 s	68.00	S2 s
18	66.82	S2 s	68.40	S2 s	80.00	S1

[a] Limiting factors for surface, sprinkler, and drip irrigation: s (Soil texture and/or Soil depth and/or calcium carbonate), t (slope) and w (drainage).

TABLE 24.3

Distribution of Surface, Sprinkler, and Drip Irrigation Suitability

	Surface Irrigation			Sprinkler Irrigation			Drip Irrigation		
Suitability	Land Unit	Area (ha)	Ratio (%)	Land Unit	Area (ha)	Ratio (%)	Land Unit	Area (ha)	Ratio (%)
S_1	3	43.71	1.32	—	—	—	3, 14, 18	238.61	7.22
S_2	2, 10, 11, 14, 15, 16, 17, 18	1213.3	36.69	1, 2, 3, 6, 7, 10, 11, 14, 16, 17, 18	1,526.9	46.17	1, 2, 5, 6, 7, 10, 11, 12, 15, 16 17	2067.18	62.51
S_3	1, 4, 5, 6, 7, 8, 12	1836.9	55.55	4, 5, 8, 12, 15	1,567.03	47.39	4, 8, 13	845.53	25.57
N_1	9, 13	69.34	2.10	9, 13	69.34	2.10	9	11.95	0.36
N_2	—	—	—	—	—	—	—	—	—
Mis. land[a]		143.71	4.35		143.71	4.35		143.71	4.35
Total		3306.98	100		3306.98	100		3306.98	100

[a] Miscellaneous land: hill, Rocks and flood way.

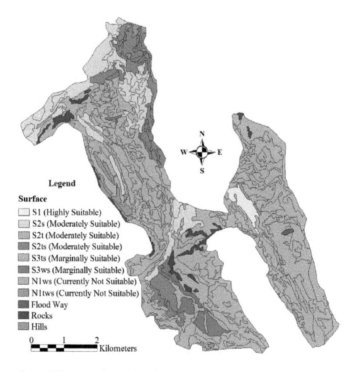

FIGURE 24.2 Land suitability map for surface irrigation.

high calcium carbonate content (26%–41% in series 1, 4, 5, 6, and 12), improper slope (2%–25%), and drainage (8 Series). The currently not suitable lands are located in the western and southern border region. They are currently not suitable lands because of silty clay texture of the soil, high calcium carbonate content (37.95% in series 13), and very poor drainage. The salinity and soil depth (except the lands of 11 soil series) aren't of limiting factors for surface irrigation in all lands of the region.

24.3.2 Land Suitability for Sprinkler Irrigation

The land suitability of the Sudjan Plain for sprinkler irrigation was investigated (Tables 24.2 and 24.3). The results showed highly suitable (S1) lands were not found for the sprinkler irrigation method. Soil series coded 1, 2, 3, 6, 7, 10, 11, 14, 16, 17, and 18 with the area of 1526.9 ha (46.17%) were classified as moderately suitable (S2), while soil series coded 4, 5, 8, 12, and 15 with the area of 1567.03 ha (47.39%) were marginally suitable (S3) for sprinkler irrigation. Soil series coded 9 and 13 with the area of 69.34 ha (2.10%) were classified as currently not suitable (N1).

Highly suitable land for sprinkler irrigation (Figure 24.3) was not found in the Sudjan Plain. Moderately suitable lands were located in the eastern, northern, and southwestern areas of this plain. These lands have clay loam and silty clay texture, high slope, and $CaCO_3$ (15%–38%). The lands that are marginally suitable are located in the central and southern parts of this plain because of the soil texture of clay and clay loam, high slope (2%–25%), $CaCO_3$ (9%–41%), and improper drainage (8 Series) in this area. The currently not suitable lands are located in the western and southern border region. They are currently not suitable lands because of the silty clay texture of the soil, high $CaCO_3$ (37.95% in the 13 series), and very poor drainage. The salinity and soil depth (except the lands of 11 soil series) are not limiting factors for sprinkler irrigation in all studied regions.

24.3.3 Land Suitability for Drip Irrigation

The land suitability of the Sudjan Plain for drip irrigation was investigated (Tables 24.2 and 24.3). The results from evaluating the suitability of lands for drip irrigation system based on parametric method showed that the soil series coded 3, 14, and 18 (238.61 ha – 7.22%) were highly suitable (S1). The soil series coded 7, 10, 11, 12, 15, 16, and 17 (2067.18 ha – 62.51%) were moderately suitable (S2), while soil series coded 4, 8, and 13 (845.53 ha – 25.57%) were classified as marginally suitable (S3). Soil series coded 9 (11.95 ha – 0.36%) were classified as currently not suitable (N1).

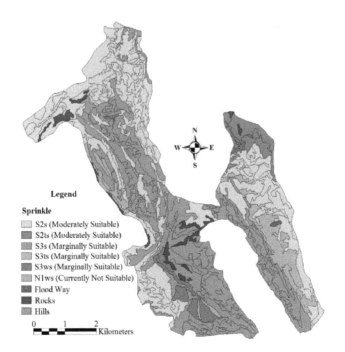

FIGURE 24.3 Land suitability map for sprinkler irrigation.

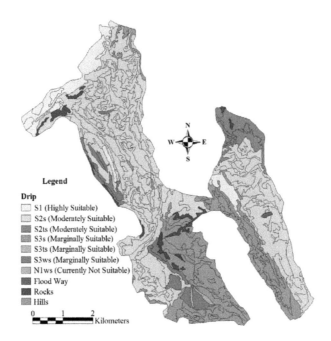

FIGURE 24.4   Land suitability map for drip irrigation.

Figure 24.4 reveals the suitability of the region for drip irrigation. Highly suitable lands for drip irrigation are located in the northwestern, eastern, and a small part in the western border lands of the studied region. In these lands, all factors such as calcium carbonate content, soil texture, salinity, soil depth, and slope are in the desired condition. The moderately suitable lands (S2) cover the eastern, central, northern, and southwestern lands. These lands have the soil texture of silty clay and clay, high slope (8%–25% in the 1 Series), $CaCO_3$ (15%–38%), and low depth of soil (11 series). The lands with marginal suitability (S3) are located in the southern and western border of the plain. The limiting factors in these areas are the soil texture (silty clay and clay), high $CaCO_3$ (9%–41%), and poor drainage (8 and 13 series). The currently not suitable lands are located in a very small part of the southern border region. They are currently not suitable lands because of silty clay texture of the soil and very poor drainage. The salinity and soil depth (except the lands of 11 soil series) are not the limiting factors for drip irrigation in the studied region.

24.3.4 SELECTION OF THE MOST SUITABLE IRRIGATION SYSTEM

The mean capability index (Ci) for surface irrigation, sprinkler irrigation, and drip irrigation are 56.73 (marginally suitable), 60.14 (moderately suitable), and 65.53 (moderately suitable), respectively. A comparison of the capability index (Ci) for irrigation systems (Tables 24.2 and 24.4) reveals that the drip irrigation is more suitable than the surface and sprinkler irrigations in all series of studied lands (except the soil series 3 and 17). In the soil series 3 and 17, the surface irrigation is more suitable than the drip and sprinkler irrigation systems. It is due to the percentage of calcium carbonate content (31%) in these lands. Figure 24.5 shows the application of the most suitable irrigation system in the lands of Sudjan Plain. As shown in the map, the major part of the Sudjan Plain was suitable for drip irrigation, and a small part of the plain was suitable for surface irrigation.

Tables 24.2 and 24.4 show that by applying drip and sprinkler irrigation systems instead of the surface irrigation method, land suitability classes of 3119.7 ha (94.34%) and 2807.82 ha (84.91%) of this plain will improve, respectively. The surface irrigation has shown more suitability in 187.28 ha (5.66%) compared to drip and sprinkler irrigations. Comparison of different irrigation methods in

TABLE 24.4
The Most Suitable Land Units for Surface, Sprinkle and Drip Irrigation Systems by Notation to Capability Index (Ci) for Different Irrigation Systems

Codes of Land Units	The Maximum Capability Index for Irrigation (Ci)	Suitability Classes	The Most Suitable Irrigation Systems	Limiting Factors[a]
1	63.53	S2 ts	Drip	t and s
2	76.15	S2 s	Drip	s
3	82.49	S1	Surface	None exist
4	57.65	S3 s	Drip	s
5	59.40	S2 s	Drip	s
6	64.12	S2 s	Drip	s
7	63.84	S2 s	Drip	s
8	56.53	S3 ws	Drip	w and s
9	40.38	N1 ws	Drip	w and s
10	67.82	S2 s	Drip	s
11	72.00	S2 s	Drip	s
12	61.20	S2 s	Drip	s
13	52.22	S3 ws	Drip	w and s
14	80.75	S1	Drip	None exist
15	68.00	S2 s	Drip	s
16	68.00	S2 s	Drip	s
17	70.12	S2 s	Surface	s
18	80.00	S1	Drip	None exist

[a] Limiting factors for surface irrigation: s (Soil texture and calcium carbonate). Limiting factors for drip irrigation: s (Soil texture and/or Soil depth and/or calcium carbonate), t (slope), and w (drainage).

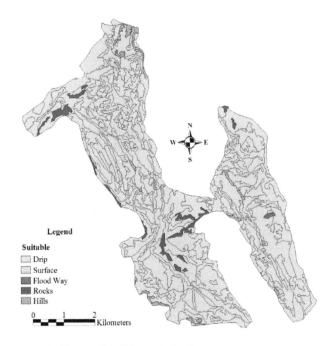

FIGURE 24.5 The most suitable map for different irrigation systems.

the studied region showed that the drip irrigation method is more efficient than surface and sprinkler irrigations (except the soil series 3 and 17), and it enhances the land suitability for irrigation goals. The second suitable option in the lands of this plain (except the soil series 3 and 17) is sprinkler irrigation enhancing the land suitability, compared to surface irrigation. The above discussion suggests that the irrigation systems suitable for the Sudjan Plain are in the order: drip > sprinkler > surface irrigation. In the studied region, the most limiting factors for surface irrigation are slope, soil texture, and drainage, and the most limiting factors for sprinkler and drip irrigations are soil texture, calcium carbonate content, and drainage.

24.4   CONCLUSIONS

In this study, an attempt has been made to analyze and to compare three irrigation systems by taking the various soils and land characteristics into account. The results obtained showed that drip and sprinkler irrigation methods are more suitable in many series of the studied soil, compared to the surface irrigation method. Due to a shortage of the surface and groundwater resources in semiarid regions, it is recommended that drip and sprinkler irrigation methods be applied instead of a surface irrigation method in order to sustain the use of water and soil resources. Drip and sprinkler irrigation systems are more suitable than the surface irrigation method in the whole region (except the soil series 3 and 17). Landi et al. (2008); Rezania et al. (2009); Boroomand Nasab et al. (2010); Jovzi et al. (2012); and Albaji et al. (2008, 2012a, 2014a, 2014b, 2015, 2016) found similar results.

The most limiting factors for drip and sprinkler irrigation methods were soil texture, calcium carbonates content, and drainage; whereas, slope, soil texture, and drainage were limiting factors for surface irrigation method.

In arid and semiarid regions, because of water scarcity, it is necessary to improve the water use efficiency in order to increase the agricultural production yield. Comparing the maps showed that sprinkler and drip irrigation methods can be considered as an optimal and useful solution in different policies for water management. Shifting from surface to drip and sprinkler irrigation methods is resulting in saving water resources considerably. In general, drip and sprinkler irrigation methods preserve the soil at the field capacity by using little amount of water, compared to the surface irrigation method. Therefore, it is necessary and more beneficial to use drip and sprinkler irrigation instead of a surface irrigation method to improve the water use efficiency and to resolve water shortage problems in the Sudjan Plain.

REFERENCES

Agricultural Jihad Organization of Chaharmahal and Bakhtiari Province. 2008. Semi-detailed Soil Study Report of Sudjan Plain: Iran. *Department of Water and Soil*, 163 p. (in Persian).

Albaji, M., Boroomand Nasab, S., Kashkuli, H.A., Naseri, A.A., Sayyad, G., Jafari, S. 2008. Comparison of different irrigation methods based on the parametric evaluation approach in North Molasani Plain, Iran. *Journal of Agronomy* 7 (2), 187–191.

Albaji, M., Boroomand Nasab, S., Hemadi, J. 2012a. Comparison of different irrigation methods based on the parametric evaluation approach in West North Ahvaz Plain. In: *Problems, Perspectives and Challenges of Agricultural Water Management*. M. Kumar, (ed.), InTech, Croatia, pp. 259–274.

Albaji, M., Papan, P., Hosseinzadeh, M., Barani, S. 2012b. Evaluation of land suitability for principal crops in the Hendijan region. *International Journal of Modern Agriculture* 1 (1), 24–32.

Albaji, M., Golabi, M., Boroomand Nasab, S., Jahanshahi. M. 2014a. Land suitability evaluation for surface, sprinkler and drip irrigation systems. *Transactions of the Royal Society of South Africa* 69 (2), 63–73.

Albaji, M., Golabi, M., Piroozfar, V.R., Egdernejad, A., Nazari Zadeh, F. 2014b. Evaluation of Agricultural Land Resources for Irrigation in the Ramhormoz Plain by using GIS. *Agriculturae Conspectus Scientificus* 79 (2), 93–102.

Albaji, M., Boroomand Nasab, S., Golabi, M., Sorkheh Nezhad, M., Ahmadee, M. 2015. Application possibilities of different irrigation methods in Hofel Plain. *YYÜ TAR BİL DERG (YYU J AGR SCI)* 25 (1), 13–23.

Albaji, M., Golabi, M., Hooshmand, A.R., Ahmadee, M. 2016. Investigation of surface, sprinkler and drip irrigation methods using GIS. *Jordan Journal of Agricultural Sciences* 12 (1), 211–222.

Albaji, M., Alboshokeh, A. 2017. Assessing agricultural land suitability in the Fakkeh region, Iran. *Outlook on Agriculture* 46 (1), 57–65.

Bagherzadeh, A., Paymard, P. 2015. Assessment of land capability for different irrigation systems by parametric and fuzzy approaches in the Mashhad Plain, northeast Iran. *Soil Water Research* 10, 90–98. https://doi.org/10.17221/139/2014-SWR.

Barberis, A., Minelli, S. 2005. Land evaluation in the Shouyang County, Shanxi Province, China. *25th Course Professional Master.* November 8th, 2004–June 23, 2005. IAO, Florence, Italy.

Behzad, M., Albaji, M., Papan, P., Boroomand Nasab, S., Naseri, A.A., Bavi, A. 2009. Qualitative evaluation of land suitability for principal crops in the Gargar Region, Khuzestan Province, Southwest Iran. *Asian Journal of Plant Sciences* 8 (1), 28.

Bienvenue, J.S., Ngardeta, M., Mamadou, K. 2002. Land evaluation in the province of Thies, Senegal. *23rd Course Prof. Master, Geometric and Nat. Resour. Eval.* 8th Nov 2002. IAO, Florence, Italy.

Boroomand Nasab, S., Albaji, M., Naseri, A.A. 2010. Investigation of different irrigation systems based on the parametric evaluation approach in Boneh Basht plain, Iran. *African Journal of Agricultural Research* 5 (5), 372–379.

Dengiz, O. 2006. Comparison of different irrigation methods based on the parametric evaluation approach. *Turkish Journal of Agriculture and Forestry* 30, 21–29.

FAO, F.A.O. of the U.N. 1976. A framework for land evaluation. *Soil Bulletin* No. 32, 72, FAO, Rome, Italy.

Fatapour, E., Eslami, H. 2014. Locating suitable areas for pressurized irrigation systems using GIS. *Bulletin of Environment, Pharmacology and Life Sciences* 3, 153–156.

Hargreaves, G.H., Merkley, G.P. 1998. Irrigation Fundamentals: An applied technology text for teaching irrigation at the intermediate level. *Water Resources Publication LLC*, xvi, 182 p.

Ikawa, H. 1992. Soil Taxonomy and land evaluation for forest establishmentIn: Conrad, Engene C.; Newell, Leonard A., Tech. Cords. *Proceedings on the Session on Tropical Forestry for People of the Pacific, XVII Pacific Science Congress*; May 27–28. Gen. Tech. Rep. PSW-129. Berkeley, CA: US Department of Agriculture, Forest Service, Pacific Southwest Forest and Range Experiment Station, pp. 56–57.

Jovzi, M., Albaji, M., Gharibzadeh, A. 2012. Investigating the suitability of lands for surface and underpressure (drip and sprinkler) irrigation in Miheh Plain. *Research Journal of Environmental Sciences* 6 (2), 51–61.

Landi, A., Boroomand-Nasab, S., Behzad, M., Tondrow, M.R., Albaji, M., Jazaieri, A. 2008. Land suitability evaluation for surface, sprinkle and drip irrigation methods in Fakkeh Plain, Iran. *Journal of Applied Sciences* 8 (20), 3646–3653.

Masoudi, M., Ebrahimi, A., Jokar, P. 2018. Comparison of different irrigation methods based on the parametric evaluation approach in Chikan and Mourzian Subbasin, Iran. *The International Journal of Agricultural Management and Development* 8, 355–364.

McBratney, A.B., Odeh, I.O.A., Bishop, T.F.A., Dunbar, M.S., Shatar, T.M. 2000. An overview of pedometric techniques for use in soil survey. *Geoderma* 97, 293–327. https://doi.org/10.1016/S0016-7061(00)00043-4.

Naseri, A.A., Albaji, M., Boroomand Nasab, S., Landi, A., Papan, P., Bavi, A. 2009a. Land suitability evaluation for principal crops in the Abbas Plain, Southwest Iran. *Journal of Food, Agriculture & Environment* 7 (1), 208–213.

Naseri, A.A., Albaji, M., Khajeh Sahoti, G.R., Sharifi, S., Sarafraz, A., Eghbali, M.R. 2009b. Investigation of soil quality for different irrigation systems in Baghe Plain, Iran. *The Journal of Food, Agriculture and Environment* 7, 713–717. https://doi.org/10.1234/4.2009.2339.

Rezania, A.R., Naseri, A.A., Albaji, M. 2009. Assessment of soil properties for irrigation methods in North Andimeshk Plain, Iran. *Journal of Food, Agriculture & Environment* 7 (3&4), 728–733.

Soil Survey Staff. 2000. *Keys to Soil Taxonomy.* U.S. Department of Agriculture, Soil Conservation Service, Washington, DC.

Sys, C., Van Ranst, E., Debaveye, J. 1991. Land evaluation, part 1, principles. In: *Land Evaluation and Crop Production Calculations. Int. Train. Cent. Post-graduate Soil Sci. Univ. Ghent*, 265 p.

25 Irrigation System Choice in Miheh Plain, Iran

Mehdi Jovzi, Hamid Zare Abyaneh, Majid Heydari,
Shahrokh Fatehi, Amin Behmanesh,
Mohammad Albaji, and Saeid Eslamian

CONTENTS

25.1 INTRODUCTION

Agriculture is the source of people's basic subsistence, which is directly related to the survival and development of the human society (Lu et al., 2019). The soil and water resources are limited, and they experience gradual degradation (Chitsaz and Azarnivand, 2017; Singh, 2018). Moreover, farm production requires to be increased using these limited resources for feeding the growing global population (Li and Zhang, 2015; Habibi Davijani et al., 2016; Liu et al., 2016; Lomba et al., 2017; Xie et al., 2018). In dry regions, given that normal rainfall in these areas is highly unreliable (Herrmann et al., 2016; Adhikari et al., 2017), the exacerbation of irrigated agriculture is required for realizing food security (Das et al., 2015; Singh et al., 2016). On the other hand, the sustainability of irrigated agriculture depends upon consistently achieving high irrigation application efficiency (Bavi et al. 2009). Therefore, land classification is necessary for application of different irrigation systems in order to protect and optimize the use of soil and water resources. According to FAO (1976) methodology, land suitability is strongly related to "land qualities" including erosion resistance, water availability, and flood hazards, which are derived from slope angle and length, rainfall, and soil texture. Sys et al. (1991) proposed a parametric evaluation system for irrigation methods that was primarily based on chemical and physical soil attributes. Factors affecting the soil suitability for irrigation are physical properties such as permeability and available water content; chemical properties such as soil salinity, alkalinity, and acidity; drainage properties such as depth of groundwater; and environmental factors such as slope. The capability index for irrigation (CI) is calculated based on the multiplication of the above factors.

Bienvenue et al. (2002), by using the parametric evaluation methods, evaluated the land suitability for drip and surface irrigation in Senegal. For drip irrigation, about 45.25% of the area was classified as suitable (S2), while 25.03% was highly suitable (S1), and a small area (5.83%) was relatively suitable (N1) or unsuitable (N2). The researchers' results showed that the limiting factors for surface irrigation were soil depth, soil texture, and drainage of the region. For surface irrigation, there was no area classified as highly suitable (S1). Only 20.24% of the study area was moderately suitable (S2, 7.73%) or marginally suitable (S3, 12.51%), and 57.66% of the study area was classified as permanently not suitable (N2). The limiting factor to this kind of land use was mainly the soil texture and drainage that was mostly sandy, while surface irrigation generally requires heavier soils. Mbodj et al. (2004), by using the parametric methods, investigated the suitability of lands for two methods of drip and surface irrigation in the north of Tunisia. Their results showed that because of slope, drainage, soil depth, and texture limitations, the suitability of drip irrigation is more than surface irrigation. Dengiz (2006) studied the different methods of irrigation including surface and drip irrigation based on the parametric evaluation south of Ankara, Turkey, and concluded that the drip irrigation resulted in the increase of suitability of the lands about 38% compared to the surface irrigation. Albaji et al. (2010) studied the surface, sprinkler, and drip irrigation methods based on the parametric evaluation system in the Elam Province in the west of Iran. Their results showed that in the investigated region, the most limiting factor in the surface and sprinkler irrigation methods was the soil texture. Whereas the limiting factors for drip irrigation were the calcium carbonate content and soil texture. Also, by applying sprinkler irrigation instead of surface and drip irrigation methods, the land capability of 21,250 ha (72.53%) in the Abbas Plain will improve. Rabia et al. (2013) investigated spatially evaluated land suitability of the Kilte Awulaelo district in Ethiopia for surface and drip irrigation methods based on GIS and remote sensing approaches. Their results showed that the GIS and remote sensing were highly efficient for modeling and developing land suitability maps together with spatially compared land suitability for deferent irrigation methods. Final suitability maps showed the irregularity of suitability classes' distribution over the study area. Results show that only 15% of the study area is suitable for the surface irrigation method. This is due to the limitation of the topography and stoniness factors for surface irrigation suitability. Albaji et al. (2015b) studied different irrigation methods based upon a parametric evaluation system in an area of 15,000 ha in the Jaizan Plain, Iran. Once the soil attributes were analyzed and evaluated by using GIS, suitability maps were generated for surface, sprinkler, and drip irrigation systems. Their results showed that by applying sprinkler irrigation instead of drip and surface irrigation methods, the arability of 13,875 ha (92.5%) in the Jaizan Plain will improve. The main limiting factor for using surface irrigation methods was drainage; for sprinkler irrigation methods, the limiting factors were soil texture, gravel, calcium carbonate, and drainage; and for drip irrigation methods, the limiting factors were drainage and calcium carbonate.

The aim of this research was to select suitable irrigation systems (surface, sprinkler, and drip irrigation) based on the parametric evaluation systems for the Miheh Plain in the Chahar Mahal and Bakhtyari Province of Iran.

25.2 MATERIALS AND METHODS

25.2.1 Ecological Conditions of the Study Area

The present study was conducted in an area of about 3011.9 ha in the Miheh Plain, in the Zayandeh Rood watershed on the Chaharmahal and Bakhtiari Province of Iran. The study area is located 60 km northwest of the city of Shahr-e Kord, 50° 14′ 43″ to 50° 24′ 13″ E and 32° 21′ 02″ to 32° 29′ 21″ N. The maximum and minimum elevations for the area are 3,282 m and 2,179 m for the eastern and northwestern parts, respectively. A review of the existing statistics for 1978–2005 shows that the average annual temperature is 9.5°C. The annual evaporation of the area is 1,865 mm. The annual rainfall is 1,409 mm, much of it rain in the autumn and winter (1,205 mm). In this period, the average temperature is low, and no plant is growing in the region in this time so no water is used. During spring and summer, when the plants are growing, there is little rainfall (204 mm) and the region is categorized as semiarid. Most often the lands are dry farmed, and a limited area of lands are irrigated by fountains or through pumping water from rivers. Currently, the irrigation systems used by farmlands in the region are border, furrow, and basin irrigation schemes (Agricultural Jihad Organization of Chaharmahal and Bakhtiari Province, 2008). In this plain, like other plains in Iran, the important major crops are wheat and barley (Behzad et al., 2009; Naseri et al., 2009; Albaji et al., 2012b; Albaji and Alboshokeh, 2017).

25.2.2 Irrigation System Selection

The land evaluation was determined based on topography and soil properties of the region. The properties included slope, soil texture, depth, drainage, salinity, and calcium carbonate content, which were taken into account. Soil characteristics such as organic matter (OM), cation exchange capacity (CEC), percentage of basic saturation (PBS), and pH were considered in terms of soil fertility. Sys et al. (1991) suggested that soil properties such as PBS and OM did not require any evaluation in arid regions, whereas the clay CEC rate usually exceeds the plant requirement without further limitation—thus, fertility properties can be excluded from land evaluation if it is done for the purpose of irrigation.

To determine soil properties, the semi-detailed soil survey report of the Miheh Plain, Kohrang, was used (Agricultural Jihad Organization of Chaharmahal and Bakhtiari Province, 2008). Based on the laboratory analysis and profile description, the soil groups that had similar properties and were located in the same physiographic unit were categorized as soil series; taxonomy was used to form a soil family as per the Soil Survey Staff (2000). Finally, four soil series and 14 series phases were chosen for the land suitability in order to choose the best irrigation method among different irrigation methods (surface, sprinkler, and drip irrigation).

To obtain the average salinity, soil texture, and $CaCO_3$ for the upper 150 cm of soil surface, the profile was subdivided into six equal sections and weighting factors of 2, 1.5, 1, 0.75, 0.50, and 0.25 were used for each section, respectively. The weight factors are used in order to give more importance to the upper part of the soil profile where the root is developing more (Sys et al., 1991). To evaluate the suitability of lands for sprinkler, drip, and surface irrigation methods, the parametric evaluation system was applied. The parametric evaluation system is based on morphology, physical, and chemical properties of soil (Sys et al., 1991).

The parametric evaluation system consisted of six parameters including slope, drainage properties, electrical conductivity of soil solution, calcium carbonate status, depth soil, and soil texture. Each of the six aforementioned parameters is scaled according to the related tables, and the capability index for irrigation (Ci) is calculated using them as the equation 24.1. The ranges of the capability index and the corresponding suitability classes were shown in Table 24.1 in the previous chapter.

The semi-detailed soil map of the Miheh Plain is shown in Figure 25.1. This soil map was used to develop land suitability maps for different irrigation methods. By using Arc GIS 9.2 software, all the data for soil properties were incorporated in the map. The digital soil map base preparation was the first step toward the presentation of a GIS module in order to identify land suitability maps for different irrigation methods. A total of 14 different land mapping units (LMUs) were determined in the base map. Soil properties were also given for each LMU. These values were used to generate the land suitability maps for different irrigation methods (surface, sprinkler, and drip irrigation) using geographic information systems (GIS).

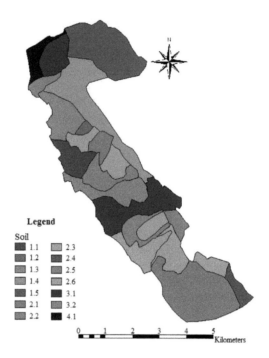

FIGURE 25.1   Soil map of the study area.

25.3   RESULTS AND DISCUSSION

The surface irrigation method is applied in most lands of the Miheh Plain. Wheat, barley, walnut, and peach are the major plants in agricultural lands and gardens (Agricultural Jihad Organization of Chaharmahal and Bakhtiari Province, 2008). The semi-detailed soil study resulted in recognition of 4 series of soil and altogether 14 series phases in the Miheh Plain. The series of soil are introduced in Figure 25.1 as a basis of evaluation. The soils of the study region are rated as Molli soil, Alfi soil, and Incepti soil. The soil moisture regime is xeric, and the soil temperature regime is mesic. Non-agricultural lands consisting of residential area (0.87%) and hills (2.14%) constitute 90.9 ha (3.01%) of all 3011.9 ha area of the region (Agricultural Jihad Organization of Chaharmahal and Bakhtiari Province, 2008).

25.3.1   LAND SUITABILITY FOR SURFACE IRRIGATION

The values of capability index (Ci), suitability classes of surface, sprinkler and drip irrigation for each land unit, and distribution of different irrigation suitability are shown in Tables 25.1 and 25.2.

As shown in Tables 25.1 and 25.2 for the surface irrigation method, just the soil series coded 4.1 (73 ha – 2.42%) were highly suitable (S1); soil series coded 1.1, 1.2, 1.3, 2.1, 2.2, 2.3, and 3.1 with the area of 1355.8 ha (45.01%) were found to be moderately suitable (S2). Soil series coded 2.4, 2.5, 2.6, and 3.2 with the area of 767.1 ha (25.47%) were classified as marginally suitable (S3). Soil series coded 1.4 and 1.5 with the area of 725.1 ha (24.07%) were found to be currently not suitable (N1).

The map analyzing land suitability for surface irrigation in Figure 25.2 shows that a small area (in the northwest) is highly suitable for the surface irrigation method. This was due to good drainage, suitable texture, deep soil, salinity, and proper slope of the area; however, the calcium carbonate content was high (27.54%). Most lands of the region are moderately suitable (S2). In general, the soil texture of these lands is silty clay loam and clay, and the lands are mildly sloped. Other factors such as soil depth, salinity, and drainage have no effect on the land suitability of the region. The map also shows that some of the lands located in western and southeastern margins are marginally suitable

TABLE 25.1

Capability Index for Irrigation (Ci) Values and Suitability Classes of Surface, Sprinkler, and Drip Irrigation for Each Land Unit

Codes of Land Units	Surface Irrigation		Sprinkler Irrigation		Drip Irrigation	
	Ci	Suitability Classes	Ci	Suitability Classes	Ci	Suitability Classes
1.1	66.43	S2 t[a]	76.71	S2 s[b]	80.75	S1[b]
1.2	66.43	S2 t	76.71	S2 s	80.75	S1
1.3	78.15	S2 t	90.25	S1	95.00	S1
1.4	42.67	N1 ts	56.71	S3 ts	57.87	S3 ts
1.5	43.88	N1 ts	56.71	S3 ts	57.87	S3 ts
2.1	78.15	S2 t	90.25	S1	95.00	S1
2.2	66.43	S2 t	76.71	S2 s	80.75	S1
2.3	68.38	S2 t	80.75	S1	85.50	S1
2.4	53.20	S3 ts	64.60	S2 s	68.40	S2 s
2.5	54.71	S3 ts	64.60	S2 s	68.40	S2 s
2.6	54.85	S3 t	70.89	S2 t	72.34	S2 t
3.1	66.84	S2 t	76.71	S2 s	80.75	S1
3.2	53.20	S3 ts	64.60	S2 s	68.40	S2 s
4.1	82.49	S1	76.00	S2 s	80.00	S2 s

[a] Limiting factors for surface irrigation: s (Soil texture) and t (slope).
[b] Limiting factors for sprinkle and drip irrigation: s (Soil texture and/or calcium carbonate) and t (slope).

TABLE 25.2

Distribution of Surface, Sprinkler, and Drip Irrigation Suitability

Suitability	Surface Irrigation			Sprinkler Irrigation			Drip Irrigation		
	Land Unit	Area (ha)	Ratio (%)	Land Unit	Area (ha)	Ratio (%)	Land Unit	Area (ha)	Ratio (%)
S_1	4.1	73	2.42	1.3, 2.1, 2.3	907.6	30.13	1.1, 1.2, 1.3, 2.1, 2.2, 2.3, 3.1	1355.8	45.01
S_2	1.1, 1.2, 1.3, 2.1, 2.2, 2.3, 3.1	1355.8	45.01	1.1, 1.2, 2.2, 2.4, 2.5, 2.6, 3.1, 3.2, 4.1	1288.3	42.77	2.4, 2.5, 2.6, 3.2, 4.1	840.1	27.89
S_3	2.4, 2.5, 2.6, 3.2	767.1	25.47	1.4, 1.5	725.1	24.07	1.4, 1.5	725.1	24.07
N_1	1.4, 1.5	725.1	24.07	—	—	—	—	—	—
N_2	—	—	—	—	—	—	—	—	—
[a]Mis. land		90.9	3.02		90.9	3.02		90.9	3.02
Total		3011.9	100		3011.9	100		3011.9	100

[a] Miscellaneous land: Hill and Residential areas.

(S3) due to silty clay loam texture of the soil, gravels (15%–35%), and improper slope (8%–25%). The currently not suitable lands (N1) are located in the eastern and northern parts of the region. They are currently not suitable lands because of sharp slope of lands (12%–25%) and silty clay loam texture of the soil with coarse gravels (15%–35%). The drainage, salinity, and $CaCO_3$ weren't limiting factors for surface irrigation method in all lands of the region.

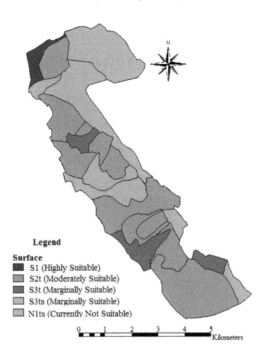

FIGURE 25.2   Land suitability map for surface irrigation method.

25.3.2   LAND SUITABILITY FOR SPRINKLER IRRIGATION

As shown in Tables 25.1 and 25.2, the land suitability of the Miheh Plain for sprinkler irrigation was studied. The results showed that the soil series coded 1.3, 2.1, and 2.3 with the area of 907.6 ha (30.13%) were highly suitable (S1) for the sprinkler irrigation method. Soil series coded 1.1, 1.2, 2.2, 2.4, 2.5, 2.6, 3.1, 3.2, and 4.1 with the area of 1288.3 ha (42.77%) were moderately suitable (S2), while soil series coded 1.4, and 1.5 with the area of 725.1 ha (24.07%) were classified as marginally suitable (S3) for sprinkler irrigation.

Figure 25.3 shows that the highly suitable (S1) land for the sprinkler irrigation method consisted of the western and southern parts and a small part of the central area of the Miheh Plain; this is due to the soil texture, soil depth, good drainage, salinity, and proper slope. Most parts of the lands investigated in the Miheh Plain are moderately suitable (S2) for sprinkler irrigation due to these lands having silty clay loam and clay texture, and a mild slope. A small part of the northwest lands was only moderately (S2) suitable for sprinkler irrigation due to a high percentage of $CaCO_3$. Also, small parts in the central, western, and southeastern areas of this studied region were moderately suitable (S2) due to the sharp slope. The lands that were marginally suitable (S3) were located in the eastern and northern parts of this plain due to the sharp slope (12%–25%) and soil texture of silty clay loam with coarse gravels (15%–35%). The salinity, drainage, and soil depth were not limiting factors for the sprinkler irrigation method in all the lands of the studied regions.

25.3.3   LAND SUITABILITY FOR DRIP IRRIGATION

The land suitability of the Miheh Plain for drip irrigation was investigated, as shown in Tables 25.1 and 25.2. The results from evaluating the suitability of lands for a drip irrigation system based on parametric method showed that the soil series coded 1.1, 1.2, 1.3, 2.1, 2.2, 2.3, and 3.1 (1355.8 ha – 45.01%) were highly suitable (S1). The soil series coded 2.4, 2.5, 2.6, 3.2, and 4.1 (840.1 ha – 27.89%) were moderately suitable (S2), while the soil series coded 1.4 and 1.5 (725.1 ha – 24.07%) were classified as marginally suitable (S3).

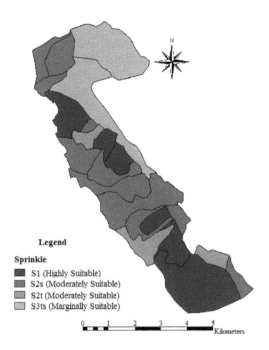

FIGURE 25.3 Land suitability map for sprinkler irrigation method.

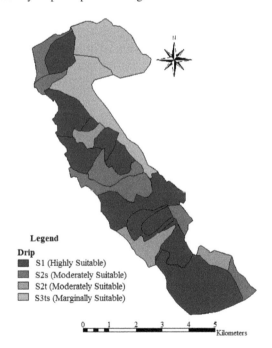

FIGURE 25.4 Land suitability map for drip irrigation method.

As shown in Figure 25.4, most lands located in the southern and central parts and a small part in the northern lands of the studied region are highly suitable for drip irrigation. In these lands, all factors such as soil depth, soil texture, calcium carbonate content, salinity, and slope were in favorable condition. The moderately suitable lands (S2) are located at the central parts and some small parts in the northwestern and southeastern lands. These lands have silty clay loam soil with medium calcium carbonate content (15.65% – soil series 2.4, 2.5, and 3.2) and coarse gravel (15%–35%).

The lands in some parts of the central, southwestern, and southeastern areas of the studied region were moderately suitable (S2) for drip irrigation due to the sharp slope. Also, a small part of the northwestern area was classified as moderately suitable (S2) because of a high percentage of $CaCo_3$. The lands identified as marginally suitable (S3) are located in the eastern and northern areas of the studied region. The limiting factors in these lands were the sharp slope and the soil texture (silty clay loam with coarse gravel 15%–35%). The salinity, soil depth, and drainage were not limiting factors for the drip irrigation method in the studied region.

25.3.4 SELECTION OF THE MOST SUITABLE IRRIGATION SYSTEM

The mean capability index (Ci) numbers for surface, sprinkler, and drip irrigation were 62.56 (moderately suitable), 73.02 (moderately suitable), and 76.56 (moderately suitable), respectively. Comparing the capability index for different irrigation methods in Tables 25.1 and 25.3 shows that the drip irrigation method was more suitable than the sprinkler and surface irrigations in all series of studied lands (except soil series 4.1). The surface irrigation method in soil series 4.1 was more suitable than the other irrigation methods. It was due to the high percentage of calcium carbonate content (27.54%) in this area.

Figure 25.5 shows the usage of the most suitable irrigation system in the Miheh Plain area. As this map shows, a small part of the Miheh Plain is suitable for the surface irrigation method, whereas the major part of the plain is suitable for the drip irrigation method.

Tables 25.1 and 25.3 show that by applying the drip irrigation method instead of the surface and sprinkler irrigation methods, land suitability classes of 2,848 ha (94.56%) of this plain will improve. The surface irrigation method has shown more suitability in 73 ha (2.42%) compared to other irrigation methods. Comparison of different irrigation methods in the Miheh Plain showed that drip irrigation is more efficient than sprinkler and surface irrigation methods (except for soil series 4.1), and it enhances the land suitability for irrigation goals. The second suitable option in the lands of this plain is the sprinkler irrigation method, which enhances the land suitability compared to the surface irrigation method. The above discussion suggests that the irrigation systems suitable for the Miheh Plain are in order: drip > sprinkler > surface irrigation methods.

TABLE 25.3

The Most Suitable Land Units for Surface, Sprinkle, and Drip Irrigation Systems by Notation to Capability Index (Ci) for Different Irrigation Systems

Codes of Land Units	The Maximum Capability Index for Irrigation (Ci)	Suitability Classes	The Most Suitable Irrigation Systems	Limiting Factors
1.1	80.75	S1	Drip	None exist
1.2	80.75	S1	Drip	None exist
1.3	95.00	S1	Drip	None exist
1.4	57.87	S3 ts	Drip	Soil texture and slope
1.5	57.87	S3 ts	Drip	Soil texture and slope
2.1	95.00	S1	Drip	None exist
2.2	80.75	S1	Drip	None exist
2.3	85.50	S1	Drip	None exist
2.4	68.40	S2 s	Drip	Soil texture
2.5	68.40	S2 s	Drip	Soil texture
2.6	72.34	S2 t	Drip	Slope
3.1	80.75	S1	Drip	None exist
3.2	68.40	S2 s	Drip	Soil texture
4.1	82.49	S1	Surface	None exist

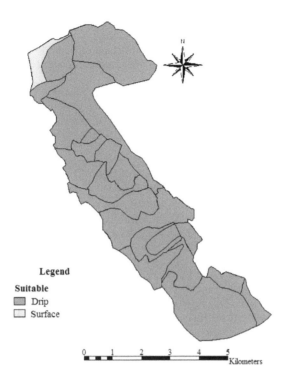

FIGURE 25.5   The most suitable map for different irrigation methods in the Miheh Plain.

In the Miheh Plain, the most limiting factors for drip and sprinkler irrigation methods are soil texture, calcium carbonate content, and slope; for the surface irrigation method, the limiting factors are soil texture and slope.

25.4   CONCLUSIONS

In this chapter, selection of the suitable irrigation system was conducted based on the properties of different soil series and other land properties. The results of this study showed that sprinkler and drip irrigation methods were more suitable compared to the surface irrigation method in many of the study lands. Landi et al. (2008); Rezania et al. (2009); Boroomand Nasab et al. (2010); and Albaji et al. (2008, 2012a, 2014a, 2014b, 2015a, 2015b, 2016) found similar results.

The most limiting factors for the surface irrigation method were slope and soil texture, whereas the most limiting factors for drip and sprinkler irrigation methods were soil texture, slope, and calcium carbonate content. In arid and semiarid areas, because of water scarcity, it is necessary to improve the water use efficiency for increasing the agricultural production yield. In this regard, the use of land suitability maps for irrigation—obtained in this study—can be useful in the Miheh Plain. Furthermore, changing the surface irrigation method to sprinkler and drip irrigation methods ensures that water resources are saved. In general, drip and sprinkler irrigation methods compared to the surface irrigation method preserve the soil at the field capacity by using less amount of water. Therefore, it would be useful to use drip and sprinkler irrigation methods in the Miheh Plain. Finally, due to water scarcity in semiarid regions, and for sustainable use of water and soil resources, it is recommended that the drip and sprinkler irrigation methods be applied instead of the surface irrigation method.

REFERENCES

Adhikari, U., Nejadhashemi, A.P., Herman, M.R., Messina, J.P. 2017. Multiscale assessment of the impacts of climate change on water resources in Tanzania. *J. Hydrol. Eng.* 22, 05016034. https://doi.org/10.1061/(ASCE)HE.1943-5584.0001467.

Agricultural Jihad Organization of Chaharmahal and Bakhtiari Province. 2008. Semi-detailed Soil Study Report of Miheh Plain, Iran. *Department of Water and Soil*, 93 p. (in Persian).

Albaji, M., Boroomand Nasab, S., Kashkuli, H.A., Naseri, A.A., Sayyad, G., Jafari, S. 2008. Comparison of different irrigation methods based on the parametric evaluation approach in North Molasani Plain, Iran. *J. Agron.* 7 (2), 187–191.

Albaji, M., Boroomand-Nasab, S., Naseri, A., Jafari, S. 2010. Comparison of different irrigation methods based on the parametric evaluation approach in Abbas Plain: Iran. *J. Irrig. Drain. Eng.* 136, 131–136. https://doi.org/10.1061/(ASCE)IR.1943-4774.0000142.

Albaji, M., Boroomand Nasab, S., Hemadi, J. 2012a. Comparison of different irrigation methods based on the parametric evaluation approach in West North Ahvaz Plain. In: *Problems, Perspectives and Challenges of Agricultural Water Management*. M. Kumar, (ed.), InTech, Croatia, pp. 259–274.

Albaji, M., Papan, P., Hosseinzadeh, M., Barani, S. 2012b. Evaluation of land suitability for principal crops in the Hendijan region. *Int. J. Mod. Agric.* 1(1), 24–32.

Albaji, M., Golabi, M., Boroomand Nasab, S., Jahanshahi. M. 2014a. Land suitability evaluation for surface, sprinkler and drip irrigation systems. *Trans. R. Soc. South Africa* 69 (2), 63–73.

Albaji, M., Golabi, M., Piroozfar, V.R., Egdernejad, A., Nazari Zadeh, F. 2014b. Evaluation of agricultural land resources for irrigation in the Ramhormoz Plain by using GIS. *Agric. Conspec. Sci.* 79 (2), 93–102.

Albaji, M., Boroomand Nasab, S., Golabi, M., Sorkheh Nezhad, M., Ahmadee, M. 2015a. Application possibilities of different irrigation methods in Hofel Plain. *YYÜ TAR BİL DERG (YYU J AGR SCI)* 25 (1), 13–23.

Albaji, M., Golabi, M., Boroomand Nasab, S., Zadeh, F.N. 2015b. Investigation of surface, sprinkler and drip irrigation methods based on the parametric evaluation approach in Jaizan Plain. *J. Saudi Soc. Agric. Sci.* 14, 1–10. https://doi.org/10.1016/j.jssas.2013.11.001.

Albaji, M., Golabi, M., Hooshmand, A.R., Ahmadee, M. 2016. Investigation of surface, sprinkler and drip irrigation methods using GIS. *Jordan J. Agric. Sci.* 12 (1), 211–222.

Albaji, M., Alboshokeh, A. 2017. Assessing agricultural land suitability in the Fakkeh region, Iran. *Outlook Agr.* 46 (1), 57–65. https://doi.org/10.3923/jas.2008.654.659.

Bavi, A., Kashkuli, H.A., Boroomand, S., Naseri, A.A., Albaji, M. 2009. Evaporation losses from sprinkler irrigation systems under various operating conditions. *J. Appl. Sci.* 9 (3), 597–600.

Behzad, M., Albaji, M., Papan, P., Boroomand Nasab, S., Naseri, A.A., Bavi, A. 2009. Qualitative evaluation of land suitability for principal crops in the Gargar Region, Khuzestan Province, Southwest Iran. *Asian J. Plant Sci.* 8 (1), 28.

Bienvenue, J.S., Ngardeta, M., Mamadou, K. 2002. Land evaluation in the province of Thies, Senegal. *23rd Course Prof. Master, Geometric and Nat. Resour. Eval.* 8th Nov 2002. IAO, Florence, Italy.

Boroomand Nasab, S., Albaji, M., Naseri, A.A. 2010. Investigation of different irrigation systems based on the parametric evaluation approach in Boneh Basht plain, Iran. *African J. Agric. Res.* 5 (5), 372–379.

Chitsaz, N., Azarnivand, A. 2017. Water scarcity management in arid regions based on an extended multiple criteria technique. *Water Resour. Manag.* 31, 233–250. https://doi.org/10.1007/s11269-016-1521-5.

Das, B., Singh, A., Panda, S.N., Yasuda, H. 2015. Optimal land and water resources allocation policies for sustainable irrigated agriculture. *Land Use Policy* 42, 527–537. https://doi.org/10.1016/j.landusepol.2014.09.012.

Dengiz, O. 2006. Comparison of different irrigation methods based on the parametric evaluation approach. *Turkish J. Agric. For.* 30, 21–29.

FAO, F.A.O. of the U.N. 1976. A framework for land evaluation. *Soil Bulletin* No. 32, 72, FAO, Rome, Italy.

Habibi Davijani, M., Banihabib, M.E., Nadjafzadeh Anvar, A., Hashemi, S.R. 2016. Multi-objective optimization model for the allocation of water resources in arid regions based on the maximization of socioeconomic efficiency. *Water Resour. Manag.* 30, 927–946. https://doi.org/10.1007/s11269-015-1200-y.

Herrmann, F., Kunkel, R., Ostermann, U., Vereecken, H., Wendland, F. 2016. Projected impact of climate change on irrigation needs and groundwater resources in the metropolitan area of Hamburg (Germany). *Environ. Earth Sci.* 75, 1104. https://doi.org/10.1007/s12665-016-5904-y.

Landi, A., Boroomand-Nasab, S., Behzad, M., Tondrow, M.R., Albaji, M., Jazaieri, A. 2008. Land suitability evaluation for surface, sprinkle and drip irrigation methods in Fakkeh Plain, Iran. *J. Appl. Sci.* 8 (20), 3646–3653.

Li, C.Y., Zhang, L. 2015. An inexact two-stage allocation model for water resources management under uncertainty. *Water Resour. Manag.* 29, 1823–1841. https://doi.org/10.1007/s11269-015-0913-2

Liu, X.M., Huang, G.H., Wang, S., Fan, Y.R. 2016. Water resources management under uncertainty: Factorial multi-stage stochastic program with chance constraints. *Stoch. Environ. Res. Risk Assess.* 30, 945–957. https://doi.org/10.1007/s00477-015-1143-0.

Lomba, A., Strohbach, M., Jerrentrup, J.S., Dauber, J., Klimek, S., McCracken, D.I. 2017. Making the best of both worlds: Can high-resolution agricultural administrative data support the assessment of High Nature Value farmlands across Europe? *Ecol. Indic.* 72, 118–130. https://doi.org/10.1016/j.ecolind.2016.08.008

Lu, S., Bai, X., Li, W., Wang, N. 2019. Impacts of climate change on water resources and grain production. *Technol. Forecast. Soc. Change* 143, 76–84. https://doi.org/10.1016/j.techfore.2019.01.015

Mbodj, C., Mahjoub, I., Sghaiev, N. 2004. Land evaluation in the oud rmel catchment, Tunisia. *Proc., 24th Course Prof. Master Geom. Nat. Resour. Eval.* 10 November, 2003 - 23 June 2004, IAO, Florence, Italy.

Naseri, A.A., Albaji, M., Boroomand Nasab, S., Landi, A., Papan, P., Bavi, A. 2009. Land suitability evaluation for principal crops in the Abbas Plain, Southwest Iran. *J. Food, Agric. Environ.* 7 (1), 208–213.

Rabia, A.H., Figueredo, H., Huong, T.L., Lopez, B.A.A., Solomon, H.W., Alessandro, V. 2013. Land suitability analysis for policy making assistance: A GIS based land suitability comparison between surface and drip irrigation systems. *Int. J. Environ. Sci. Dev.* 4 (1), 1–6.

Rezania, A.R., Naseri, A.A., Albaji, M. 2009. Assessment of soil properties for irrigation methods in North Andimeshk Plain, Iran. *J. Food, Agric. Environ.* 7 (3&4), 728–733.

Singh, A. 2018. Assessment of different strategies for managing the water resources problems of irrigated agriculture. *Agric. Water Manag.* 208, 187–192. https://doi.org/10.1016/j.agwat.2018.06.021

Singh, A., Panda, S.N., Saxena, C.K., Verma, C.L., Uzokwe, V.N.E., Krause, P., Gupta, S.K. 2016. Optimization modeling for conjunctive use planning of surface water and groundwater for irrigation. *J. Irrig. Drain. Eng.* 142, 04015060. https://doi.org/10.1061/(ASCE)IR.1943-4774.0000977

Soil Survey Staff. 2000. *Keys to Soil Taxonomy.* U.S. Department of Agriculture Natural Resources Conservation Service, Washington, DC.

Sys, C., Van Ranst, E., Debaveye, J. 1991. Land evaluation, part 1, principles in land evaluation and crop production calculations. *Int. Train. Cent. Post-graduate Soil Sci. Univ. Ghent*, 265 p.

Xie, Y.L., Xia, D.X., Ji, L., Huang, G.H. 2018. An inexact stochastic-fuzzy optimization model for agricultural water allocation and land resources utilization management under considering effective rainfall. *Ecol. Indic.* 92, 301–311. https://doi.org/10.1016/j.ecolind.2017.09.026

Index

9 780367 518776